Results and Problems in Cell Differentiation

47

Dietmar Richter
Center for Molecular Neurobiology
University Medical Center Hamburg-Eppendorf (UKE)
University of Hamburg
Martinistrasse 52
20246 Hamburg
Germany
richter@uke.uni-hamburg.de

Henri Tiedge
The Robert F. Furchgott Center for Neural and Behavioral Science
Department of Physiology and Pharmacology
Department of Neurology
SUNY Health Science Center at Brooklyn
Brooklyn, New York 11203
USA
htiedge@downstate.edu

Series Editors
D. Richter, H. Tiedge

Wolfgang Meyerhof, Sigrun Korsching (eds.)

Chemosensory Systems in Mammals, Fishes, and Insects

Springer

Editors
Prof. Dr. Wolfgang Meyerhof
Deutsches Institut für
Ernährungsforschu
Potsdam-Rehbrücke (DIFE)
Abt. Molekulare Genetik
Arthur-Scheunert-Allee
114-116
14558 Nuthetal
Germany
meyerhof@dife.de

Prof. Sigrun Korsching
Universität Köln
Institut für Genetik
Zülpicher Str. 47
50674 Köln
Germany
sigrun.korsching@uni-koeln.de

ISBN 978-3-540-69918-7   e-ISBN 978-3-540-69919-4
DOI 10.1007/978-3-540-69919-4

Results and Problems in Cell Differentiation ISSN 0080-1844

Library of Congress Control Number: 2009921136

© Springer-Verlag Berlin Heidelberg 2009
This work is subject to copyright. All rights are reserved, whether the whole or part of the material is concerned, specifically the rights of translation, reprinting, reuse of illustrations, recitation, broadcasting, reproduction on microfilm or in any other way, and storage in data banks. Duplication of this publication or parts thereof is permitted only under the provisions of the German Copyright Law of September 9, 1965, in its current version, and permission for use must always be obtained from Springer. Violations are liable for prosecution under the German Copyright Law.
The use of registered names, trademarks, etc. in this publication does not imply, even in the absence of a specific statement, that such names are exempt from the relevant protective laws and regulations and therefore free for general use.

Cover design: WMXDesign GmbH, Heidelberg

Printed on acid-free paper

Springer is part of Springer Science+Business Media (www.springer.com)

Preface

The sense of smell has an essential role in locating food, detecting predators, navigating, and communicating social information. Accordingly, the olfactory system has evolved complex repertoires of receptors to face these problems. Although the sense of taste has less far-reaching tasks, they are every bit as essential for the animals well-being, allowing it to reject toxic materials and to select nutritionally valuable food. The last decade has seen a massive advance in understanding the molecular logic of chemosensory information processing, beyond that already achieved in the first few years following Linda Bucks discovery of odorant receptors. Shortly afterwards, the major principles of olfactory representation had been established in mammals as the one neuron/one receptor rule and the convergence of neurons, which express the same receptor, onto individual modules in the olfactory bulb. In recent years, such studies have been extended to lower vertebrates, including fishes and other phyla, i.e., arthropods, worms, and insects, showing both the general validity of these concepts and some exceptions to the rule. In parallel, hallmarks of the molecular logic of taste sensation have been deciphered and found to differ in interesting ways from those of smell sensation. In keeping with the emphasis of the taste system on decision making vs the strength of the olfactory system in complex distinction and recognition tasks, taste receptor cells are specific for taste qualities, not necessarily for single taste receptors, and are linked to stereotyped behavioral outputs. We consider it timely to present the current state of the art in gustatory and olfactory research, as seen by leading researchers in the field. In total 12 contributions are presented, about half of them from each field that cover our current knowledge in mammalian, fish and insect models.

Shi and Zhang start out by presenting an overview of olfactory and gustatory receptor gene families in vertebrates and discuss evolutionary rates, species-specific gene expansions and pseudogenization as factors shaping receptor gene repertoires. Four olfactory receptor families, odorant receptors (ORs), vomeronasal receptors type I (V1Rs), vomeronasal receptors type II (V2Rs), and trace amine-associated receptors (TAARs), first described in mammals, have orthologs in teleost fish. All of them are G protein-coupled receptors (GPCRs). Shi and Zhang describe how ORs are both less numerous and more diverse in teleost fish compared to tetrapods. A loss of the entire V2R repertoire has occurred at least three times during tetrapod evolution, leading to the complete absence of V2Rs in several species, whereas rodent V2R, as well as V2R-related olfC genes of some fish species, show large scale gene amplification. Extreme variation of family size also characterizes the T2R taste receptor family in vertebrates. Some gene losses, but nearly no gene gains are observed in the small vertebrate T1R repertoire. The authors present the case for adaptive evolution of chemosensory receptor families to reflect changing requirements for chemical senses as species evolve to fill different ecological niches.

Zhang and Firestein discuss the genomics of olfactory receptors and the accuracy of genomic datamining. Through computational analysis of genomic databases, OR repertoires of multiple species were identified, revealing an exceedingly large OR gene family of over 1,000 genes in rodents, and a surprisingly large, unrelated family of about 800 chemosensory receptor genes in nematodes. Evolutionary fluctuation is prominent between different species. Pseudogenization is a leading cause for decreases in repertoire sizes, with pseudogenes representing two-thirds of the OR repertoire in primates such as humans and chimpanzees, and six-sevenths of an avian species, chicken. The characteristics of OR genes were explored through computational and experimental methods, showing a complicated gene structure and particular genomic distribution. Phylogenetically, OR genes may be divided into class I and class II ORs, the latter showing a massive gene expansion in tetrapods compared to teleosts. Class I genes form a single large cluster, but class II genes exist in several clusters, often of closely related OR genes, as well as isolated OR genes. Utilizing high-throughput OR microarrays, expression profiles of the mouse and human OR repertoires were examined, their olfactory functions verified, and their zonal, ectopic and developmental expression determined. Class I genes occupy a particular, and molecularly distinct zone within the olfactory epithelium (dorsalmost, zone 1), and a correspondingly segregated target region in the dorsal olfactory bulb. Variation in human smelling abilities results from different functional OR repertoires, variable expression levels and polymorphisms in the copy number of OR genes.

Korsching presents a comprehensive review of teleost olfactory receptor repertoires focusing on evolutionary history, phylogenomic properties and similarities as well as differences to the corresponding mammalian families. Representatives of all four families, the OR, vomeronasal V1R-related ORA, V2R-related OlfC, and TAAR receptors are found in cartilaginous fish and/or jawless fish, indicating an evolutionary origin before the segregation between cartilaginous and bony fish or cartilaginous and jawless fish, respectively. Gene repertoires of teleost olfactory receptors are smaller in size (OR, ORA), comparable (OlfC), or even larger (TAAR) than the corresponding mammalian gene repertoires, but all teleost families show much larger divergence than their mammalian counterparts. Evolutionary rates vary greatly between families, with evidence for positive selection in teleost OR genes, whereas the *ora* genes are unusually conserved among all teleost species. With one exception, ligands are not known for any of the four teleost olfactory receptor gene families so far. ORs are expressed stochastically within expression domains, similar to the stochastic expression of mammalian ORs within expression zones. The range of odors relevant to fish is rather well known, and contains amino acids, bile acids, nucleotides, steroid and prostaglandin hormones and metabolites, with different groups of chemicals being processed in different subpopulations of olfactory receptor neurons.

Imai and Sakano describe odorant receptor gene choice and axonal projection in the mouse olfactory system. Each olfactory sensory neuron (OSN or ORN) expresses a single type of odorant receptor (OR) out of about 1,000 different genes. In fact, from the two alleles of an OR gene, only one is chosen for expression (monoallelic expression). Furthermore, the axons of olfactory receptor neurons expressing the same OR converge onto a specific pair of glomeruli in the olfactory bulb. For unknown reasons the mammalian (but not the fish) olfactory bulb contains a duplicated, mirror-symmetrical map of glomeruli, hence the pair of target glomeruli. These two basic principles are fundamental to the peripheral olfactory system, and are regulated by the expressed OR protein itself. Somatic recombination or gene conversion play no role in guiding OR expression, since mice cloned from single olfactory receptor neurons contain the full set of expressed OR genes. Singular OR gene choice is ensured by a two-step mechanism, the first being a stochastic enhancer-promoter interaction with a single enhancer element neighboring a cluster of OR genes. In a second step, negative feedback regulation by OR proteins occurs, which blocks transcription from other clusters of OR genes. In the axonal projection, OR-derived cAMP signals and neuronal activity determine the expression levels of axon guidance/sorting molecules, and thereby direct glomerular positioning and axon sorting.

Rodriguez and Boehm discuss pheromone sensing in mice. Among other strategies mice employ urine investigation as a tool to discriminate between individuals. The authors summarize the available information about the chemical nature of rodent pheromones, their physiological sources, and biological function. Pheromones turn out to belong to structurally diverse classes of chemicals, including peptides secreted by some exocrine lacrimal glands, small volatile molecules and protein fragments present in urine. Most pheromones activate both vomeronasal and main olfactory sensory neurons, contrary to the initial hypothesis of a neat segregation of the main and the accessory or vomeronasal olfactory system. In fact, besides the VR genes, some OR genes are expressed in the vomeronasal organ, and even MHC molecules may play a role in odor detection here. Selective gene-targeting of the main and accessory olfactory systems in mice has shown that both systems can converge and synergize to express the complex array of stereotyped behaviors and hormonal changes triggered by pheromones. Moreover, rodent noses house at least two other distinct chemosensory epithelia: the Grüneberg ganglion and the septal organ; their functions are currently examined.

Yoshihara presents a molecular genetic dissection of the zebrafish olfactory system. His contribution details the advantages of zebrafish as a vertebrate model system, which includes external fertilization, large clutch sizes, rapid development, transparency of embryos, and the availability of various genetic engineering technologies such as transgenesis, mutagenesis, gene knockdown, and transposon-mediated gene transfer. Yoshihara shows that the 'one neuron/one receptor rule' established in mammalian olfaction mostly holds true for zebrafish, and that 'convergence of axons to target glomeruli' is preserved as well. Using a transgenic approach the author showed the existence of two segregated neural circuits, one originating from ciliated and the other from microvillous olfactory sensory neurons in the olfactory epithelium to distinct regions of the olfactory bulb. These two segregated pathways are likely to convey different types of olfactory information (e.g. pheromones and odorants) to the higher olfactory centers. The chemotopic odor map present in the olfactory bulb is partially retained in the forebrain, but an integration of different input channels begins to be visible as well. A discussion of the chemical nature of fish odor stimuli is included, together with an evaluation of three zebrafish mutants showing defects in olfactory axonal path-finding and smell-guided behavior.

Sato and Touhara present a review of the functional anatomy of the insect olfactory system and discuss some remarkable similarities to the vertebrate system despite the evolutionary independent origins. The authors describe the complete olfactory receptor gene repertoire in the fruitfly and compare it to that of several other insect species. As many as 62, 79, 131, 157, 48, and 265 ORs have been identified in *Drosophila*, *Anopheles*, *Aedes*, *Apis*, *Bombyx*, and *Tribolium*, respectively. In adult *Drosophila* about 1,300 olfactory receptor neurons are housed in about 500 sensilla of three different subtypes expressing 40 different ORs, whereas larvae possess just 21 olfactory receptor neurons expressing 25 ORs. The majority of receptor neurons express a single OR (some do express up to three different ORs), together with the ubiquitous OR 83b, which forms a heterodimer with many of the unique ORs and is involved in signal transduction. Even co-expression of OR and a taste receptor are observed for some olfactory sensory neurons. The tuning-curves of olfactory receptor neurons (ORN) are discussed, the most obvious difference to vertebrate ORN being a notable spontaneous activity allowing for activation as well as inhibition as odor response. Unlike vertebrate OR, insect OR form directly gated ion channels for signal transduction. Olfactory-guided behavior is discussed and technical applications in pest control are represented with the example of the insect repellent N,N-diethyl-3-methylbenzamide (DEET) and its behavioral and molecular mechanism of action.

Gerber, Stocker, Tanimura and Thum use *Drosophila* to elucidate the generation of behavior from olfactory and gustatory sensation. The functional anatomy of *Drosophila* olfactory receptor neurons is described both for mature flies and larvae, which emerge as simpler model system with fewer olfactory receptors and with attraction and repulsion as easily testable, behavioral outcomes.

Although insect taste cells are neurons, unlike their vertebrate counterparts, their responses can be categorized in the same modalities vertebrates possess, plus an additional sensitivity to water. Gerber et al. give a detailed description of the regulation of distinct behaviors by multiple taste organs distributed over the fly's body. The authors point out that the largest differences between olfaction and gustation do not lie in the peripheral sensation but in the central processing leading to selection of appropriate motor behaviors. Central olfactory pathways in mushroom body and lateral horn, i.e., beyond the antennal lobe, the insect equivalent of the olfactory bulb of vertebrates, are characterized by a certain segregation of pheromone processing vs normal food odors. The role of mushroom body Kenyon cells as coincidence detectors is explored. The existence of a discrete CNS pathway for encoding experience-dependent changes in olfactory behavior is shown. In contrast, gustatory information seems to bypass the brain proper, being received by the subesophageal ganglion, from which premotor commands likely can be triggered directly. A discussion of olfactory and gustatory learning includes conclusions from mutant studies and examines convergence of olfactory and gustatory information in the brain.

Vigues, Dotson and Munger discuss the molecular mechanism of sweet taste in mammals. Due to its distinct hedonic value, and the associated flip-side, overingestion of sugars associated with obesity and obesity-related diseases, sweet taste is of large interest to neuroscientists, dieticians and others, including the general public. The sweet taste receptor, a heterodimer of two class C G protein-coupled receptors, T1R2 and T1R3, responds to a vast array of chemically diverse natural and artificial sweeteners. Natural sweeteners come from several chemical classes, including sugars, sugar alcohols, proteins and amino acids, whereas synthetic sweeteners include sulfamates, dipeptides, halogenated sugars and sulfonyl amides. Mammalian species vary strongly in their sweetener preference, similarly, polymorphisms within species lead to large differences in sensitivity to sweeteners. Such polymorphisms in inbred mice strains have in fact led to the molecular identification of the sweet taste receptors. Receptor chimeras have identified the extracellular domain, the cysteine-rich domain and the transmembrane domain as sites of interaction with different sweeteners. Modeling the T1R heterodimer structure has provided evidence for allosteric interactions being involved in sweetener action and allowed further insights in the location of the ligand binding sites.

Behrens and Meyerhof discuss mammalian bitter taste perception and summarize our current knowledge. The authors describe results of taste cell and taste fiber responses to tastants, and compare them to the large array of data obtained for heterologously expressed taste receptors. Multiple TAS2R (synonym T2R) genes are coexpressed in individual bitter taste receptor cells, thus creating taste cells with broader agonist spectra than any given receptor responds to. The heterotrimeric G protein composition for bitter taste transduction is given as Gα-gustducin, Gβ3, and Gγ13, with the beta/gamma subunits activating phospholipase Cβ2. PLCβ2 is an essential molecule in taste signal transduction causing increases in IP_3 levels, which in turn lead to rising cytosolic calcium concentrations, and eventually to activation of the transient receptor potential channel M5. Structure-function investigations have shed some light on the tuning curves of several TAS2R, which range between highly promiscuous and specific for particular chemical substructures. The genetic variability of taste receptors is explored, which generates a heterogenous human population that contains tasters and non-tasters for several compounds, phenylthiocarbamide (PTC) being most famous among them.

Passilly-Degrace, Gaillard and Besnard expand the taste world beyond the established categories of sweet, sour, bitter, and salt perception to explore the sensation of lipids. Lipid-rich food is spontaneously preferred by both rodents and humans. Although a necessity in times of food deficiency as the optimal source of energy, lipid-rich food is also preferred in times of affluence, unforeseen as a stable state by Nature. Thus, overconsumption of energy-dense fats is a major cause of obesity in industrialized countries. Fats appear to be sensed as fatty acids that are liberated from triglycerides by lingual lipase. Among three candidates for a long chain fatty acid receptor, the authors make the case for the receptor-like glycoprotein CD36, whereas another candidate, the delayed-rectifying potassium channel Kv1.5, appears less likely to be involved in lipid sensing. A third candidate, the G protein-coupled receptor, GPR120, which is a receptor for unsaturated fatty acids, requires further analysis to confirm its status as lipid sensor. The authors discuss the orosensory mechanisms of fat detection with emphasis on CD36 signal transduction.

In the last chapter, Yasuoka and Abe present a summary of the taste system in fish, which in some species have much higher sensitivity than that of mammals. Taste buds are distributed over nearly the entire body of fish and are innervated by three cranial nerves. The authors discuss the evolution of V2R-related taste receptors (T1Rs) in several fish species including a fish-specific expansion of subfamily T1R2, and the comparatively small fish T2R repertoire, together with molecules involved in signal transduction such as phospholipase Cβ2 and the transient receptor potential channel TRPM5. Mutually exclusive expression of T1R and T2R receptors, as well as heterodimer formation within families mimics the situation in mammals. The authors continue to describe the ligand spectra of fish taste receptors, and point out that the fish orthologs of mammalian sweet taste receptors (T1R) recognize amino acids. Despite the differences in agonist spectra, T1R activation appears to be similarly linked to attractive behaviors in both mammals and fishes. Similarly, in both phyla activation of T2R bitter receptors lead to aversion, i.e., rejection of food.

The current knowledge of the genetics, molecular biology, and neurobiology of the several distinct chemosensory systems along with the insight into the molecular architecture of the various chemoreceptor molecules and the functional connectivity of the cells processing chemosensory information summarized in this edition has formed a solid basis for identifying challenging research topics for the period to come. Such challenges will include cracking the neural codes and understanding how chemosensory information triggers behavioral outputs. To this end, new experimental tools need to be developed such as novel genetically engineered strains of mice, fish and insects, molecules for neuroanatomical tracing, in vivo imaging systems, genetic reporters of neuronal activity, and in silico computation. In this sense, the editors wish that the present book serves researchers who are new in the field as a guide to our current knowledge and inspire those already involved to design future research activities.

December 2008 Wolfgang Meyerhof
Sigrun Korsching

Contents

**Extraordinary Diversity of Chemosensory
Receptor Gene Repertoires Among Vertebrates**.................................. 1
P. Shi and J. Zhang

Genomics of Olfactory Receptors 25
Xiaohong Zhang and Stuart Firestein

**The Molecular Evolution of Teleost Olfactory
Receptor Gene Families**.. 37
Sigrun Korsching

**Odorant Receptor Gene Choice and Axonal Projection
in the Mouse Olfactory System** 57
T. Imai and H. Sakano

Pheromone Sensing in Mice.. 77
I. Rodriguez and U. Boehm

**Molecular Genetic Dissection of the Zebrafish
Olfactory System**... 97
Y. Yoshihara

**Insect Olfaction: Receptors, Signal Transduction,
and Behavior**.. 121
K. Sato and K. Touhara

Smelling, Tasting, Learning: *Drosophila* as a Study Case 139
B. Gerber, R.F. Stocker, T. Tanimura, and A.S. Thum

The Receptor Basis of Sweet Taste in Mammals.................................. 187
S. Vigues, C.D. Dotson, and S.D. Munger

Mammalian Bitter Taste Perception 203
M. Behrens and W. Meyerhof

Orosensory Perception of Dietary Lipids in Mammals 221
P. Passilly-Degrace, D. Gaillard, and P. Besnard

Gustation in Fish: Search for Prototype of Taste Perception.............. 239
A. Yasuoka and K. Abe

Index... 257

Contributors

Keiko Abe
Graduate School of Agricultural and Life Sciences, Department of Applied Biological Chemistry, The University of Tokyo, 1-1-1 Yayoi, Bunkyo-ku, Tokyo 113-8657, Japan
aka7308@mail.ecc.u-tokyo.ac.jp

Maik Behrens
German Institute of Human Nutrition Potsdam-Rehbruecke, Arthur-Scheunert-Allee 114–116, 14558 Nuthetal, Germany
behrens@dife.de

Philippe Besnard
Physiologie de la Nutrition, UMR INSERM U 866, Ecole Nationale Supérieure de Biologie Appliquée à la Nutrition et à l'Alimentation (ENSBANA), Université de Bourgogne, 1, Esplanade Erasme, 21000 Dijon, France
pbesnard@u-bourgogne.fr

Ulrich Boehm
Center for Molecular Neurobiology, Institute for Neural Signal Transduction, Falkenried 94, 20251 Hamburg, Germany
ulrich.boehm@zmnh.uni-hamburg.de

Cedrick D. Dotson
Department of Anatomy and Neurobiology, University of Maryland School of Medicine, Baltimore, MD 21210, USA
cdots001@umaryland.edu

Stuart Firestein
Department of Biological Sciences, Columbia University, New York, NY, USA

Dany Gaillard
Physiologie de la Nutrition, UMR INSERM U 866, Ecole Nationale Supérieure de Biologie Appliquée à la Nutrition et à l'Alimentation (ENSBANA), Université de Bourgogne, 1, Esplanade Erasme, 21000 Dijon, France

Bertram Gerber
Universität Würzburg, Lehrstuhl für Genetik und Neurobiologie, Biozentrum,
Am Hubland, 97074 Würzburg, Germany
bertram.gerber@biozentrum.uni-wuerzburg.de

Takeshi Imai
Graduate School of Science, Department of Biophysics and Biochemistry,
The University of Tokyo, Tokyo 113-0032, Japan

Sigrun Korsching
Institut für Genetik, Universität zu Köln, Germany
sigrun.korsching@uni-koeln.de

Wolfgang Meyerhof
German Institute of Human Nutrition Potsdam-Rehbruecke, Arthur-Scheunert-Allee 114–116, 14558 Nuthetal, Germany
meyerhof@dife.de

Steven D. Munger
Department of Anatomy and Neurobiology, University of Maryland School of Medicine, Baltimore, MD 21210, USA
smung001@umaryland.edu

Patricia Passilly-Degrace
Physiologie de la Nutrition, UMR INSERM U 866, Ecole Nationale Supérieure de Biologie Appliquée à la Nutrition et à l'Alimentation (ENSBANA), Université de Bourgogne, 1, Esplanade Erasme, 21000 Dijon, France

Ivan Rodriguez
Department of Zoology and Animal Biology, University of Geneva,
30 Quai Ernest Ansermet, 1211 Geneva, Switzerland
Ivan.Rodriguez@zoo.unige.ch

Hitoshi Sakano
Department of Biophysics and Biochemistry, Graduate School of Science, University of Tokyo, 2-11-16 Yayoi, Bunkyo-ku, Tokyo 113-0032, Japan
sakano@mail.ecc.u-tokyo.ac.jp

Koji Sato
Department of Integrated Biosciences, The University of Tokyo, 5-1-5 Kashiwanoha, Kashiwa, Chiba 277-8562, Japan
satou-kouji@k.u-tokyo.ac.jp

Peng Shi
State Key Laboratory of Genetic Resources and Evolution, Kunming Institute of Zoology, The Chinese Academy of Sciences, Kunming 650223, China
ship@mail.kiz.ac.cn

Contributors

Reinhard F. Stocker
Department of Biology, University of Fribourg, 10, Chemin du Musée,
1700 Fribourg, Switzerland
reinhard.stocker@unifr.ch

Teiichi Tanimura
Graduate School of Sciences, Department of Biology, Kyushu University,
Ropponmatsu, Fukuoka 810-8560, Japan
tanimura@rc.kyushu-u.ac.jp

Andreas S. Thum
Department of Biology, University of Fribourg, 10, Chemin du Musée,
1700 Fribourg, Switzerland
andreas.thum@unifr.ch

Kazushige Touhara
Department of Integrated Biosciences, The University of Tokyo,
5-1-5 Kashiwanoha, Kashiwa, Chiba 277-8562, Japan
touhara@k.u-tokyo.ac.jp

Stephan Vigues
Department of Anatomy and Neurobiology, University of Maryland School
of Medicine, Baltimore, MD 21210, USA
svigu001@umaryland.edu

Akihito Yasuoka
Department of Biological Engineering, Maebashi Institute of Technology,
Gunma 371-0816, Japan

Yoshihiro Yoshihara
Laboratory for Neurobiology of Synapse, RIKEN Brain Science Institute,
2-1 Hirosawa, Wako, Saitama 351-0198, Japan
yoshihara@brain.riken.jp

Jianzhi Zhang
Department of Ecology and Evolutionary Biology, University of Michigan,
1075 Natural Science Building, 830 North University Avenue, Ann Arbor,
MI 48109, USA
jianzhi@umich.edu

Xiaohong Zhang
Department of Biological Sciences, Columbia University, New York, NY, USA

Extraordinary Diversity of Chemosensory Receptor Gene Repertoires Among Vertebrates

P. Shi and J. Zhang

Abstract Chemosensation (smell and taste) is important to the survival and reproduction of vertebrates and is mediated by specific bindings of odorants, pheromones, and tastants by chemoreceptors that are encoded by several large gene families. This review summarizes recent comparative genomic and evolutionary studies of vertebrate chemoreceptor genes. It focuses on the remarkable diversity of chemoreceptor gene repertoires in terms of gene number and gene sequence across vertebrates and the evolutionary mechanisms that are responsible for generating this diversity. We argue that the great among-species variation of chemoreceptor gene repertoires is a result of adaptations of individual species to their environments and diets.

1 Introduction

Chemosensation is responsible for the detection of chemicals in the external environment and is essential for an organism's survival and reproduction (Prasad and Reed 1999). Chemosensation originated very early in evolution, as even bacteria can respond to chemical changes in the environment. This type of chemoreception is known as the general chemical sense and is universal among organisms (Smith 2000). In this review, however, we will not study this general chemical sense. Instead, we will focus on two types of chemoreception that are animal-specific: olfaction (detection of odorants and pheromones) and gustation (detection of tastants). Owing to space limitations, we will only discuss vertebrates. It is widely thought that chemoreception plays multiple important roles in a vertebrate's daily life, including food detection and discrimination, toxin and predator avoidance, mating, and territoriality (Prasad and Reed 1999). Vertebrate chemosensory

---

P. Shi (✉)
State Key Laboratory of Genetic Resources and Evolution, Kunming Institute of Zoology,
The Chinese Academy of Sciences, Kunming 650223, China
email: ship@mail.kiz.ac.cn

systems include the olfactory system, which detects odorants and pheromones in the nasal cavity, and the gustatory system, which perceives different tastants with the tongue. Within the olfactory system there are two anatomically distinct organs: the main olfactory epithelium (MOE) and the vomeronasal organ (VNO) (Dulac and Torello 2003). It was initially thought that the MOE and the VNO have distinct functions, as the MOE is largely responsible for the detection of ordinary odorants, while the VNO detects pheromones (Dulac 1997; Buck 2000), although the current view is that the two systems can both detect odorants and pheromones. For the gustatory system, the tongue can perceive five basic tastes: sour, salty, bitter, sweet, and umami (Kinnamon and Margolskee 1996; Lindemann 2001). Among them, the sweet and umami tastes can influence appetitive reactions and generally reflect the identification of nutrients, whereas the bitter taste may result in aversion and therefore is a defensive mechanism against ingestion of toxins (Herness and Gilbertson 1999).

The ability of the chemosensory system to detect a diverse array of chemicals is mediated by the distinct chemoreceptors encoded by several gene families. The characterization of chemoreceptor genes started in 1991 with the Nobel-prize-winning discovery of 18 rat odorant receptor (OR) genes (Buck and Axel 1991). ORs have seven transmembrane domains and belong to the G-protein-coupled receptor (GPCR) family A. It has been proposed that the potential odorant-binding pocket is formed by the third, fifth, and sixth transmembrane domains (Emes et al. 2004). OR genes have no introns and the coding region of each gene has about 1,000 nucleotides. They are mainly expressed in sensory neurons of MOEs. It is widely accepted that a single OR allele is expressed in each olfactory sensory neuron, known as the "one neuron—one gene" hypothesis (Mombaerts 2004). Recently, trace amine-associated receptors (TAARs) were demonstrated to be the second class of chemosensory receptors in the MOE (Liberles and Buck 2006). TAARs and ORs share many features, including the gene structure and expression profile (Liberles and Buck 2006), but TAARs and ORs are not coexpressed in any neurons (Liberles and Buck 2006).

Two distinct superfamilies of GPCRs, V1Rs and V2Rs, have been identified as vomeronasal receptors (Dulac and Axel 1995; Herrada and Dulac 1997; Matsunami and Buck 1997; Ryba and Tirindelli 1997). Like ORs and TAARs, V1R genes have intronless coding regions. They are coexpressed with the G-protein subunit $G_{\alpha i2}$ in sensory neurons whose cell bodies are located in the apical part of the vomeronasal epithelium (Dulac and Torello 2003; Mombaerts 2004). In contrast, V2Rs are characterized by the presence of a long, highly variable N-terminal domain. They are encoded by multiexon genes expressed in $G_{\alpha o}$-positive neurons whose cell bodies are located basally in the vomeronasal epithelium (Dulac and Torello 2003; Mombaerts 2004). Neurons expressing V1R and V2R receptors project to the anterior and posterior accessory olfactory bulb, respectively, where they form multiple glomeruli in spatially conserved domains (Dulac and Torello 2003). Interestingly, the four olfactory-system-related gene families (OR, TAAR, V1R, and V2R) are not evolutionarily related, although all of them belong to GPCRs.

In mammals, T1Rs and T2Rs have been identified as sweet/umami and bitter taste receptors, respectively. In additional to the distinct physiological functions, T1Rs and T2Rs also differ in expression pattern and molecular structure. Multiple T2Rs are coexpressed in individual cells that also express α-gustducin, a G-protein subunit (Adler et al. 2000; Nelson et al. 2001; Behrens et al. 2007). T1Rs, on the other hand, are not coexpressed with α-gustducin. T1R1 and T1R2 are expressed in distinct taste receptor cells, but they are always coexpressed with T1R3. Consistent with the expression feature, T1R3 forms a heteromeric receptor with T1R1 to detect L-amino acids and monosodium L-glutamate, which is the taste of umami, or combines with T1R2 to broadly respond to sweet tastants (Nelson et al. 2001; Li et al. 2002). Different from T1R genes, which contain multiple introns, T2R genes are intronless in the coding region. T1R proteins are characterized by a long N-terminal extracellular domain, whereas T2Rs have a short N-terminal domain (Hoon et al. 1999; Adler et al. 2000; Meyerhof 2005).

Chemoreception varies substantially among vertebrates, probably because of the tremendous diversity of chemical stimuli in the external environments of various species. We summarize in this review recent comparative genomic studies on the variation of gene number and gene sequence of chemoreceptor gene families among vertebrates. We also discuss the genetic mechanisms and evolutionary consequences of these features with an emphasis on the adaptive diversification of chemosensory receptor genes.

2 Chemoreceptor Gene Families in the Main Olfactory System

2.1 The OR Gene Family – the Largest Gene Family in Mammals

The OR gene family is known to be the largest gene family in the vertebrate genome. Since the original discovery in 1991 (Buck and Axel 1991), OR genes have been partially cloned from many vertebrates (Mombaerts 1999). The first near-complete OR gene repertoire was not unveiled until 2001, when the draft human genome sequence became available (Glusman et al. 2001; Zozulya et al. 2001). Since then, the complete OR gene repertoires have been characterized in several major vertebrate lineages for which the genome sequences are available, including teleosts (pufferfish, fugu, and zebrafish), amphibians (frog), birds (chicken), and mammals (human, mouse, rat, dog, cow, opossum, and platypus) (Alioto and Ngai 2005; Niimura and Nei 2005, 2006, 2007; Grus et al. 2007). As shown in Table 1, the number of functional OR genes varies greatly among species, ranging from 44 genes in fugu to over 1,200 genes in rat (Alioto and Ngai 2005; Niimura and Nei 2005, 2007). The gene number is substantially smaller in fishes than in birds and mammals, while that of the amphibian frog appears to be in the middle. The largest known fish functional OR gene repertoire is in zebrafish, with at least 102 genes (Alioto and Ngai 2005; Niimura and Nei 2005).

However, this number is still much lower than that in mammals, even when we consider only human and platypus, two mammals that are believed to have reduced main olfactory sensitivity (Niimura and Nei 2006; Grus et al. 2007). Variation in the number of functional OR genes also exists among species of the same class. In mammals, the smallest numbers are 387 and 262, found in human and platypus, respectively (Young and Trask 2002; Niimura and Nei 2003, 2007 Malnic et al. 2004; Grus et al. 2007), while the largest numbers are 1,207 and 1,188, for rat and opossum, respectively (Niimura and Nei 2007).

Phylogenetic analysis helps us understand the evolutionary history and mechanism of the extraordinary diversity of the vertebrate OR gene family. The phylogenetic tree shows that the OR gene family can be classified into two groups, type 1 and type 2 (Fig. 1a). The divergence of these two types predated the split between jawed vertebrates and jawless vertebrates (Fig. 1b). The phylogenetic analysis revealed at least

Table 1 Sizes of chemosensory receptor gene repertoires in vertebrates

Species	OR	TAAR	V1R	V2R	T1R	T2R
Human	387(415)[a]	6(3)[b]	5(115)[c]	0(20)[d,e]	3[f]	25(11)[f]
Mouse	1,035(356)[g]	15(1)[b]	191(117)[h]	121(158)[d,e]	3[f]	35(6)[f]
Rat	1,207(560)[g]	17(2)[b]	117(72)[h]	79(142)[d,e]	3[f]	37(5)[f]
Dog	811(289)[g]	2(2)[b]	8(33)[c,i]	0(9)[d,e]	3[f]	15(5)[f]
Cow	970(1,159)[g]	17(9)[b]	40(45)[c,i,d]	0(16)[d,e]	3[f]	12(15)[f]
Opossum	1,188(295)[g]	22(0)[b]	98(30)[d,i]	86(79)[d,e]	3[f]	26(5)[f]
Platypus	262(315)[g,i]	4(1)[b]	270(579)[b]	15(112)[b]	ND	ND
Chicken	82(476)[j,k]	3(0)[b]	0(0)[d]	0(0)[d]	2[f]	3(0)[f]
Frog	410(478)[j,k]	2(1)[b]	21(2)[d]	249(448)[d]	0[f]	49(12)[f]
Fugu fish	44(13)[l]	13(6)[m]	5(0)[n]	18(29)[d]	4(1)[f]	4(0)[f]
Pufferfish	44(54)[j,k]	ND	5(0)[n]	4(21)[d]	5(1)[f]	6(0)[f]
Zebrafish	102(35)[j,k]	109(10)[m]	6(0)[n]	44(8)[d]	1[f]	4(0)[f]

The number of nonintact genes, containing truncated genes and pseudogenes, is shown in *parentheses*.
OR odorant receptor, *TAAR* trace amine-associated receptor, *ND* not determined
[a]From Niimura and Nei (2003)
[b]From Grus et al. (2007)
[c]From Young et al. (2005)
[d]From Shi and Zhang (2007)
[e]From Young and Trask (2007)
[f]From Shi and Zhang (2006)
[h]From Zhang et al. (2007)
[i]From Grus et al. (2005)
[j]From Niimura and Nei (2005)
[k]From Niimura and Nei (2006)
[l]From Alioto and Ngai (2005)
[m]From Hashiguchi and Nishida (2007)
[n]From Saraiva and Korsching (2007)

nine ancestral OR genes (or gene lineages) in the most recent common ancestor (MRCA) of fishes and tetrapods (Niimura and Nei 2005) (Fig. 1b). Eight of the nine ancestral gene lineages have been maintained in fishes (Fig. 1b), probably because the extant teleosts share a similar environment with the MRCA of fishes and tetrapods. By contrast, only two ancestral gene lineages have been retained in mammals or birds, but these gene lineages have expanded extensively in mammals and birds, giving rise to the largest gene family in the mammalian genome. By contrast, these two gene lineages, although present in today's fishes, have not expanded in evolution (Fig. 1b). On the other hand, four gene lineages that are present in fishes have been lost completely in mammals and birds (Fig. 1a,b). Taken together, these observations show that the long-term evolutionary dynamics of OR genes follows the "birth-and-death" process, characterized by frequent gene duplication and gene loss (Nei et al. 1997). The large amount of turnover of OR genes in vertebrate evolution probably reflects the functional requirement for different olfactory abilities in different evolutionary lineages. This view is further supported by the fact that frogs have both mammal-like and fish-like OR genes (Niimura and Nei 2005).

Many OR gene gains and losses have also been observed within mammals. Comparative genomic and phylogenetic analyses show that gene family expansions occurred independently in monotremes, marsupials, and placental mammals. Consequently, many lineage-specific genes are observed in today's mammalian genomes (Fig. 1c). The largest gene family expansion occurred in the marsupial lineage, with at least 750 gene gains. Similarly, more than 400 genes were gained in the cetartiodactyl and rodent lineages (Fig. 1c). On the other hand, the number of gene losses in the primate lineage is much greater than that in other lineages (Niimura and Nei 2007). As shown in Fig. 1c, since the human—mouse split, 385 ancestral genes have been inactivated in the human lineage, while only 55 new OR genes have been acquired. In the mouse lineage, the two numbers are 277 and 623, respectively. These gene gains and losses explain the dramatic difference in OR gene family size between human and mouse. Furthermore, even when the gene number is similar between two species, the gene content may differ, owing to rapid gene gains and losses (Niimura and Nei 2007).

OR gene number variation also exists among closely related species. Rouquier et al. (2000) found a higher fraction of OR pseudogenes in ten primates than in mice. A subsequent study by Gilad et al. (2004), based on an analysis of 100 orthologous OR genes in 19 nonhuman primate species, found that the percentage of OR pseudogenes is significantly higher in humans, apes, and Old World monkeys than in most New World monkeys and mice (Gilad et al. 2004). Interestingly, in the howler monkey, the only New World monkey with a full trichromatic vision as humans, apes, and Old World monkeys have, approximately 30% of the genes are pseudogenes, similar to the number in Old World monkeys. On the basis of this result, the authors suggested that the loss of OR genes in humans, apes, and Old World monkeys is a result of the acquisition of trichromatic vision. On the basis of the analysis of 50 orthologous OR genes, Gilad et al. (2003) suggested that humans also have fewer functional genes and more pseudogenes than chimpanzees, which was later substantiated by a genome-wide comparison of human and chimpanzee

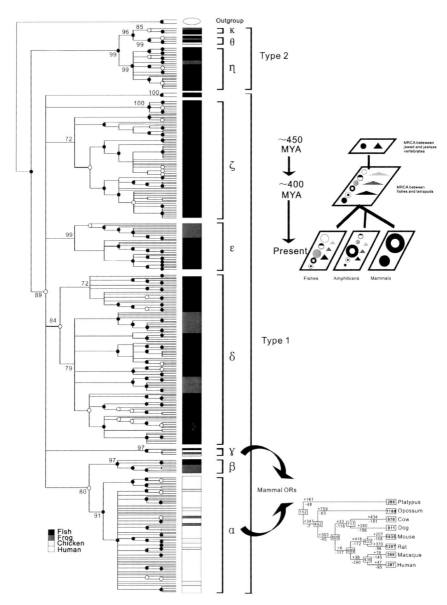

Fig. 1 Evolution of vertebrate odorant receptors (*ORs*). **a** Condensed phylogenetic tree for 310 functional OR genes from fishes, frogs, chickens, and humans at the 70% bootstrap level. The nine major clades are labeled α, β, γ, δ, ε, ζ, η, θ, and κ. Human class I genes are within the α clade and human class II genes are within the γ clade. *Open circles* and *closed circles* at nodes represent branches with bootstrap values greater than 90% and greater than 80%, respectively. **b** Evolutionary dynamics of vertebrate OR genes. There are at least nine ancestral genes in the most recent common ancestor (MRCA) of fishes and tetrapods. Fishes maintain eight of nine ancestral genes, whereas mammals contain only two of them. c OR gene gains and losses in mammals. The *numbers with plus and minus signs* for each branch indicate the numbers of gene gains and losses, respectively, and the *numbers in rectangular boxes* denote the functional OR genes for the extant or ancestral species. *MYA* million years ago. (**a, c** Modified from Niimura and Nei 2003. **b** Modified from Niimura and Nei 2007)

OR repertoires (Gilad et al. 2005). More recently, the variation of OR gene number was also identified among different human individuals, which is known as the copy number variation (Trask et al. 1998; Wong et al. 2007). Interestingly, the level of interspecific divergence relative to that of intraspecific variation in OR gene copy number is not significantly different between functional genes and pseudogenes, suggesting that human intraspecific and human—chimpanzee interspecific OR gene number variations may not have any fitness consequence (Nozawa et al. 2007; Zhang 2007).

In addition to the vast variation in gene family size among vertebrates, the remarkable diversity of ORs is also reflected in the high sequence (and potentially functional) variation among alleles found within species. For example, in humans, pygmy populations tend to have higher frequencies of intact alleles than Caucasians in 32 OR genes examined (Gilad and Lancet 2003). A further study of 51 human OR loci in 189 ethnically diverse individuals reached the same conclusion and suggested that different evolutionary forces may have shaped the OR repertoire in different human populations (Menashe et al. 2003). Similarly, great allelic diversity was found in different mouse strains and dog breeds, respectively (Zhang et al. 2004; Tacher et al. 2005). In mice, single nucleotide polymorphisms (SNPs) were counted by comparing two mouse genome sequences, which were derived from different strains (Zhang et al. 2004). It was estimated that there are 2.68 SNPs per OR gene coding region, about twice that in other mouse GPCR genes (Zhang et al. 2004). In dogs, a high level of allelic variation among 20 different breeds was observed in a survey of 16 OR genes among 95 individuals. All genes were found to have SNPs and 50% of SNPs are nonsynonymous. More interestingly, some SNPs are breed-specific and they may be the basis of breed-specific olfactory sensitivity (Tacher et al. 2005).

2.2 The TAAR Gene Family – the Second Class of Olfactory Receptor Gene Family

TAAR genes were initially identified to respond to trace amines in rodents and were later shown to be chemosensory receptors in the MOE (Borowsky et al. 2001; Liberles and Buck 2006). Gloriam et al. (2005) performed the first genome-wide investigation in zebrafish and identified 57 intact TAAR genes, which is almost 10 times the number in human. More recently, a comprehensive scan of ten vertebrate genomes found a large variation in the size of this gene family among vertebrates (Hashiguchi and Nishida 2007). In sharp contrast to the OR gene family, which is larger in tetrapods than in fishes, the TAAR gene family is smaller in tetrapods than in some fishes such as zebrafish and stickleback (Table 1). The largest TAAR gene repertoire, found in zebrafish, has 102 intact genes, whereas the smallest repertoire, in chicken, has only three intact genes (Hashiguchi and Nishida 2007). The comparative genomic and phylogenetic analyses suggested that the large gene repertoires in some fishes are attributable to the genome duplication in teleosts followed by additional gene duplications (Hashiguchi and Nishida

2007). By contrast, in frog, chicken, and the majority of mammals, the TAAR family lost several ancestral genes but gained virtually no new members. Opossum, cow, mouse, and rat are exceptions, with some gene duplications (Fig. 2). These findings suggest that the TAAR gene family is also subject to the birth-and-death evolutionary process observed in the OR gene family and that biogenic amine odorants are more important for fish than for tetrapods (Hashiguchi and Nishida 2007).

3 Vomeronasal Receptor Gene Families

The vomeronasal system is present in most tetrapods (amphibians, reptiles, and mammals), but is absent in fishes (Dulac and Torello 2003). Thus, the nomenclature of the fish chemosensory receptors that are homologous to mammalian vomeronasal receptors has been confusing. On the basis of sequence homology, some authors termed them "V1R-like" and "V2R-like" genes (Hashiguchi and Nishida 2005, 2006; Pfister and Rodriguez 2005; Pfister et al. 2007), while some authors separately designated them as "olfactory receptor A family GPCR" and "olfactory receptor C family GPCR" by considering their expression pattern and phylogenetic position in the GPCR family (Alioto and Ngai 2006; Saraiva and Korsching 2007). Here we use the former terminology for three reasons. First, most, if not all, VR-like genes have been identified by sequence homology and their expression pattern and biological function are usually unknown. Second, although teleost fishes do not have a morphologically distinct vomeronasal organ, they may have a primordial vomeronasal system (Grus and Zhang 2006). Third, the latter nomenclature is also confusing and undistinguishable from that for other chemosensory receptors expressed in the MOE.

3.1 The V1R Gene Family – the Family with the Highest Among-Species Variation in Gene Family Size

In 1995, V1R genes were first identified in rats by comparative hybridization of complementary DNA libraries from individual VSNs (Dulac and Axel 1995). The first complete V1R repertoire was described for mouse by Rodriguez et al. (2002). In this work, they identified 137 functional V1R genes from the mouse draft genome sequence and subsequently classified them into 12 subfamilies according to protein sequence identity (Rodriguez et al. 2002). In human, the entire V1R repertoire, including functional genes and pseudogenes, has approximately 200 members. However, the functional V1R repertoire is small, with only four open reading frames in most individuals (Rodriguez and Mombaerts 2002; Zhang and Webb 2003). Grus et al. (2005) identified functional V1R genes from five orders of placental and marsupial mammals (Table 1). The intact V1R repertoire size varies by at least 23-fold among mammals with functional VNOs and this size ratio represents the greatest among-species variation in gene family size of all

Extraordinary Diversity of Chemosensory Receptor Gene Repertoires

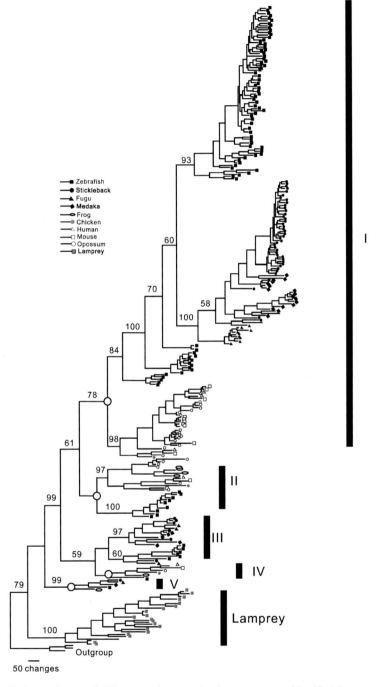

Fig. 2 Phylogenetic tree of 268 trace amine-associated receptor genes identified from ten vertebrates. The tree was reconstructed by the maximum-likelihood method using nucleotide sequences. Bootstrap percentages greater than 50 are shown on interior branches. *White dots* on nodes indicate the MRCA of fishes and tetrapods. (Modified from Hashiguchi and Nishida 2007)

mammalian gene families (Grus et al. 2005). Young et al. (2005) made similar findings. A more recent study found that the platypus, a semiaquatic monotreme, has the largest V1R repertoire characterized to date, with 270 potentially functional genes and 579 pseudogenes. Thus, the functional V1R repertoire size varies by at least 34-fold among mammals with functional VNOs (Grus et al. 2007). This large variation in V1R repertoire size is also observed between closely related species. For example, mouse has 191 functional V1R genes, about 50% more than the number in rat (117) (Shi et al. 2005; Young et al. 2005; Zhang et al. 2007). The number of V1R genes also varies tremendously among nonmammalian vertebrates (Shi and Zhang 2007). No V1R genes were found in chicken, consistent with the fact that birds have neither VNO nor VNO-mediated olfaction (Keverne 1999). A total of 21 functional genes and two pseudogenes was found in the western clawed frog (Shi and Zhang 2007). In contrast to mammals, fishes have highly conserved V1R-like repertoires, containing four genes in two pufferfish species and five genes in zebrafishes, stickleback, and medaka, respectively (Hashiguchi and Nishida 2006; Saraiva and Korsching 2007; Shi and Zhang 2007). Interestingly, the number of intact V1R genes is positively correlated with the morphological complexity of the VNO, suggesting that VNO morphology is a good indicator of vomeronasal sensitivity (Grus et al. 2005).

A phylogenetic analysis of all vertebrate V1Rs suggests that V1R genes can be divided into at least three major clades that diverged from one another before the separation of tetrapods and teleosts (Shi and Zhang 2007) (Fig. 3a). Clade 1 now contains genes from frog and mammals, but was lost in fishes. Clade 2 and clade 3 include frog and fish genes, which are absent in mammals. Major expansions of the V1R gene repertoire occurred in some mammals (clade 1), whereas minor expansions occurred in frog (clades 1 and 2) (Fig. 3a) (Shi and Zhang 2007). Very recently, two very divergent V1R-like genes were found in teleost fishes (Saraiva and Korsching 2007). It appears that the evolutionary diversity of V1R genes in fishes is much larger than that in tetrapods, although the gene family size is smaller in fishes than in tetrapods.

In mammals, both gene duplicate and pseudogenization have played important roles in generating the remarkable among-species variation in V1R gene repertoire (Grus et al. 2005; Young et al. 2005). On one hand, substantial numbers of gene duplication events occurred independently in monotremes, marsupials, and placentals, giving rise to platypus-specific, opossum-specific, and placental-specific gene clusters (Grus et al. 2005, 2007) (Fig. 3a). Following the initial gene duplications in the MRCA of placental mammals, additional expansions occurred most prominently in rodents, in which the evolution of the V1R repertoire is characterized by rapid gene turnover and species-specific phylogenetic clustering (Grus and Zhang 2004; Lane et al. 2004; Grus et al. 2005; Shi et al. 2005; Young et al. 2005) (Fig. 3a). Extreme examples include two subfamilies of V1R genes that appear in mouse but not in rat (Grus and Zhang 2004; Shi et al. 2005). On the other hand, human, cow, and dog lost many ancestral V1R genes. In humans, only three of the 12 ancestral family groups were observed. This is also the case in cow and dog, where at least four ancestral family groups are missing (Grus et al. 2005; Young et al. 2005).

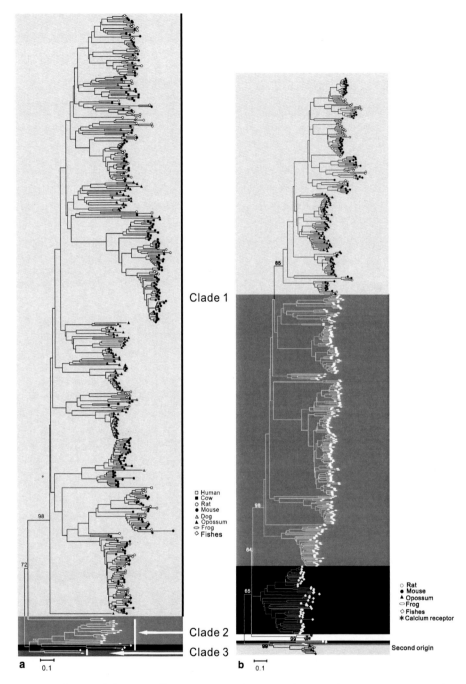

Fig. 3 Neighbor-joining trees of intact vomeronasal receptors from vertebrates. **a** The V1R tree. **b** The V2R tree. The trees were reconstructed with protein Poisson distances. Bootstrap percentages for some major groups are presented. Vomeronasal receptors from the mammals, frogs, and fishes are shown by *light grey background, dark grey background,* and *black background,* respectively. (Modified from Shi and Zhang 2007)

3.2 The V2R Gene Family – Independent Origins of Two Types of V2Rs

The first extensive characterization of any V2R gene repertoire was conducted in 2005 (Yang et al. 2005), 8 years after the initial identification of the gene family (Dulac 1997; Herrada and Dulac 1997; Ryba and Tirindelli 1997; Yang et al. 2005). Comparative genomic studies showed that the across-vertebrate variation in gene number for V2R genes is not lower than that for V1R genes. The largest V2R repertoire is unexpectedly found in frog, with 249 intact genes and 408 disrupted genes. By contrast, no intact V2R genes are found in chicken, cow, dog, and human. In addition, 4, 18, 44, 90, 120, and 70 intact V2R genes are present in the green spotted pufferfish, fugu, zebrafish, opossum, mouse, and rat, respectively (Shi and Zhang 2007; Young and Trask 2007) (Table 1).

In comparison with the V1R gene repertoire, the V2R repertoire is frequently lost in terrestrial vertebrates. There are at least three independent losses of the entire V2R repertoire in chicken, human, and cow/dog, respectively, which is consistent with the loss of certain morphological features of the VNO (Shi and Zhang 2007). By contrast, the V2R gene repertoire expanded in teleosts with prominent patterns of lineage-specific gene amplifications (Alioto and Ngai 2006; Hashiguchi and Nishida 2006; Shi and Zhang 2007) (Fig. 3b). The V2R repertoire also expanded in rodents and opossums and exhibit the characteristics of rapid gene turnover and species-specific gene clustering (Yang et al. 2005; Young and Trask 2007), as seen in V1Rs. Interestingly, the V2R gene family tree has an unique branching pattern, where not all V2R genes cluster in one monogenetic clade. The phylogenetic analysis shows that V2Rs of family C (also termed V2R2 subfamily) are quite different from those of families A and B and are evolutionarily closer to Ca^{2+}-sensing receptors than to V2Rs of families A and B. This observation suggests that family C and families A and B had independent origins (Yang et al. 2005) (Fig. 3b). This evolutionary history may explain the differences in expression pattern and transport mechanism between the two types of V2Rs and suggests that family C V2Rs may be functionally distinct from those of families A and B (Yang et al. 2005; Young and Trask 2007).

3.3 Diversity of Protein Families Interacting with Vomeronasal Receptors

In mice, there are two gene families that are known to function in concert with V2Rs. One of them is the M10 family of major histocompatibility class Ib molecules, which appear to function as escort molecules in the transport of some V2Rs to the cell membrane of vomeronasal sensory neurons (Loconto et al. 2003). The second is the exocrine gland peptide (ESP) family, which can activate the V2R-expressing vomeronasal sensory neurons and have been suggested to be ligands of some V2Rs (Kimoto et al. 2005). A recent study compared these two gene families in 11 vertebrates and found them to have been coevolving with V2Rs (Shi and Zhang 2007). Consistent with the absence of V2R genes in dog, cow, and human, neither M10 nor ESP genes are found in these species. Unexpectedly, however, M10 and ESP genes are not found in the opossum genome, despite the presence of numerous intact V2R genes, suggesting that the requirement of M10 molecules for the transport of some V2Rs to cell membranes is probably a rodent-specific phenomenon and the use of ESPs as potentially V2R-recognizing pheromones is also rodent-specific (Shi and Zhang 2007). Interestingly, these two gene families share with the V2R family the rapid birth-and-death evolutionary pattern. Very recently, major urinary proteins were identified as V2R-recognizing pheromones in mice and the major urinary protein family size was found to covary with the V2R family size across vertebrates (Chamero et al. 2007).

A similar story can be told for the transient receptor potential channel C2 (TRPC2) gene, which encodes an ion channel indispensable for vomeronasal signal transduction. TRPC2 is absent in the catarrhine primates (humans, apes, and Old Word monkeys), which possess only vestigial VNOs and have no or significantly reduced ability of pheromone detection (Liman and Innan 2003; Zhang and Webb 2003). Consistently, the majority of V1R genes and all V2R genes have disrupted open reading frames in catarrhine primates (Zhang and Webb 2003; Shi and Zhang 2007; Young and Trask 2007). Similarly, the lack of the TRPC2 gene and V1R and V2R genes is observed in chicken, reflecting the ancient loss of the VNO in birds. Conversely, the TRPC2 open reading frame is maintained in all vertebrates known to have functional vomeronasal receptors (Grus and Zhang 2006).

4 Taste Receptor Gene Families

4.1 T2R Gene Family – the More Variable Group of Taste Receptors

Taste receptor genes were the last chemoreceptor genes to be isolated. In 2000, the T2R gene family (also known as TRBs or Tas2Rs) was identified and two mouse T2R genes were shown to be bitter taste receptors (Adler et al. 2000; Chandrashekar et al. 2000; Matsunami et al. 2000). To date, the complete T2R gene repertoires have been described in mammals, birds, amphibians, and some fishes (Conte et al. 2002, 2003; Go 2006; Shi and Zhang 2006). In addition, a small number of T2R genes have also been described in several nonhuman primates (Parry et al. 2004; Wang et al. 2004; Fischer et al. 2005; Go et al. 2005). As shown in Table 1, the T2R gene repertoire varies extremely among vertebrates, ranging from three genes in chicken to 50 genes in amphibians (Go 2006; Shi and Zhang 2006). This observation is consistent with the fact that bitter taste perception, as a mechanism of guarding against the ingestion of toxins, varies enormously among vertebrates that have different diets and environments. Most interestingly, the comparative genomic analysis shows that the size of the gene family appears to be positively correlated with the number of bitter toxins that an organism is likely to encounter (Shi and Zhang 2006). Omnivorous mammals tend to have the largest T2R gene repertoires and the lowest fractions of T2R pseudogenes, probably because they consume both animal and plant tissues and consequently encounter more toxic compounds than herbivorous and carnivorous mammals do. By contrast, carnivores have a small number of functional T2R genes than herbivores, because animal tissues contain fewer toxins than plant tissues do (Shi and Zhang 2006). Cow was found to have the highest proportion of T2R pseudogenes (44%), suggesting that detecting poisons in the diet is not as important in ruminants as in other animals, probably owing to the detoxification role of cow's rumen microbes (Shi and Zhang 2006). These hypotheses need to be scrutinized in more mammalian species.

A phylogenetic analysis of all vertebrate T2R genes suggests that there were multiple T2R genes in the common ancestor of tetropods and teleosts, because T2R genes from teleost fishes do not cluster into one monophyletic clade (Fig. 4a). In addition, the overall evolutionary pattern of vertebrate T2R genes follows the birth-and-death process, similar to that observed in several other chemoreceptor gene families. Specifically, the T2R gene repertoire expanded considerably in the common ancestor of tetrapods, followed by additional independent expansions in frogs and mammals and contraction in chicken (Go 2006; Shi and Zhang 2006). The comparative genomic analysis of human and mouse T2R genes shows that some T2R genes exhibit one-to-one orthologous pairing, whereas others form species (lineage) specific clusters, in which the genes from the same species cluster together in the phylogenetic tree. These species-specific genes are the results of tandem gene duplications and are probably used for detecting species-specific bitter tastants (Shi et al. 2003). One-to-one orthologous genes were found to be subject to stronger selective constraints than species-specific genes, suggesting that each of the one-to-one orthologous genes is possibly detecting one or several distinct bitter compounds that are encountered by a wide range of animals (Shi et al. 2003). This still requires further verification by functional analysis of the receptors, although two recent evolutionary studies (Go 2006; Shi and Zhang 2006) that extended the study of T2Rs to nine additional vertebrate species supported the hypothesis.

Comparative analysis of the T2R gene family between several species of primates and rodents revealed that the genes were under reduced selective constraints in primates compared with rodents (Parry et al. 2004; Wang et al. 2004; Fischer et al. 2005; Go et al. 2005). The proportion of pseudogenes in the T2R repertoire is lower in mice (15%) than in apes (21–28%), which is in turn lower than that in humans (31%) (Fischer et al. 2005; Go et al. 2005). The prevalence of lineage- or species-specific pseudogenes in primates further supports this conclusion (Go et al. 2005). In addition, the ratio of nonsynonymous to synonymous substitution rates for T2R genes is lower in rodents than in primates (Wang et al. 2004; Fischer et al. 2005; Go et al. 2005). The most likely explanation is that primates have reduced bitter taste needs owing to changes in the environment and diet (Go et al. 2005). Actually, some ecological studies support this explanation. For instance, meat accounts for 2–13% of diet in chimpanzees, whereas it has never been found in the diet of other apes (Wang et al. 2004). Furthermore, there were significant changes in human diet, such as decreased intake of plant tissues and the controlled use of fire to detoxify food (Wang et al. 2004). Both factors may have caused a reduction in the importance of bitter taste and consequently triggered a functional relaxation on T2Rs in humans, as has been observed (Wang et al. 2004).

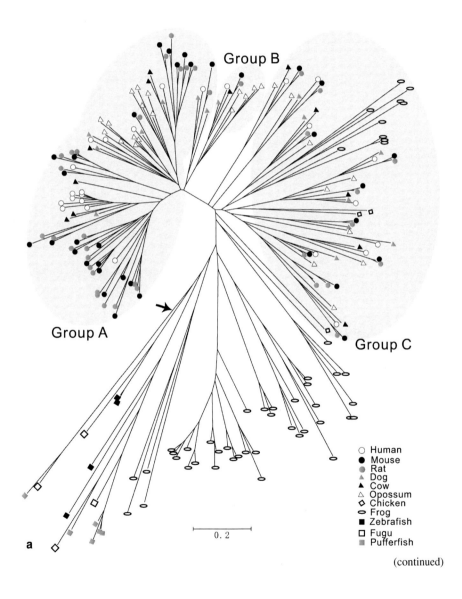

(continued)

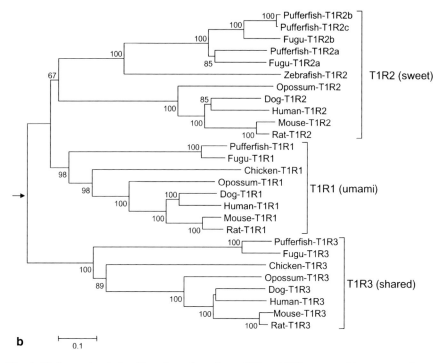

Fig. 4 Phylogenetic relationships of vertebrate intact T1R and T2R genes. **a** The T1R tree. **b** The T2R tree. The tree was reconstructed by the neighbor-joining method with protein Poisson distances. Bootstrap percentages greater than 50 are shown on interior branches. Groups A, B, and C in the T2R tree were previously defined on the basis of the tree of human and mouse T2R genes (Shi et al. 2003). The *arrow* indicates the root of the tree, which was determined by using vertebrate V1R genes as outgroups for T2R genes, and using V2R genes as outgroups in the T1R tree. (Reprinted from Shi and Zhang 2006, copyright 2006, with permission from Oxford University Press)

4.2 The T1R Gene Family – the More Conserved Group of Taste Receptors

In 1999, two GPCR genes that are now named T1R1 and T1R2 were described in subsets of taste receptor cells (Hoon et al. 1999). The third T1R gene, T1R3, was identified almost simultaneously by six groups in 2001 (Kitagawa et al. 2001; Max et al. 2001; Montmayeur et al. 2001; Nelson et al. 2001; Sainz et al. 2001; Zhao et al. 2003). In contrast to the T2R gene family, the T1R family is evolutionary conserved in both gene family size and gene sequence (Shi and Zhang 2006). In terms of the family size, the number of T1R genes is virtually constant across mammals, which might reflect the necessity of both sweet and umami tastes among mammals (Shi and Zhang 2006). But the number of T1R genes varies in some nonmammalian vertebrates, including gene duplications observed in pufferfish and stickleback and gene losses found in western clawed frog and chicken (Shi and Zhang 2006; Hashiguchi et al. 2007). Interestingly, the western clawed frog does not have any T1R genes, although it has many T2R genes. In chicken, the T1R2 gene is missing (Shi and Zhang 2006) (Fig. 4b). In addition, cats and closely related carnivores are also known to miss the T1R2 genes, which is likely the cause of the insensitivity to sweet tastants in these species (Li et al. 2005). Thus, pseudogenization of T1R2 happened multiple times in evolution.

At the protein sequence level, T1R genes evolve more slowly than T2R genes at both interspecific and intraspecific levels. For interspecific comparison, sequence divergence among orthologs is significantly lower for T1R genes than for T2R genes when human, mouse, rat, and opossum genes were compared (Shi and Zhang 2006). Within human populations, the mean pairwise difference per nucleotide between sequences of T1Rs is also lower than that for T2Rs (Kim et al. 2005, 2006).

In sum, the contrasting evolutionary modes between T1R and T2R gene families suggest the relative constancy in the number and type of sweet and umami tastants encountered by various vertebrates or low binding specificities of T1Rs but a large variation in the number and type of bitter compounds detected by different species.

5 Adaptive Diversification of Chemoreceptor Gene Repertoires

Why do the chemosensory receptor gene families vary so much among vertebrates? One potential answer is the variable functional requirements for different species to adapt to their specific environments. Here we summarize evidence for the adaptive hypothesis at three levels: (1) gene family, (2) newly duplicated paralogous genes, and (3) intraspecific variation.

Adaptive evolution at the gene family level may be detected by comparing the gene repertoires of nasal chemosensory receptors in terrestrial vertebrates with those in aquatic vertebrates because terrestrial vertebrates tend to encounter volatile chemicals, while aquatic vertebrates encounter water-soluble chemicals. A recent analysis of V1R, V2R, and OR gene families in several vertebrate genomes showed that the ratio of the number of intact V1R genes to that of intact V2R genes increased by approximately 50 fold in the evolutionary transition from water to land. Note that circumstantial evidence suggests that V1Rs tend to recognize airborne molecules, while V2Rs tend to recognize water-soluble ligands. The comparison of the number of class II ORs to that of class I ORs, which have been suggested to bind to volatile and water-solvable molecules, respectively, also shows a similar pattern of change during the evolutionary transition of vertebrates from aquatic to terrestrial environments. By contrast, a comparison of pairs of randomly chosen gene families from the zebrafish and mouse genomes does not show such dramatic changes, indicating that the observation made in the nasal chemoreceptor genes is unlikely caused by random gene turnovers (Shi and Zhang 2007). Rather, the two nasal chemosensory systems appear to show a consistent pattern of a shift from receptors for water-soluble molecules to those for volatiles in the vertebrate transition from water to land, reflecting a rare case of adaptation to terrestrial life at the gene family level.

Gene duplication is believed to be the primary source of new genes with novel functions (Zhang 2003). In chemoreceptor gene families, gene duplication occurs frequently and the newly generated genes may acquire the ability to recognize new ligands, which could increase an organism's fitness. As expected, analysis of ORs, V1Rs, V2Rs, T1Rs, and T2Rs revealed positive selection acting on newly duplicated genes in most vertebrate lineages (Hughes and Hughes 1993; Mundy and Cook 2003; Shi et al. 2003, 2005; Emes et al. 2004; Alioto and Ngai 2005, 2006; Yang et al. 2005; Shi and Zhang 2006). More interestingly, positive selection tends to happen in potential ligand-binding regions. For instance, the analysis of eight closely related human T2R genes suggested positive selection in extracellular domains, while purifying selection in transmembrane and intracellular domains (Shi et al. 2003). Although no crystal structure of T2Rs has been solved, existing functional data suggest that extracellular domains of T2Rs are involved in binding to ligands (Soranzo et al. 2005). Positive selection was also detected in extracellular domains in V1Rs, which are most closely related to T2Rs in sequence (Mundy and Cook 2003; Shi et al. 2005). These results contrast the majority of positively selected sites in ORs, which are located in transmembrane domains that are thought to be the binding pocket in ORs (Emes et al. 2004). For the V2R and T1R gene families, most positively selected sites are mapped to the long N-terminus, which is believed to be involved in heterodimerization or homodimerization and ligand binding (Yang et al. 2005; Shi and Zhang 2006). Together, these positive selection analyses suggest that newly generated chemoreceptor genes tend to be subject to diversifying selection, probably because of the ability to recognize a diverse array of chemicals that the animals encounter in exploring new habitats and foods.

As described above, sequence variation of chemoreceptor genes is prevalent even within species. If these polymorphisms affect an individual's fitness such as mate selection and sibling sustenance, some of them may be under positive selection. There is now evidence supporting this possibility. For example, Zhang et al. (2004) compared V1R gene sequences from two mouse draft sequences which were derived from different inbred mouse strains. They found a high ratio of nonsynonymous to synonymous substitutions, a possible result of positive selection on these genes. In humans, most chemoreceptor genes are under relaxed selective constraints. However, positive selection has been detected in the human T2R16 gene, which encodes a β-glucopyranoside receptor (Bufe et al. 2002). By analyzing the sequences from different human populations, Soranzo et al. (2005) detected signatures of positive selection on a derived allele, which was found in all human populations except Africans. Compared with the ancestral allele, the derived allele exhibits increased sensitivity in detecting -glucopyranoside. It was suggested that the derived allele may provide better protection against harmful cyanogenic plant foods and natural toxins (Soranzo et al. 2005). Another interesting case is the T2R38 gene, which is largely responsible for the human polymorphism in tasting phenylthiocarbamide (Kim et al. 2003). Balancing selection was suggested to maintain both taster and nontaster T2R38 alleles in human populations (Wooding et al. 2004), although a subsequent analysis found the evidence for balancing selection unconvincing (Wang et al. 2004). Interestingly, chimpanzees are also known to have tasters and nontasters of phenylthiocarbamide, but the nontaster allele is apparently a null allele (Wooding et al. 2006).

6 Conclusions

The hallmark of vertebrate chemoreceptor gene family evolution is the extremely high diversity of gene family size and gene sequence among species. The general genetic mechanisms involved in generating this pattern include frequent gene duplication and pseudogenization, conforming to the "birth-and-death" process (Nei et al. 1997). Adaptation to changing environments and diets is likely the major selective force behind this evolutionary process, in addition to the random factor of genomic drift (Nei 2007). Although evolutionary and genomic studies have resulted in enormous advances in this field in the last several years, many fundamental questions are yet to be answered (Mombaerts 2004; Meyerhof 2005). It is expected that evolutionary analysis, coupled with functional assays of chemoreceptors, will yield useful information on the molecular mechanisms and selective forces behind vertebrate chemoreceptor gene diversification.

Acknowledgement This work was supported by a start-up fund of "Hundreds-Talent Program" from Chinese Academy of Sciences to P.S. and research grants from the National Institutes of Health to J.Z.

References

Adler E, Hoon MA, Mueller KL, Chandrashekar J, Ryba NJ, Zuker CS (2000) A novel family of mammalian taste receptors. Cell 100:693–702

Alioto TS, Ngai J (2005) The odorant receptor repertoire of teleost fish. BMC Genomics 6:173

Alioto TS, Ngai J (2006) The repertoire of olfactory C family G protein-coupled receptors in zebrafish: candidate chemosensory receptors for amino acids. BMC Genomics 7:309

Behrens M, Foerster S, Staehler F, Raguse JD, Meyerhof W (2007) Gustatory expression pattern of the human TAS2R bitter receptor gene family reveals a heterogenous population of bitter responsive taste receptor cells. J Neurosci 27:12630–12640

Borowsky B, Adham N, Jones KA, Raddatz R, Artymyshyn R, Ogozalek KL, Durkin MM, Lakhlani PP, Bonini JA, Pathirana S, Boyle N, Pu X, Kouranova E, Lichtblau H, Ochoa FY, Branchek TA, Gerald C (2001) Trace amines: identification of a family of mammalian G protein-coupled receptors. Proc Natl Acad Sci USA 98:8966–8971

Buck LB (2000) The molecular architecture of odor and pheromone sensing in mammals. Cell 100:611–618

Buck L, Axel R (1991) A novel multigene family may encode odorant receptors: a molecular basis for odor recognition. Cell 65:175–187

Bufe B, Hofmann T, Krautwurst D, Raguse JD, Meyerhof W (2002) The human TAS2R16 receptor mediates bitter taste in response to beta-glucopyranosides. Nat Genet 32:397–401

Chamero P, Marton TF, Logan DW, Flanagan K, Cruz JR, Saghatelian A, Cravatt BF, Stowers L (2007) Identification of protein pheromones that promote aggressive behaviour. Nature 450:899–902

Chandrashekar J, Mueller KL, Hoon MA, Adler E, Feng L, Guo W, Zuker CS, Ryba NJ (2000) T2Rs function as bitter taste receptors. Cell 100:703–711

Conte C, Ebeling M, Marcuz A, Nef P, Andres-Barquin PJ (2002) Identification and characterization of human taste receptor genes belonging to the TAS2R family. Cytogenet Genome Res 98:45–53

Conte C, Ebeling M, Marcuz A, Nef P, Andres-Barquin PJ (2003) Evolutionary relationships of the Tas2r receptor gene families in mouse and human. Physiol Genomics 14:73–82

Dulac C (1997) Molecular biology of pheromone perception in mammals. Semin Cell Dev Biol 8:197–205

Dulac C, Axel R (1995) A novel family of genes encoding putative pheromone receptors in mammals. Cell 83:195–206

Dulac C, Torello AT (2003) Molecular detection of pheromone signals in mammals: from genes to behaviour. Nat Rev Neurosci 4:551–562

Emes RD, Beatson SA, Ponting CP, Goodstadt L (2004) Evolution and comparative genomics of odorant- and pheromone-associated genes in rodents. Genome Res 14:591–602

Fischer A, Gilad Y, Man O, Paabo S (2005) Evolution of bitter taste receptors in humans and apes. Mol Biol Evol 22:432–436

Gilad Y, Lancet D (2003) Population differences in the human functional olfactory repertoire. Mol Biol Evol 20:307–314

Gilad Y, Wiebe V, Przeworski M, Lancet D, Paabo S (2004) Loss of olfactory receptor genes coincides with the acquisition of full trichromatic vision in primates. PLoS Biol 2:E5

Gilad Y, Man O, Glusman G (2005) A comparison of the human and chimpanzee olfactory receptor gene repertoires. Genome Res 15:224–230

Gilad Y, Man O, Paabo S, Lancet D (2003) Human specific loss of olfactory receptor genes. Proc Natl Acad Sci U S A 100:3324–3327

Gloriam DE, Bjarnadottir TK, Yan YL, Postlethwait JH, Schioth HB, Fredriksson R (2005) The repertoire of trace amine G-protein-coupled receptors: large expansion in zebrafish. Mol Phylogenet Evol 35:470–482

Glusman G, Yanai I, Rubin I, Lancet D (2001) The complete human olfactory subgenome. Genome Res 11:685–702

Go Y (2006) Proceedings of the SMBE Tri-National Young Investigators' Workshop 2005. Lineage-specific expansions and contractions of the bitter taste receptor gene repertoire in vertebrates. Mol Biol Evol 23:964–972

Go Y, Satta Y, Takenaka O, Takahata N (2005) Lineage-specific loss of function of bitter taste receptor genes in humans and nonhuman primates. Genetics 170:313–326

Grus WE, Zhang J (2004) Rapid turnover and species-specificity of vomeronasal pheromone receptor genes in mice and rats. Gene 340:303–312

Grus WE, Zhang J (2006) Origin and evolution of the vertebrate vomeronasal system viewed through system-specific genes. Bioessays 28:709–718

Grus WE, Shi P, Zhang YP, Zhang JZ (2005) Dramatic variation of the vomeronasal pheromone receptor gene repertoire among five orders of placental and marsupial mammals. Proc Natl Acad Sci USA 102:5767–5772

Grus WE, Shi P, Zhang JZ (2007) Largest vertebrate vomeronasal type 1 receptor gene repertoire in the semiaquatic platypus. Mol Biol Evol 24:2153–2157

Hashiguchi Y, Nishida M (2005) Evolution of vomeronasal-type odorant receptor genes in the zebrafish genome. Gene 362:19–28

Hashiguchi Y, Nishida M (2006) Evolution and origin of vomeronasal-type odorant receptor gene repertoire in fishes. BMC Evol Biol 6:76

Hashiguchi Y, Nishida M (2007) Evolution of trace amine associated receptor (TAAR) gene family in vertebrates: lineage-specific expansions and degradations of a second class of vertebrate chemosensory receptors expressed in the olfactory epithelium. Mol Biol Evol 24:2099–2107

Hashiguchi Y, Furuta Y, Kawahara R, Nishida M (2007) Diversification and adaptive evolution of putative sweet taste receptors in threespine stickleback. Gene 396:170–179

Herness MS, Gilbertson TA (1999) Cellular mechanisms of taste transduction. Annu Rev Physiol 61:873–900

Herrada G, Dulac C (1997) A novel family of putative pheromone receptors in mammals with a topographically organized and sexually dimorphic distribution. Cell 90:763–773

Hoon MA, Adler E, Lindemeier J, Battey JF, Ryba NJ, Zuker CS (1999) Putative mammalian taste receptors: a class of taste-specific GPCRs with distinct topographic selectivity. Cell 96:541–551

Hughes AL, Hughes MK (1993) Adaptive evolution in the rat olfactory receptor gene family. J Mol Evol 36:249–254

Keverne EB (1999) The vomeronasal organ. Science 286:716–720

Kim UK, Jorgenson E, Coon H, Leppert M, Risch N, Drayna D (2003) Positional cloning of the human quantitative trait locus underlying taste sensitivity to phenylthiocarbamide. Science 299:1221–1225

Kim UK, Wooding S, Ricci D, Jorde LB, Drayna D (2005) Worldwide haplotype diversity and coding sequence variation at human bitter taste receptor loci. Hum Mutat 26:199–204

Kim UK, Wooding S, Riaz N, Jorde LB, Drayna D (2006) Variation in the human TAS1R taste receptor genes. Chem Senses 31:599–611

Kimoto H, Haga S, Sato K, Touhara K (2005) Sex-specific peptides from exocrine glands stimulate mouse vomeronasal sensory neurons. Nature 437:898–901

Kinnamon SC, Margolskee RF (1996) Mechanisms of taste transduction. Curr Opin Neurobiol 6:506–513

Kitagawa M, Kusakabe Y, Miura H, Ninomiya Y, Hino A (2001) Molecular genetic identification of a candidate receptor gene for sweet taste. Biochem Biophys Res Commun 283:236–242

Lane RP, Young J, Newman T, Trask BJ (2004) Species specificity in rodent pheromone receptor repertoires. Genome Res 14:603–608

Li X, Staszewski L, Xu H, Durick K, Zoller M, Adler E (2002) Human receptors for sweet and umami taste. Proc Natl Acad Sci USA 99:4692–4696

Li X, Li W, Wang H, Cao J, Maehashi K, Huang L, Bachmanov AA, Reed DR, Legrand-Defretin V, Beauchamp GK, Brand JG (2005) Pseudogenization of a sweet-receptor gene accounts for cats' indifference toward sugar. PLoS Genet 1:e3

Liberles SD, Buck LB (2006) A second class of chemosensory receptors in the olfactory epithelium. Nature 442:645–650

Liman ER, Innan H (2003) Relaxed selective pressure on an essential component of pheromone transduction in primate evolution. Proc Natl Acad Sci USA 100:3328–3332

Lindemann B (2001) Receptors and transduction in taste. Nature 413:219–225

Loconto J, Papes F, Chang E, Stowers L, Jones EP, Takada T, Kumanovics A, Fischer Lindahl K, Dulac C (2003) Functional expression of murine V2R pheromone receptors involves selective association with the M10 and M1 families of MHC class Ib molecules. Cell 112:607–618

Malnic B, Godfrey PA, Buck LB (2004) The human olfactory receptor gene family. Proc Natl Acad Sci USA 101:2584–2589

Matsunami H, Buck LB (1997) A multigene family encoding a diverse array of putative pheromone receptors in mammals. Cell 90:775–784

Matsunami H, Montmayeur JP, Buck LB (2000) A family of candidate taste receptors in human and mouse. Nature 404:601–604

Max M, Shanker YG, Huang L, Rong M, Liu Z, Campagne F, Weinstein H, Damak S, Margolskee RF (2001) Tas1r3, encoding a new candidate taste receptor, is allelic to the sweet responsiveness locus Sac. Nat Genet 28:58–63

Menashe I, Man O, Lancet D, Gilad Y (2003) Different noses for different people. Nat Genet 34:143–144

Meyerhof W (2005) Elucidation of mammalian bitter taste. Rev Physiol Biochem Pharmacol 154:37–72

Mombaerts P (1999) Molecular biology of odorant receptors in vertebrates. Annu Rev Neurosci 22:487–509

Mombaerts P (2004) Genes and ligands for odorant, vomeronasal and taste receptors. Nat Rev Neurosci 5:263–278

Montmayeur JP, Liberles SD, Matsunami H, Buck LB (2001) A candidate taste receptor gene near a sweet taste locus. Nat Neurosci 4:492–498

Mundy NI, Cook S (2003) Positive selection during the diversification of class I vomeronasal receptor-like (V1RL) genes, putative pheromone receptor genes, in human and primate evolution. Mol Biol Evol 20:1805–1810

Nei M (2007) The new mutation theory of phenotypic evolution. Proc Natl Acad Sci USA 104:12235–12242

Nei M, Gu X, Sitnikova T (1997) Evolution by the birth-and-death process in multigene families of the vertebrate immune system. Proc Natl Acad Sci USA 94:7799–7806

Nelson G, Hoon MA, Chandrashekar J, Zhang Y, Ryba NJ, Zuker CS (2001) Mammalian sweet taste receptors. Cell 106:381–390

Niimura Y, Nei M (2003) Evolution of olfactory receptor genes in the human genome. Proc Natl Acad Sci USA 100:12235–12240

Niimura Y, Nei M (2005) Evolutionary dynamics of olfactory receptor genes in fishes and tetrapods. Proc Natl Acad Sci USA 102:6039–6044

Niimura Y, Nei M (2006) Evolutionary dynamics of olfactory and other chemosensory receptor genes in vertebrates. J Hum Genet 51:505–517

Niimura Y, Nei M (2007) Extensive gains and losses of olfactory receptor genes in Mammalian evolution. PLoS ONE 2:e708

Nozawa M, Kawahara Y, Nei M (2007) Genomic drift and copy number variation of sensory receptor genes in humans. Proc Natl Acad Sci USA 104:20421–20426

Parry CM, Erkner A, le Coutre J (2004) Divergence of T2R chemosensory receptor families in humans, bonobos, and chimpanzees. Proc Natl Acad Sci USA 101:14830–14834

Pfister P, Rodriguez I (2005) Olfactory expression of a single and highly variable V1r pheromone receptor-like gene in fish species. Proc Natl Acad Sci USA 102:5489–5494

Pfister P, Randall J, Montoya-Burgos JI, Rodriguez I (2007) Divergent evolution among teleost V1r receptor genes. PLoS ONE 2:e379

Prasad BC, Reed RR (1999) Chemosensation: molecular mechanisms in worms and mammals. Trends Genet 15:150–153

Rodriguez I, Mombaerts P (2002) Novel human vomeronasal receptor-like genes reveal species—specific families. Curr Biol 12:R409–411

Rodriguez I, Del Punta K, Rothman A, Ishii T, Mombaerts P (2002) Multiple new and isolated families within the mouse superfamily of V1r vomeronasal receptors. Nat Neurosci 5:134–140

Rouquier S, Blancher A, Giorgi D (2000) The olfactory receptor gene repertoire in primates and mouse: evidence for reduction of the functional fraction in primates. Proc Natl Acad Sci USA 97:2870–2874

Ryba NJ, Tirindelli R (1997) A new multigene family of putative pheromone receptors. Neuron 19:371–379

Sainz E, Korley JN, Battey JF, Sullivan SL (2001) Identification of a novel member of the T1R family of putative taste receptors. J Neurochem 77:896–903

Saraiva LR, Korsching SI (2007) A novel olfactory receptor gene family in teleost fish. Genome Res 17:1448–1457

Shi P, Zhang J (2006) Contrasting modes of evolution between vertebrate sweet/umami receptor genes and bitter receptor genes. Mol Biol Evol 23:292–300

Shi P, Zhang J (2007) Comparative genomic analysis identifies an evolutionary shift of vomeronasal receptor gene repertoires in the vertebrate transition from water to land. Genome Res 17:166–174

Shi P, Zhang J, Yang H, Zhang YP (2003) Adaptive diversification of bitter taste receptor genes in mammalian evolution. Mol Biol Evol 20:805–814

Shi P, Bielawski JP, Yang H, Zhang YP (2005) Adaptive diversification of vomeronasal receptor 1 genes in rodents. J Mol Evol 60:566–576

Smith CUM (2000) Biology of sensory systems. Wiley, New York

Soranzo N, Bufe B, Sabeti PC, Wilson JF, Weale ME, Marguerie R, Meyerhof W, Goldstein DB (2005) Positive selection on a high-sensitivity allele of the human bitter-taste receptor TAS2R16. Curr Biol 15:1257–1265

Tacher S, Quignon P, Rimbault M, Dreano S, Andre C, Galibert F. (2005) Olfactory receptor sequence polymorphism within and between breeds of dogs. J Hered 96: 812–816

Trask BJ, Friedman C, Martin-Gallardo A, Rowen L, Akinbami C, Blankenship J, Collins C, Giorgi D, Iadonato S, Johnson F, Kuo WL, Massa H, Morrish T, Naylor S, Nguyen OT, Rouquier S, Smith T, Wong DJ, Youngblom J, van den Engh G. (1998) Members of the olfactory receptor gene family are contained in large blocks of DNA duplicated polymorphically near the ends of human chromosomes. Hum Mol Genet 7: 13–26

Wang X, Thomas SD, Zhang J. (2004) Relaxation of selective constraint and loss of function in the evolution of human bitter taste receptor genes. Hum Mol Genet 13: 2671–2678

Wong KK, deLeeuw RJ, Dosanjh NS, Kimm LR, Cheng Z, Horsman DE, MacAulay C, Ng RT, Brown CJ, Eichler EE, Lam WL. (2007) A comprehensive analysis of common copy-number variations in the human genome. Am J Hum Genet 80: 91–104

Wooding S, Kim UK, Bamshad MJ, Larsen J, Jorde LB, Drayna D. (2004) Natural selection and molecular evolution in PTC, a bitter-taste receptor gene. Am J Hum Genet 74: 637–646

Wooding S, Bufe B, Grassi C, Howard MT, Stone AC, Vazquez M, Dunn DM, Meyerhof W, Weiss RB, Bamshad MJ. (2006) Independent evolution of bitter-taste sensitivity in humans and chimpanzees. Nature 440: 930–934

Yang H, Shi P, Zhang YP, Zhang JZ. (2005) Composition and evolution of the V2r vorneronasal receptor gene repertoire in mice and rats. Genomics 86: 306–315

Young JM, Trask BJ. (2002) The sense of smell: genomics of vertebrate odorant receptors. Human Molecular Genetics 11: 1153–1160

Young JM, Trask BJ. (2007) V2R gene families degenerated in primates, dog and cow, but expanded in opossum. Trends Genet 23: 212–215

Young JM, Kambere M, Trask BJ, Lane RP. (2005) Divergent V1R repertoires in five species: Amplification in rodents, decimation in primates, and a surprisingly small repertoire in dogs. Genome Res 15: 231–240

Zhang J. (2003) Evolution by gene duplication: an update. Trends Ecol Evol 18: 292–298

Zhang J. (2007) The drifting human genome. Proc Natl Acad Sci U S A 104: 20147–20148

Zhang J, Webb DM. (2003) Evolutionary deterioration of the vomeronasal pheromone transduction pathway in catarrhine primates. Proc Natl Acad Sci U S A 100: 8337–8341

Zhang X, Rodriguez I, Mombaerts P, Firestein S. (2004) Odorant and vomeronasal receptor genes in two mouse genome assemblies. Genomics 83: 802–811

Zhang X, Zhang X, Firestein S. (2007) Comparative genomics of odorant and pheromone receptor genes in rodents. Genomics 89: 441–450

Zhao GQ, Zhang Y, Hoon MA, Chandrashekar J, Erlenbach I, Ryba NJ, Zuker CS. (2003) The receptors for mammalian sweet and umami taste. Cell 115: 255–266

Zozulya S, Echeverri F, Nguyen T. (2001) The human olfactory receptor repertoire. Genome Biol 2: RESEARCH0018

Genomics of Olfactory Receptors

Xiaohong Zhang and Stuart Firestein

Abstract In many species, the sense of smell plays important roles in locating food, detecting predators, navigating, and communicating social information. The olfactory system has evolved complex repertoires of odor receptors (ORs) to fulfill these functions. Through computational data mining, OR repertoires of multiple species were identified, revealing a surprisingly large OR gene family in rodents and evolutionary fluctuation among different organisms. Characteristics of OR genes were explored through computational and experimental methods, showing a complicated gene structure and special genomic distribution. Utilizing high-throughput OR microarrays, expression profiles of the mouse and human OR repertoire were examined, their olfactory functions verified, and their zonal, ectopic and developmental expression determined. Variation in human smelling abilities results from different functional OR repertoires, variable expressional levels and polymorphisms in the copy number of the OR genes. These genomic approaches have both provided new data and generated new questions.

1 Introduction

The molecular era in olfaction began in 1991 with the landmark discovery of a large, multigene family of odor receptors in rat by Buck and Axel (Buck and Axel 1991). The first few olfactory receptors were cloned based on the assumption that olfactory receptors would comprise a diverse repertoire of G-protein coupled receptors (GPCRs) with seven-transmembrane topology, and they would be expressed exclusively in the olfactory epithelium. Later combined with the availability of numerous completely sequenced genomes, this pioneering discovery opened the way for the characterization of the OR gene family through exhaustive computational data mining.

X. Zhang and S. Firestein
Department of Biological Sciences, Columbia University, New York, NY, USA

2 Data Mining of OR Repertoire

The coding region of OR gene sequences is encoded by a single exon with conserved amino acid motifs that distinguish them from other non OR seven-transmembrane proteins (Glusman et al. 2001). This feature facilitates genomic screening for putative OR genes in any species with known genome sequences. Different laboratories performed data mining for OR repertoires in multiple species using similar strategies (Glusman et al. 2001; Young et al. 2002; Young and Trask 2002; Zhang and Firestein 2002).

The general strategy comprises the following steps (see Fig. 1): First, known OR sequences which have been cloned and examined through classical molecular methods were compiled from gene databases, such as Genbank. Redundant sequences were removed to keep a representative group of known ORs. This group should be as diverse as possible to represent the width of the full repertoire. For example, 30 OR sequences from 30 different families will have better coverage than 30 sequences from a single family. Secondly, a well-assembled genome is searched with the representative sequences as query. The sequence quality of the genome would of course have an effect on the completeness of the OR sequences identified. For example, mistakenly assembled scaffolds will result in partial or pseudo genes if adjacent sites reside in OR genes. This

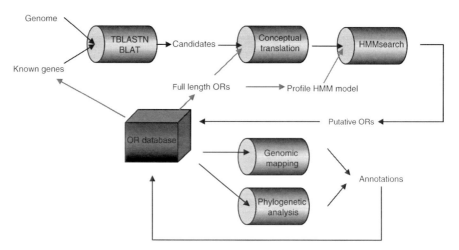

Fig. 1 *Reiterative Genomic Data Mining Pipeline Optimized for Large Gene Families.* To ensure the search is fast and exhaustive, a large number of BLAT searches or low stringent TBLASTN searches using representative genes was utilized in the first step. The output sequences from conceptual translation were matched to pHMM model which was built with the HMMER package and known ORs to achieve high specificity. Sequences with low p-values were selected as putative ORs. These low p-value OR candidates, combined with the original known ORs, were used as queries for the second-round search. The reiteration is stopped after no new sequences with p-values lower than the threshold are discovered

screening is a rough alignment between the known OR genes and the genome to identify the likely OR gene coordinates on the DNA. Thirdly, a conceptual translation is performed using a model of known ORs as the template and the DNA sequences obtained from the second step. This model is necessary in that it instructs the translation process by adding necessary frame-shifts within the coding region. Finally, we compare the similarity between the conceptually translated sequences and the model built with known ORs. Sequences that have high similarity with known ORs are selected and subject to additional filtering to avoid redundancy, followed by assigning them as "OR candidates". These putative ORs share the conserved amino acid motifs that distinguish ORs from other non OR seven-transmembrane proteins. Receptors that also function in odor recognition but don't share any sequence similarity with known ORs will not be identified in this searching. For example, the TAARs were missed in the data mining performed for ORs.

2.1 Olfactory Receptor Repertoire

The size of OR repertoire varies dramatically among different species, but the reasons for this are unknown. While many primates have a reduced repertoire of functional receptor genes, family size is not in general coordinated with the apparent dependence on olfaction in the species niche; for example dogs have many fewer than rats. Exhaustive data with updated genomes has revealed that humans have 851 olfactory receptors genes, but only 384 among them are functionally intact (Glusman et al. 2001; Aloni et al. 2006) (details in Table 1). Another primate, the chimpanzee, has a similar number of OR genes as humans. In rodents, the mouse has fewer OR genes than the rat, with a number of 1,375 versus 1,576 (Zhang et al. 2004a, 2007a). The olfactory subgenomes of other mammals were also explored as their genome sequences became available. For example, dog, being well-known for its excellent smelling capability has 971 ORs (July 2004 assembly of the boxer breed) (Olender et al. 2004). The chicken's OR repertoire consists of 554 genes,

Table 1 A comprehensive collection of OR genes in complete mammalian genomes

Organism	Species name	Genome assembly	Number of OR genes	Number of Intact OR genes	Number of OR pseudogenes
Human	*Homo sapiens*	hg17	851	384	467
Chimp	*Pan troglodytes*	PCAP1O26	899	353	546
Dog	*Canis familiaris*	canFam1	971	713	258
Mouse	*Mus musculus*	mm5	1,375	1,194	181
Rat	*Rattus norvegicus*	rn3	1,576	1,284	292
Opossum	*Monodelphis domestica*	mon Dom1	1,516	899	617
Chicken	*Gallus gallus*	galGal2	554	78	476

while only 78 among them are intact (Aloni et al. 2006). Lower organisms, such as fishes and insects, have dozens of OR genes and they are more distantly related with each other than those of mammals. The Pufferfish has 44 ORs while the Zebrafish has 98 (Niimura and Nei 2005). Insects have similar sized OR gene families. For example, the fruit fly and mosquito have 62 and 79 ORs respectively (Hill et al. 2002; Robertson et al. 2003). Nematode worms have around 800 functional chemoreceptor genes, which are thought to have arisen independently compared to insects and vertebrates (Bargmann 1998).

The size of the OR repertoire is not the only feature to vary across species, the proportion of intact and pseudo genes are also significantly different. A startlingly high fraction of the human OR (around 55%) repertoire has degenerated to pseudogenes. Other primates, such as the chimpanzee, also have a similar percentage (Gilad et al. 2005). In contrast, the proportion is much lower in rodents. In the most current version of the rat genome, pseudogenes in the OR repertoire only constitute 18.5% of the total and in mouse they account for only 13.1% (Zhang et al. 2007a). This has resulted in the moue effectively possessing over three times as many intact genes as humans. However, we do not know if this translates into a wider range of detectable odorants or a superior discriminatory ability in mouse over human. Intact human OR genes are found in most of mouse OR subfamilies through phylogenetic analysis (Zhang and Firestein 2002) suggesting that human receptors are well represented within the mouse repertoire. Unexpectedly the decline of the OR gene family in some primates has been found to coincide with the acquisition of trichromatic vision, suggesting that better visual capability may make olfaction partially redundant (Gilad et al. 2004).

Although it is often thought that a genomics is a relatively static inquiry, this is untrue and in fact this research is quite dynamic. Genome sequences are regularly updated and significant revisions occur. For example, mouse OR repertoire was reported to consist of 1,296 ORs and about 20% of them are pseudogenes by Zhang et al. in 2002 using the Celera first draft of mouse genome (Zhang and Firestein 2002). Through the same computational methods and thresholds, 1,375 mouse ORs were identified in mouse genome version mm5 (http://hgdownload.cse.ucsc.edu/), decreasing the fraction of pseudogenes to 13.1% and about 100 pseudogenes were corrected to be intact genes (Zhang et al. 2007a). As the assembly has been optimizing, we expected to see a few pseudogenes to be annotated as intact while the repertoire size are relatively similar. Results from different research groups also have small discrepancies because of using slightly variable filtering criteria or searching methods.

Additionally the curated database that includes functions likely for particular genes is occasionally discovered to be inaccurate. These errors can propagate through the database, as a gene incorrectly identified by its initial discoverer is used as a model in data mining, and all similar sequences are classified as having a similar function. If the initial identification is incorrect so will be the subsequent classifications.

Finally new bioinformatics tools are regularly introduced allowing new data to be derived from existing databases.

2.2 OR Gene Sequences

Mammalian ORs can be separated into two broad classes by phylogenetic analysis. Class I receptors resemble the family that was first found in fish and frog, but had been considered an evolutionary relic in mammals (Fig. 2). As identified in the human genome, at first, 102 Class I ORs constitute about one-tenth of the human OR repertoire (Glusman et al. 2001). In rodents, the mouse possesses 158 Class I ORs, and the rat has 153, also comprising about one-tenth of their repertoires respectively. In both human and rodents, the Class I ORs contain a lower fraction of pseudogenes than the Class II ORs. Class I ORs are also more conserved in terms of both genomic location and gene sequences. Class II ORs are dispersed on numerous chromosomes, while Class I ORs are tightly clustered on one chromosomal segment. For example, all human Class I ORs are located in one super cluster, 11p15; and the mouse Class I ORs are located in a single cluster on chromosome 7. Comparative genomic analysis revealed that sequences of Class I ORs are even more conserved between mouse and rat than those of Class II ORs (Zhang et al. 2007a).

Gene expression data showed that Class I ORs are restricted to the dorsal zone of the mouse olfactory epithelium (Zhang et al. 2004b). These unique features lead us to suspect that Class I ORs may have some special and important functions, driving positive evolutionary pressure on mammals to maintain a high level of conservation.

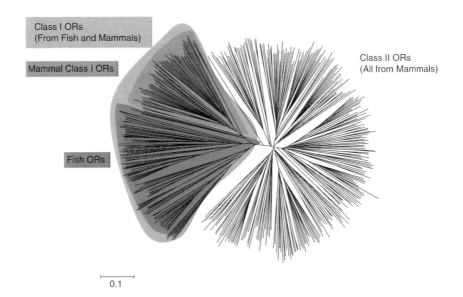

Fig. 2 *Classification of Fish and Mammalian ORs.* OR genes can be separated into two classes based on their sequences. Mammals have both Class I and Class II ORs, whereas all fish ORs belong to Class I. Insect ORs are quite different from each other and also different from fish and mammal ORs. All Class I ORs are shaded in blue. Class I ORs from mammals are shaded in orange and fish ORs are shaded in green

OR gene structures are complicated by their multiple splicing alternatives. Although the coding regions are encoded by a single exon about 1 kb long, the entire gene has multiple exons up and downstream of the coding region (Sosinsky et al. 2000). It is the coding regions of OR sequences that have been identified through computational exploration, leaving the upstream exons and downstream UTRs of most ORs yet to be identified. Through cDNA library screening, Young et al. disclosed that many OR genes have several transcriptional isoforms showing alternative splicing of their 5′ untranslated exons or utilizing more than one polyadenylation site (Young et al. 2003). The mechanisms of transcriptional control of OR genes are yet to be discovered. Whether OR transcriptional isoforms have functional significance is also not determined.

OR proteins are composed of highly conserved and variable residues fulfilling corresponding functions. First of all, each OR has seven transmembrane helices (TM1–TM7) interconnected by extracellular (EC) and intracellular (IC) loops. Additionally there is an extracellular N-terminal chain and an intracellular C-terminal chain (Buck and Axel 1991; Pilpel and Lancet 1999). In analogy to other Class A GPCRs, ligand-binding activities are thought to take place in a pocket formed by the TM helices. The helix bundles appear to form two pockets. One is formed by TM1, TM2, TM3, and TM7 and a second one is formed by TM3 through TM7 (Liu et al. 2003). Secondly, OR proteins are characterized by conserved amino acid motifs that distinguish them from other seven-transmembrane proteins. ORs share some highly conserved motifs with other non OR GPCRs, however, they do have specific motifs that only occur in ORs. These OR signature sequences are likely to participate in critical OR-specific functional activities. For example, motif "KALSTCASHLLVV" positioned partly in IC3, is expected to bind with Golf upon activation. Furthermore, different positions of OR proteins have variable degrees of conservation. Computational analysis showed that TM4, TM5, the central region of TM3, and the last segments of the N- and C-terminals are highly variable. Since ORs can bind with a variety of ligands, those globally variable residues with high specificity are considered likely to participate in ligand-binding (Liu et al. 2003).

2.3 Genomic Organization and Gene Regulation

OR genes appear haphazardly spread over dozens of loci in the genome, as singletons or in tight clusters. In mouse and rat, the isolated genes occur quite rarely, occupying only 1.5% of the OR repertoire. These isolated genes have higher fraction of pseudogenes than genes in clusters. In the mouse genome version mm5 (http://hgdownload.cse.ucsc.edu/), OR genes are distributed in 43 clusters, with an intergenic distance within the clusters of 19–45 kb. These clusters are distributed on almost all chromosomes except chromosome 18 and the Y chromosome. The largest clusters are localized on chromosome 2 and 7, which harbor 344 and 267 ORs respectively. The second large mouse cluster, which consists entirely of all the

Class I ORs, is one of the densest clusters with an average distance of 19.1 kb between neighboring genes (Zhang et al. 2007a).

The size of the repertoire and the peculiar genomic organization pose a formidable challenge to OR gene regulation. Under the "one neuron, one receptor" hypothesis, a single olfactory sensory neuron (OSN) expresses only one OR of the repertoire (Mombaerts 2004). OR genes are expressed in a monoallelic fashion, with transcripts derived from either the paternal or maternal allele in different OSNs within an individual (Chess et al. 1994). The physical proximity between clustered genes reflects a likely evolutionary proximity, whereas their global regulatory networks must still be verified by further experimental evidence. The "H region", a 2 kb DNA element upstream of the mouse MOR28 OR cluster and conserved between rodents and humans was identified bioinformatically by Sakano's group as a possible regulatory element. Experimentally it has been shown that this region of DNA loops back and interacts with the downstream OR genes acting to promote gene choice. Negative feedback regulation has been proposed in this model to insure the one receptor-one olfactory neuron rule in mouse (Serizawa et al. 2003). While positioning the H element in different positions relative to other genes can alter expression levels. The H element can also interact with active OR gene promoters from different clusters, indicating the non cluster specific 'avidity' of H elements (Lomvardas et al. 2006). However, deletion of the H element does not effect the expression of homologous genes located in trans, nor other genes located on different chromosomes (Fuss et al. 2007). These data indicate that the H element functions as an enhancer, perhaps interacting with a single promoter of OR genes to promote OR gene choice.

2.4 Expression of OR Genes

OR repertoires were initially identified through computational methods solely based on sequence similarity with known ORs. Using a high-throughput custom microarray, Zhang et al. showed that a majority of OR genes are specifically enriched in the olfactory epithelium, substantiating their function in olfaction (Zhang et al. 2004b, 2008) (see Fig. 3). Certain OR genes are exclusively or predominantly expressed in non olfactory tissues, such as testis (Spehr et al. 2004), prostate (Yuan et al. 2001) and heart (Drutel et al. 1995). It may be that these were misidentified as ORs based on sequence similarity, but should be more properly considered as OR-like GPCRs of other functions. Since there are no ligands associated with these receptors they are effectively a group of new orphan GPCRs and therefore possible targets for therapeutic drugs. One of these receptors, hOR17-4, mediates directed chemotactic movement of human sperm in vitro (Spehr et al. 2003), and also functions in olfactory sensory neurons to detect the odor bourgeonal (Spehr et al. 2004). Using the VNO or OE as the baseline for comparison, 30–100 OR genes were found to be enriched in each of several different tissues, including heart, liver, lung, and kidney (Zhang et al. 2004b). Ectopic expression of

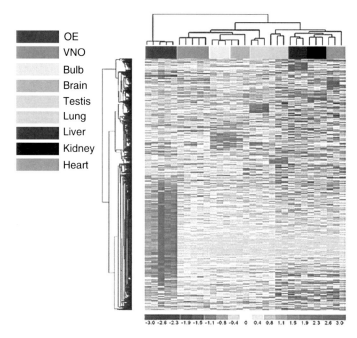

Fig. 3 *Expression profiles of OR genes across tissues.* These data show OE-specific and ectopic expression of OR genes. All tissues are from 2-month-old adult mice. Five sample replicates were collected for OE and VNO, and three sample replicates for other tissues. The gene expression values are standardized such that the mean is 0 and standard deviation is 1 for each gene. The *color* represents expression values as shown in the scale bar, with *red* corresponding to higher-than-mean expression values and *blue* corresponding to lower-than-mean values. The dendrogram on the *left* shows the clustering of genes, and the *top* dendrogram shows clustering of the samples based on the expression data. All genes showing expression in any tissue are chosen for the clustering analysis and are shown in the figure. 1,383 probe sets representing 1,095 OR genes are shown

olfactory receptor genes was also explored via EST and previously available microarray data in mouse and human (Feldmesser et al. 2006). Each different tissue was found to have a specific relatively small subset of OR genes. Human–mouse orthologous pairs did not show any correlation in the expressional level. The function or functions of these ectopically expressed ORs remain to be discovered.

In the olfactory epithelium, a given OR gene is expressed in a very small subset of OSNs within one of four parallel stripes or 'zones' that run rostral to caudal across the turbinates within the olfactory epithelium (OE). This spatial expression pattern has been determined by in situ investigation and microarray analysis (Sullivan et al. 1995; Qasba and Reed 1998; Zhang et al. 2004b). Notably, OR genes expressed in different zones often appear segregated on the chromosomes. Mouse class I ORs, which are located in one tight cluster on chromosome 7, are solely expressed in the most dorsal regions (zone 1) of the OE (Zhang et al. 2004b).

The biological importance of this compartmentalization remains unclear. Considering that zone 1 ORs cover more than one-third of the OR repertoire and the specific expression of class I ORs in zone 1, the separation between zone 1and other zones might be especially significant.

The developmental course of wide scale OR gene expression has remained largely undocumented until custom MOR arrays were utilized. At embryonic day 13 (E13), OR gene expression in the epithelium does not show a significant difference with non OE tissues, mostly because there are few OR-positive cells, making microarray detection difficult. OR expression in a relatively larger number of cells is thought to occur by about day E15-16 (Sullivan et al. 1995). Nonetheless, a large number of OR genes appear to be detected only after birth. At around postnatal day 20 (P20), the number of detected OR genes reaches a peak and remains high. Some ORs are no longer expressed as mice age, for example, at age 18 months. Even more intriguingly, the expression level of ORs can be classified into different patterns. For example, some ORs reach a peak of expression between P10 and P20 and then reduce to a low level. Other ORs reach a peak at P10, then continue to express at this high level until sometime between 7 and 18 months. These patterns may correlate with their functions; for example, the first pattern may be related to nursing, the latter one to the reproductive cycle (Zhang et al. 2008). One caveat that should be noted is that all these levels refer to mRNA, not protein, which cannot currently be measured directly.

2.5 Population Variation of Human ORs

There is an enormous diversity in the repertoire of functional OR genes among different people. Roughly 60% of human odorant receptor genes have mutated into nonfunctional pseudogenes in a relatively recent genomic process; thus a substantial fraction of human odor receptors might be expected to segregate between an intact and a pseudogene form in different individuals. Menashe et al. genotyped 51 odor receptor loci in 189 individuals of several ethnic origins to screen for SNPs that distinguish the intact and pseudogenic forms (Menashe et al. 2003). Remarkably, of the 189 individuals, 178 functionally different genomes were found. Additional variation in the population many come from differences in gene expression. Experiments with custom microarrays specialized for detecting human odorant receptor genes have found that the expressed receptor repertoire of any pair of individuals differs by at least 14% (Zhang et al. 2007b) (see Fig. 4), suggesting that polymorphisms exist not only in coding regions but also in promoter and other regulatory regions. Additionally, the copy number variation of human olfactory receptor genes also contributes significantly to individual differences in olfactory abilities (Young et al. 2008).

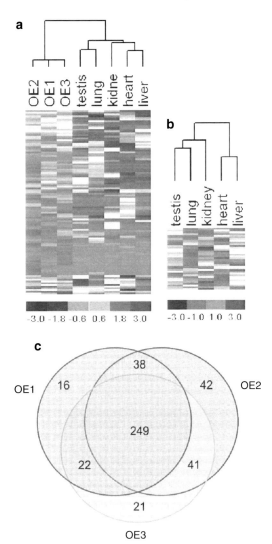

Fig. 4 *Expression profile of human OR genes across tissues and variation in three human samples.* The log transformed detection P values for OR genes in all tissues were standardized to have a mean of 0 and a standard deviation of 1 and are color coded (*red* and *blue* shades indicate values above and below the mean, respectively). The dendrograms on *top* of each panel illustrate the clustering of tissue samples based on the profile of OR gene expression. (**a**) All 578 predicted OR genes are included in a comparison between olfactory epithelium (OE) and non olfactory tissues. (**b**) Shown are the data for only the 147 OR genes with significantly elevated expression in non olfactory tissues. (**c**) The number of predicted human OR genes whose expression was detected (at $P < 0.05$) in one or more of the three olfactory epithelium (OE) samples. As can be seen, there is a substantial difference in the expressed OR gene repertoire of each of the three OE samples

2.6 Summary and Perspective

With the cloning of the superfamily of odor receptors it became clear that one of the great challenges in olfaction would be handling the large numbers of receptor genes and potential ligands. Bioinformatics and microarray technologies are both well suited for managing large data sets with high throughput, their application here has advanced the understanding of the dynamics and organization of this largest of gene families in the nervous system, and has provided new insights into chemical sensing, sensory coding, gene regulation and nervous system development. It has also opened up many new questions in evolution, individual variation, gene choice, and numerous questions about the entire gene locus beyond the coding region. Importantly, these analyses have brought the olfactory system to the attention of the wider community of neuroscientists and have demonstrated its value as a model system in which questions of general neurobiological significance can be profitably investigated.

References

Aloni R, Olender T, et al. (2006) Ancient genomic architecture for mammalian olfactory receptor clusters. Genome Biol 7(10):R88
Bargmann CI (1998) Neurobiology of the Caenorhabditis elegans genome. Science 282(5396):2028–2033
Buck L, Axel R (1991) A novel multigene family may encode odorant receptors: a molecular basis for odor recognition. Cell 65(1):175–187
Chess A, Simon I, et al. (1994) Allelic inactivation regulates olfactory receptor gene expression. Cell 78(5):823–834
Drutel G, Arrang JM, et al. (1995) Cloning of OL1, a putative olfactory receptor and its expression in the developing rat heart. Receptors Channels 3(1):33–40
Feldmesser E, Olender T, et al. (2006) Widespread ectopic expression of olfactory receptor genes. BMC Genom 7:121
Fuss SH, Omura M, et al. (2007) Local and cis effects of the H element on expression of odorant receptor genes in mouse. Cell 130(2):373–384
Gilad Y, Przeworski M, et al. (2004) Loss of olfactory receptor genes coincides with the acquisition of full trichromatic vision in primates. PLoS Biol 2(1):E5
Gilad Y, Man O, et al. (2005) A comparison of the human and chimpanzee olfactory receptor gene repertoires. Genome Res 15(2):224–230
Glusman G, Yanai I, et al. (2001) The complete human olfactory subgenome. Genome Res 11(5):685–702
Hill CA, Fox AN, et al. (2002) G protein-coupled receptors in Anopheles gambiae. Science 298(5591):176–178
Liu AH, Zhang X, et al. (2003) Motif-based construction of a functional map for mammalian olfactory receptors. Genomics 81(5):443–456
Lomvardas S, Barnea G, et al. (2006) Interchromosomal interactions and olfactory receptor choice. Cell 126(2):403–413
Menashe I, Man O, et al. (2003) Different noses for different people. Nat Genet 34(2):143–4
Mombaerts P (2004) Odorant receptor gene choice in olfactory sensory neurons: the one receptor-one neuron hypothesis revisited. Curr Opin Neurobiol 14(1):31–36
Niimura Y, Nei M (2005) Evolutionary dynamics of olfactory receptor genes in fishes and tetrapods. Proc Natl Acad Sci U S A 102(17):6039–6044
Olender T, Fuchs T, et al. (2004) The canine olfactory subgenome. Genomics 83(3):361–372
Pilpel Y, Lancet D (1999) The variable and conserved interfaces of modeled olfactory receptor proteins. Protein Sci 8(5):969–977
Qasba P, Reed RR (1998) Tissue and zonal-specific expression of an olfactory receptor transgene. J Neurosci 18(1):227–236
Robertson HM, Warr CG, et al. (2003) Molecular evolution of the insect chemoreceptor gene superfamily in Drosophila melanogaster. Proc Natl Acad Sci U S A 100(Suppl 2):14537–14542
Serizawa S, Miyamichi K, et al. (2003) Negative feedback regulation ensures the one receptor-one olfactory neuron rule in mouse. Science 302(5653):2088–2094
Sosinsky A, Glusman G, et al. (2000) The genomic structure of human olfactory receptor genes. Genomics 70(1):49–61

Spehr M, Gisselmann G, et al. (2003) Identification of a testicular odorant receptor mediating human sperm chemotaxis. Science 299(5615):2054–2058

Spehr M, Schwane K, et al. (2004) Dual capacity of a human olfactory receptor. Curr Biol 14(19):R832–R833

Sullivan SL, Bohm S, et al. (1995) Target-independent pattern specification in the olfactory epithelium. Neuron 15(4):779–789

Young JM, Trask BJ (2002) The sense of smell: genomics of vertebrate odorant receptors. Hum Mol Genet 11(10):1153–1160

Young JM, Friedman C, et al. (2002) Different evolutionary processes shaped the mouse and human olfactory receptor gene families. Hum Mol Genet 11(5):535–546

Young JM, Shykind BM, et al. (2003) Odorant receptor expressed sequence tags demonstrate olfactory expression of over 400 genes, extensive alternate splicing and unequal expression levels. Genome Biol 4(11):R71

Young JM, Endicott RM, et al. (2008) Extensive copy-number variation of the human olfactory receptor gene family. Am J Hum Genet 83(2):228–242

Yuan TT, Toy P, et al. (2001) Cloning and genetic characterization of an evolutionarily conserved human olfactory receptor that is differentially expressed across species. Gene 278(1–2):41–51

Zhang X, Firestein S (2002) The olfactory receptor gene superfamily of the mouse. Nat Neurosci 5(2):124–133

Zhang X, Rodriguez I, et al. (2004a) Odorant and vomeronasal receptor genes in two mouse genome assemblies. Genomics 83(5):802–811

Zhang X, Rogers M, et al. (2004b) High-throughput microarray detection of olfactory receptor gene expression in the mouse. Proc Natl Acad Sci U S A 101(39):14168–14173

Zhang X, Zhang X, et al. (2007a) Comparative genomics of odorant and pheromone receptor genes in rodents. Genomics 89(4):441–450

Zhang X, De la Cruz O, et al. (2007b) Characterizing the expression of the human olfactory receptor gene family using a novel DNA microarray. Genome Biol 8(5):R86

Zhang X, Macucci F, Greer C, Firestein SJ (2008) High-throughput microarray expression of olfactory receptors in rodents. In press

The Molecular Evolution of Teleost Olfactory Receptor Gene Families

Sigrun Korsching

Abstract Four olfactory receptor gene families, all of them G protein-coupled receptors, have been identified and characterized in mammals – the odorant (OR), vomeronasal (V1R and V2R) and trace amine-associated (TAARs) receptors. Much less attention has been directed towards non-mammalian members of these families. Since a hallmark of mammalian olfactory receptors is their remarkable species specificity, an evaluation of the non-mammalian olfactory receptors is instructive both for comparative purposes and in its own right. In this review I have compiled the results currently available for all four olfactory gene families and discuss their phylogenomic properties in relation to their mammalian counterparts. Representatives of all four families are found in cartilaginous fish and/or jawless fish, allowing a minimal estimate for the evolutionary origin as preceding the segregation between cartilaginous and bony fish or cartilaginous and jawless fish, respectively. Gene repertoires of teleost olfactory receptors are smaller in size (OR, ORA), comparable (olfC), or even larger (TAAR) than the corresponding mammalian gene repertoires. Despite their smaller repertoire size, the teleost OR and ORA families show much larger divergence than their mammalian counterparts. Evolutionary rates vary greatly between families, with evidence for positive selection in teleost OR genes, whereas the *ora* genes are subject to strong negative selection, and in fact are being conserved among all teleost species investigated. With one exception, ligands are unknown for any of the four teleost olfactory receptor gene families, and so the considerable knowledge about the odor responses of the olfactory epithelium and the olfactory bulb can only be linked indirectly to the receptor repertoires.

S. Korsching (✉)
Institut für Genetik, Universität zu Köln
Germany
sigrun.korsching@uni-koeln.de

1 Background

Information about the environment is to a large extent carried by the chemical senses, and in particular the olfactory sense. Thousands of structurally diverse odor molecules perceived and discriminated by vertebrates supply them with a wide range of essential information, ranging from prey localization, predator avoidance, social communication to mating behavior. The receptoire of olfactory receptor genes currently comprises four different gene families, the odorant receptors proper (ORs), vomeronasal receptor genes (V1Rs and V2Rs), and trace amine-associated receptor genes (TAARs), all of them G protein-coupled receptors (GPCRs) (Fig. 1). ORs and TAARs belong to the rhodopsin-like subclass of GPCRs, class A, with short N- and C-termini outside the seven-transmembrane domain, whereas V2Rs belong to class C, and are similar in structure to the metabotropic glutamate receptor, with an additional large N-terminal, extracellular domain. V1Rs have not been formally classified, but are closest to class A receptors.

Olfactory receptor gene families in mammals can be rather large, around 1000 OR genes in rodents (Buck and Axel 1991; Mombaerts 2004), and over 100 genes for rodent V1R and V2R genes (Dulac and Axel 1995; Matsunami and Buck 1997). A hallmark of these families is their high species specificity and rapid evolution. Species-specific expansion and loss of genes and even whole subfamilies is a recurrent theme in the mammalian receptor families (Grus et al. 2005; Lane et al. 2004; Zhang et al. 2004). It has been hypothesized that these features provide for efficient adaptation of the olfactory sense to changing environmental conditions. Several recent publications have established the respective properties of the corresponding fish

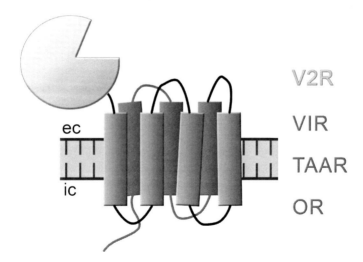

Fig. 1 The four olfactory receptor gene families. ORs, ORAs, and TAARs belong to the class A of GPCRs, with a short N-terminus and a ligand binding site within the TM domains, whereas OlfCs are class C GPCRs, similar to the metabotropic glutamate receptor, with the ligand binding pocket in the large N-terminal extracellular domain

receptor gene families. Here I delineate the four fish receptor gene repertoires and compare their evolutionary properties. Currently the genomes of five teleost species are available (zebrafish, *Danio rerio,* http://www.sanger.ac.uk; three-spined stickleback, *Gasterosteus aculeatus,* http://www.broad.mit.edu; medaka, *Oryzias latipes,* http://medaka.utgenome.org/, http://dolphin.lab.nig.ac.jp/medaka/; tetraodon, *Tetraodon nigroviridis,* http://www.broad.mit.edu/, http://www.genoscope.cns.fr/spip/; fugu, *Takifugu rubripes,* http://www.fugu-sg.org/project/info.html), as well as unfinished versions of shark and lamprey genomes (http://esharkgenome.imcb.a-star.edu.sg/, http://genome.wustl.edu/, respectively) (Fig. 2).

Zebrafish is the most popular aquatic model organism for molecular studies (Sprague et al. 2008) because of its easy (and cost-efficient) handling, short reproductive cycle, small diploid genome (less than two billion base pairs) and established genetic and genomic methods (which include forward genetic screens, transgenesis, mutagenesis-based knock-out and transient knock-down techniques, but unfortunately no homologous recombination-based modification of endogenous genes). A fast ontogenesis (only five days from fertilization to onset of feeding behavior) has turned out to be a major advantage and not only for developmental studies. Medaka has most of these advantages as well, but is not as widely studied so far. Also, zebrafish does not belong to the neoteleosts like the other four species and, as a more primitive fish, it may be considered a better model organism for tetrapods. The advantage of stickleback lies in a wealth of behavioral observations made in this species. The initial advantage of the two pufferfish species, a several-fold smaller genome, has lost its significance due to recent advances in sequencing techniques.

It will be seen that generally the divergence of the teleost receptor repertoires is larger than that of the corresponding tetrapod receptor repertoires. Repertoire size, however, can be both larger and smaller in teleost vs the corresponding tetrapod receptor families. With one exception, rapid evolution is seen in teleost receptor families, but again, the rate of evolution may be both larger and smaller in the teleost gene families compared to the corresponding tetrapod families.

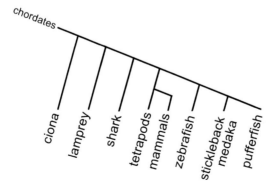

Fig. 2 Phylogenetic tree of the species discussed in this review. The five teleost species, for which completed genomes are available, are shown together with the nodes for divergence from tetrapods, cartilaginous, and jawless fish (human, shark, and lamprey, respectively)

The olfactory receptor genes can be considered the first layer of olfactory information processing and in fact they define the nature of odorants, since any molecule becomes an odorant solely by virtue of its interaction with an olfactory receptor. However, not many olfactory receptor genes are currently deorphanized, due to the sheer complexity of the task, and because heterologous expression is inefficient for many olfactory receptors. Consequently, detailed ligand response spectra so far only exist for a handful of mammalian and a single fish olfactory receptor (Luu et al. 2004).

Like the opsins, another sensory GPCR family, mammalian OR, TAAR, V1R, and V2R genes are expressed in a monogenic fashion, i.e., a particular receptor neuron expresses only a single gene from a single receptor family (Liberles and Buck 2006; Mombaerts 2004). Initial data from the teleost olfactory system seem to indicate essentially the same, if somewhat relaxed, monogenic expression (Sato et al. 2007). In mammals the neurons expressing the same receptor converge into a single glomerulus per hemisphere (main olfactory bulb, ORs, (Mombaerts 2004) expected for TAARs) or a handful of microglomeruli (accessory olfactory bulb, V1Rs, V2Rs (Mombaerts 2004)). Thus, each receptor gene specifies a separate input channel of the olfactory system and the olfactory bulb constitutes a receptotopic map of odor sensitivities, an odor map (Fried et al. 2002). Both genetic and imaging studies are consistent with such a receptotopic organization in the teleost olfactory bulb (Friedrich and Korsching 1998; Fuss and Korsching 2001; Sato et al. 2005, 2007); for a more detailed discussion see Yoshihara, this volume. Individual odorants generally bind to several receptors with different affinities and individual receptors generally bind more than one odorant (Buck 2000; Kajiya et al. 2001) – the result is a combinatorial representation by unique, albeit partly overlapping subsets of receptors, ensuring near limitless coding capability of the system. However, highly specific and possibly unique receptors may be expected for pheromones (cf. Friedrich and Korsching 1998; Kajiya et al. 2001).

2 Phylogenomic Properties of Four Olfactory Receptor Gene Families

2.1 Teleost OR Repertoires: Higher Divergence and Smaller Extent than in Mammals

In initial studies a much higher divergence of teleost OR genes as compared to their mammalian counterparts had already been noted, with homologies as low as 20% in pairwise comparisons (Ngai et al. 1993; Weth et al. 1996). Since acceptable quality genomic databases have became available, two groups have made the effort to establish the complete OR repertoire in several fish species, among them zebrafish. Niimura and Nei (2005) have identified 102 intact zebrafish OR genes (plus 35 pseudogenes), which they subdivided by phylogeny into nine groups (Fig. 3).

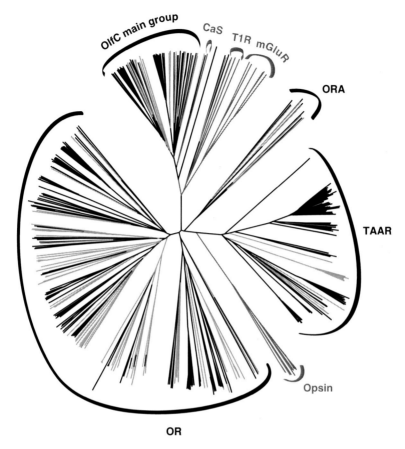

Fig. 3 Phylogenetic tree of four olfactory receptor gene families in zebrafish and fugu. Sequences were retrieved from Niimura and Nei (2005), zebrafish and fugu ORs; Saraiva and Korsching (2007), zebrafish and fugu ORAs; Hashiguchi and Nishida (2007), zebrafish and fugu TAARs; Alioto and Ngai (2006), zebrafish OlfCs; Hashiguchi and Nishida (2006), fugu OlfCs, and aligned using MAFFT version 6 (Katoh and Toh 2008). Outgroups used are metabotropic glutamate receptors, zebrafish opsins, calcium sensor, and T1R receptors. A handful of genes segregating with the outgroups were removed. The tree was constructed using the NJ algorithm (Thompson et al. 1997) and displayed using Hypertree (Bingham and Sudarsanam 2000). *Black lines*, zebrafish genes; *light grey lines*, fugu; *dark grey lines*, outgroups

With respect to the relatively large number of pseudogenes it should be noted that the reported prevalence of pseudogenes depends very much on the quality of the databases, i.e. many ostensible pseudogenes are due to errors in the databases. Indeed, in the second large study by Alioto and Ngai only 4 pseudogenes were found, but 143 intact genes plus at least 7 additional OR genes, for which only partial sequence was available (Alioto and Ngai 2005). In this publication, six of the groups from the first study were confirmed; one, not being monophyletic, was split into two groups, and two were considered as outside the scope of the OR family. Traditionally, OR genes have been subdivided into two classes, a so-called fish-specific class I, for which some mammalian family members were also later identified, and a class II, which contains most of the mammalian OR genes. Mammals have lost all but two of the eight to nine groups present in teleost fish, but large gene expansions in these two groups generated OR repertoires an order of magnitude larger than those found in teleost fish.

The zebrafish OR repertoire turns out to be several-fold larger than that of two pufferfish species, which have less than 50 OR genes (Alioto and Ngai 2005; Niimura and Nei 2005). Cloning efforts have been made for catfish, medaka, and loach OR genes (Irie-Kushiyama et al. 2004; Ngai et al. 1993; Yasuoka et al. 1999), but no genome-wide search has been performed for OR genes from other fish species and so it remains to be seen whether these differences in repertoire size reflect the difference between early and late diverging teleost fish, the genome compaction in the pufferfish genus, or, conceivably, the relative importance of the olfactory sense in these species.

The evolutionary origin of the OR gene family was elucidated by a comparison of teleost fish, amphibian, and mammalian OR repertoires. It appears that most, if not all, of the eight or nine teleost OR groups were already present in the common ancestor of teleosts and tetrapods (Alioto and Ngai 2005; Niimura and Nei 2005), and some OR genes even go back to the common ancestor of jawed and jawless fish (Freitag et al. 1999). The latter may be the ancestral genes of class I and II. However, a more thorough analysis of lamprey receptor genes will be necessary to obtain sustainable information about the evolutionary origin of ORs, considering that in a first attempt (Berghard and Dryer 1998) aminergic receptor lamprey genes had been missassigned as ORs.

Despite a slower rate of evolution compared to tetrapod OR families, teleost OR genes do show signs of positive selection (Alioto and Ngai 2005; Ngai et al. 1993). The selective pressure acting on a gene is quantifiable by the ratio of nonsynonymous to synonymous base exchanges (dN/dS), which can be calculated for a group of related genes, either as a global value or for each sequence position individually. Positive selection refers to the feature that amino acid-changing mutations are retained to a larger extent than neutral mutations, which results in a dN/dS value larger than one. The frequency of positive selection in the genome is controversial (Studer et al. 2008), but it is generally assumed to occur in transcription factors and some receptor families, including olfactory receptors (Bustamante et al. 2005). In a comparison of the whole zebrafish OR family, four sites showing positive selection were found, two of them within the transmembrane regions expected to contain the ligand binding site (Alioto and Ngai 2005).

The genomic location of teleost OR genes, like their mammalian counterparts, is characterized by the presence of several gene clusters, as well as some isolated genes (Alioto and Ngai 2005). Within the gene clusters, subfamilies are largely contiguous and subfamily members usually exhibit the same transcriptional orientation, suggesting tandem duplication as a mechanism of gene expansion. Teleost OR genes are generally expected to be monoexonic, at least for the coding region

(Alioto and Ngai 2005; Niimura and Nei 2005) like their mammalian counterparts, although no dedicated analyses have been performed to substantiate this point.

2.2 The Teleost ORA Family: A Small, Invariant Gene Repertoire

The *ora* genes are the teleost homologs of the mammalian V1R genes. The *ora* designation stands for olfactory receptor gene related to class A of G protein-coupled receptors (the mammalian designation VR stands for vomeronasal receptor, referring to the expression of VRs in a specialized olfactory organ of tetrapods not present in fish). *Ora* genes have been the latest of the four teleost olfactory receptor families to be found. The first member of this family was uncovered as late as 2005 (Pfister and Rodriguez 2005), followed by the identification of a second gene (Shi and Zhang 2007) and only in 2007 did the full extent of the family become known (Saraiva and Korsching 2007).

With respect to other teleost chemosensory receptor gene families, all fish *ORA* genes form a monophyletic clade, supporting their identification as a single family separate from the other chemosensory receptor families. The ORA clade includes all mammalian V1R receptors; thus the ORA family can be considered paraphyletic, with the mammalian V1Rs originating as a single subclade within the ORA family.

The *ora* receptor gene family is very small (Fig. 3), with only 6 members compared to over 100 genes in the corresponding rodent V1R gene family. Compared to the other four families, *ora* genes exhibit several peculiar characteristics, all of them unique for olfactory receptor gene families both in teleosts and tetrapods. Most strikingly, *ora* genes are highly conserved between all teleost species analyzed so far, such that individual orthologs for all six genes can be detected in all five teleost species analyzed so far (bar a single gene loss in the pufferfish genus) (Saraiva and Korsching 2007). Ortholog *ora* genes (closest homologs *between* species) are without exception more closely related to each other than any paralog *ora* genes (closest homologs *within* species), indicating that all six family members are evolutionarily much older than the speciation events in the teleost lineage. In fact, some *ora* genes can even be found in lamprey, a jawless fish (Saraiva and Korsching 2007).

Consistent with this very slow evolution, *ora* genes show no evidence for positive selection, in contrast to the other olfactory receptor families including the mammalian V1R family. The global dN/dS values for all six *ora* genes are very small, around 0.2, indicative of strong selective pressure on these genes. When dN/dS values are determined for each codon individually, no single instance of positive selection is found (Saraiva and Korsching 2007). For one of the *ora* genes an analysis of sequence variation was performed in a group of closely related fish species within the *Danio* genus; of several algorithms used all but one did not find evidence of positive selection (Pfister et al. 2007).

An unexpected feature of *ora* genes concerns their genomic location. Olfactory receptor genes often occur in clusters presumably arising from repeated local gene duplication and are often arranged in head-to-tail fashion. In contrast, four of the six *ora* genes are arranged in closely linked gene pairs across all fish species studied. These gene pairs are asymmetrical, head-to-head for *ora1/ora2* and tail-to-tail for *ora3/ora4*. A pairwise configuration in the phylogenetic tree suggests the existence of three ancestral ORA subclades, all of which are present in lamprey (Saraiva and Korsching 2007), one of which has been lost in amphibia, and a further one in mammals. Two subclades correspond to the above-mentioned gene pairs, the third one consisting of two isolated genes, *ora5* and *ora6*.

Another unexpected feature of the *ora* genes is the presence of introns in two of six genes. The ancestral genomic structure appears to be monoexonic for ORA1, ORA2, ORA5, and ORA6. This structure is maintained in the mammalian relatives of the ORA1–ORA2 clade (Dulac and Axel 1995; Grus et al. 2005). In marked contrast, ORA4 possesses two exons, and ORA3 four approximately equal-sized exons. ORA3 and ORA4 intron/exon borders are exactly conserved between teleost species, and the sole intron/exon border in ORA4 does not correspond to any intron/exon border in ORA3. Thus, the corresponding intron gains must have occurred after the genesis of the three subclades and indeed after the genomic rearrangements giving rise to the gene pairs, i.e., after the complete *ora* gene family was established.

All six *ora* genes are expressed specifically in the olfactory organ of zebrafish, in sparse cells within the sensory surface (Saraiva and Korsching 2007), consistent with the expectation for olfactory receptors and similar to the expression of the tetrapod subclade V1R. Taken together, the high conservation of the *ora* gene repertoire across teleosts, in striking contrast to the frequent species-specific expansions observed in tetrapods, especially mammalian V1Rs, possibly reflects a major shift in gene regulation as well as gene function upon the transition to tetrapods.

2.3 The Teleost TAAR Family Evolves Rapidly

Trace amine-associated receptors (TAARs) are related to G protein-coupled aminergic neurotransmitter receptors such as dopamine and serotonine receptors and recognize derivatives of the classical monoamines such as ß-phenylethylamine, octopamine, tryptamine, and tyramine. Initially, TAARs had been considered neurotransmitter receptors (Borowsky et al. 2001), but recently an expression in olfactory sensory neurons was shown for several mammalian *taar genes*, with expression characteristics very similar to odorant receptors (Liberles and Buck 2006). Thus the *taar* genes were recognized as a fourth GPCR family of olfactory receptors (Buck 2000).

Following the cloning of the first TAAR receptors in mammals (Borowsky et al. 2001), *taar* genes have been found in several genomes from lower vertebrates (Gloriam et al. 2005) including lamprey (Hashiguchi and Nishida 2007), but not in invertebrates. However, the delineation from classical aminergic neurotransmitter receptors has not been investigated thoroughly so far, and indeed the lamprey genes appear to represent aminergic receptors, not TAARs (Korsching, unpublished observation). The first study evaluating teleost *taar* genes by Gloriam et al. (2005) made use of very incomplete databases, and thus many of its conclusions, including the size of the family, the phylogenetic reconstruction, the genomic location, the frequency of pseudogenes, the absence of introns, and the suggested nomenclature are now outdated. Still valid are its observations that the TAAR gene family exhibits rapid evolution and correspondingly remarkably species-specific repertoires. A follow-up study confirmed these observations using a more complete data set (Hashiguchi and Nishida 2007). Particularly remarkable is the large *taar* gene repertoire of zebrafish (Fig. 3); over 100 genes are found, about 5 times the number of genes in the largest mammalian family, and double the number of *taar* genes found in stickleback (Hashiguchi and Nishida 2007). It will be interesting to study whether the selective pressure acting on teleost *taar* genes takes the form of positive selection, of which incidences have been observed in the OR, V1R, and V2R families. Currently, *taar* gene repertoires have been established for fugu, stickleback, medaka, and zebrafish. Fugu has the smallest repertoire, less than 20 genes, followed by medaka with 25 genes, stickleback with 49 genes, and zebrafish with 109 genes (Hashiguchi and Nishida 2007). Ligands for teleost TAARs have not been identified so far, but may include polyamines, which are specifically and sensitively detected by teleost fish (Michel et al. 2003; Rolen et al. 2003).

Taar genes occur in a single cluster in tetrapods, evidence of a genesis from local gene duplications, possibly via illegitimate crossover during meiotic recombination. In teleosts, *taar* genes form two large clusters (Hashiguchi and Nishida 2007), presumably resulting from the whole genome duplication occurring early in the teleost lineage (Nakatani et al. 2007). Additionally, several isolated genes and small groups are found; however, due to the still unfinished genome build in zebrafish, this may not be the final distribution. The most recent common ancestor of tetrapods and teleosts (of lobe-finned and ray-finned fishes) presumably already had a small cluster of *taar* genes.

Whereas all mammalian and all zebrafish *taar* genes are monoexonic, an intron was found in many medaka, fugu, and stickleback genes (Hashiguchi and Nishida 2007), consistent with an intron gain early in the evolution of neoteleosts, i.e., relatively late in vertebrate evolution. This is rather remarkable since several whole genome scanning studies found very little evidence for any intron gains during all of vertebrate evolution (Coulombe-Huntington and Majewski 2007; Loh et al. 2008) and may be related to the apparently low selective pressure in the *taar* gene family.

2.4 The Teleost OlfC Family is Paraphyletic

OLfC receptors belong to the class C of GPCRs like the mammalian V2Rs, which are their closest chemosensory relatives. Receptors of this class are characterized by their large N-terminal extracellular region, and their similarity to the metobotropic glutamate receptor. Unlike the other three olfactory receptor gene families they are not monophyletic, as three distinct clades have been lumped together under the olfC heading (cf. Fig. 3), following the lead of the mammalian nomenclature (Alioto and Ngai 2006). However, by far the largest clade, group I (Alioto and Ngai 2006), is monophyletic, and the closest phylogenetic neighbor of the largest mammalian V2R clade, albeit both clades are strictly segregated. Its repertoire size varies several-fold between teleost species, but is well within the range of mammalian V2R repertoires – extreme species-specific specialization has led to the complete loss of this family in several mammalian species (Young and Trask 2007). Again, zebrafish exhibits the largest of all teleost OlfC repertoires (Alioto and Ngai 2006; Hashiguchi and Nishida 2006). Two small clades of one to two genes per species appear to be more closely related to the T1R taste receptors and the calcium sensor in some phylogenetic reconstructions (Alioto and Ngai 2006). In contrast to genes from the large clade, genes from the small clades are expressed broadly in the olfactory epithelium (Sato et al. 2005; Speca et al. 1999), conceivably suggesting a modulatory or chaperone function for these genes (cf. Larsson et al. 2004; Silvotti et al. 2005).

OlfC genes exhibit five conserved intron/exon borders resulting in six exons in a characteristic short-short-long-short-short-long arrangement (Alioto and Ngai 2006). None of these intron/exon borders occur in the metabotropic glutamate receptors, but all are shared with the mammalian V2Rs and the closest outgroups, the calcium sensor and the T1R receptors (Korsching, unpublished observation), confirming a common evolutionary origin. To pinpoint the evolutionary origin, data mining in cartilaginous and jawless fish genomes will be required. However, the high species specificity suggests that OlfC/V2R genes constitute an evolutionary recent family, and accordingly may not have been present in the common ancestor of jawed and jawless fish.

Whole subfamilies of OlfC genes are present in zebrafish, but not in neoteleosts, and many instances of species-specific gene expansions are observed. The evolution of the OlfC family appears to be driven to a large extent by local gene duplication, as suggested by the arrangement of most OlfC genes in clusters of phylogenetically related genes (Alioto and Ngai 2006; Hashiguchi and Nishida 2006). Nevertheless, albeit relaxed negative selection is observed at distal ligand binding sites, there is no evidence of positive selection, and in particular the core residues of the amino acid binding motif characteristic for this family are under negative selection (Alioto and Ngai 2006). Although currently no ligands are known for any member of the largest group of OlfC genes (group 1), modelling suggests that many of them have amino acids as ligands like the one well-investigated OlfC member from one of the small groups, OlfC a1 (Luu et al. 2004). Thus, OlfC receptors may constitute the molecular basis to explain odor response studies, which predict many independent receptors for amino acids (Fuss and Korsching 2001).

3 A Comparison of Teleost and Tetrapod Olfactory Receptor Repertoires

3.1 Evolutionary Origin of the Four Olfactory Receptor Gene Families

The estimates for the evolutionary origin of the four olfactory receptor gene families OR, ORA, OlfC, and TAAR have been discussed in the preceding paragraphs. Notably there is no evidence for any of them occurring outside the chordate phylum and, in fact, so far members have only been found in vertebrates. Thus the chemosensory receptors of other phyla such as arthropods or nematodes should have an independent origin within or outside (cf. Sato et al. 2008) the G protein-coupled heptahelical receptor family – which itself dates back to the earliest eucaryotes.

The *ora* gene family appears to be the oldest family, close to the final repertoire size in lamprey, while the OR family occurs in lamprey but apparently with a much smaller gene repertoire than either later diverging teleost or tetrapod species. Neither OlfC nor TAAR receptor families have been detected in lamprey. It should be noted that some publications concerning lamprey receptor gene assignment have been performed without sufficient information about phylogenetically neighboring gene families, which has led to the erroneous assignment of whole receptor gene families (Berghard and Dryer 1998; Hashiguchi and Nishida 2007). Thus two of the four vertebrate-specific olfactory receptor gene families have their origin in jawless vertebrates, while two families appear to have emerged later in cartilaginous vertebrates.

3.2 Repertoire Size Varies Considerably Within and Between Teleost Species

Between different teleost species, pufferfish generally have the smallest olfactory receptor gene family sizes, followed by medaka (*Oryzias latipes*), whereas stickleback (*Gasterosteus aculeatus*) and zebrafish (*Danio rerio*) have several-fold larger repertoires than the two pufferfish species analyzed (*Takifugu rubripes* and *Tetraodon nigroviridis*). Apart from zebrafish, all these fish species belong to the *neoteleostei*, a modern group of teleosts, while zebrafish is a more primitive teleost, which belongs to the *ostariophysii*. Interestingly, it is the zebrafish which always has the largest olfactory receptor gene repertoires. The minimal repertoire sizes found in the pufferfish (Fig. 3) may conceivably be caused by gene loss related to the extreme genome compaction characteristic for this genus (Elgar 1996). Generally, lineage-specific gene gains as well as losses have shaped the teleost OR, TAAR, and V2R-related OlfC repertoires. On the other hand, the V1R-related *ora* genes have remained a small, rigidly conserved family throughout teleost evolution. Apart from a single gene loss in the pufferfish genus, not a single gene loss or gain event appears to have occurred throughout teleost evolution (Fig. 3).

3.3 Opposing Shifts in Family Characteristics upon Teleost/Tetrapod Transition in Two Olfactory Receptor Gene Families

Until a short time ago, conventional wisdom held that mammalian species possessed much larger olfactory receptor gene repertoires than those found in earlier diverged vertebrates such as fish. The recent assignment of TAARs as an olfactory receptor gene family (Liberles and Buck 2006) changed this narrative. By far the largest gene family in mammals is constituted by the OR gene repertoire, followed by the V1R and V2R families, and finally the rather limited TAAR gene repertoire. The evolutionary path followed in bony fish development has been distinctly different, with massive diversification in the TAAR family towards sizes close to those observed for teleost OR repertoires. As a result the teleost *taar* gene repertoires far surpass the mammalian repertoires. On the other hand, the lack of any expansion in the teleost ORA family results in the most extreme contrast in teleost vs tetrapod family size of all four olfactory receptor families.

3.4 Selective Pressure Among the Four Teleost Olfactory Gene Repertoires

A first indication of selective pressure or absence thereof is given by the species specificity of gene families, which is high for OR, OlfC, and TAARs. Quantitative determination of the selective pressure using dN/dS analysis has shown some incidences of positive selection in OR genes, and relaxed negative selection in OlfC genes, while TAARs still need to be analyzed in that respect. In contrast, the ORA genes show the lowest dN/dS values and a complete absence of positively selected sites (but see Pfister et al. 2007). A rapid evolution in chemosensory receptor gene families is consistent with either pheromonal functions which require species specificity or rapid adaptation to changing environmental specializations.

4 Beyond the Phylogenomic Analysis: Current Status of Expression Patterns, Connectivity, and Ligands

From an evolutionary point of view it is interesting to what extent the molecular and cellular logic of the mammalian system is similar to the corresponding teleost receptor gene families. For example, do the same cell types express the teleost receptors? Are the signal transduction pathways conserved? Is the logic of the axonal projection similar to the mammalian situation? With respect to the ligands, major differences between mammalian and teleost olfactory receptors might be expected due to the shift from aquatic to terrestrial life style and a corresponding shift in the biological relevance of many or most odors.

4.1 Expression Frequency and Spatial Expression Patterns of Teleost Olfactory Receptor Genes

Limited data are available concerning the tissue specificity of olfactory receptor gene expression in teleosts. For mammalian OR an extra-olfactory expression of quite a few receptor genes is known, and for two receptors a function in sperm cell chemotaxis has been postulated (Fukuda et al. 2004; Spehr et al. 2003).

Odorant receptor genes are expressed according to the one receptor neuron – one receptor gene rule, often referred to as monogenic expression. This has generally been true for the other three olfactory receptor families as well, with the exception of a few very broadly expressed receptor genes which may function as chaperones for other receptor genes. The statistics of teleost OR, ORA, and TAAR gene expression appears to conform to the monogenic expression, as most receptor genes, for which reasonably specific probes can be obtained, are expressed in very sparse cell populations, e.g., Weth et al. (1996). Double-labeling studies mostly show exclusive expression of different olfactory receptor gene combinations (Barth et al. 1997), with the exception of a few very closely related receptor gene pairs (Sato et al. 2007).

Within the main olfactory epithelium of mammals, receptor neurons expressing the same OR gene or the same TAAR gene are found at apparently random positions within medial-to-lateral subdivisions of the sensory surface, so-called expression zones (cf. Mombaerts 2004). A small subgroup of genes is expressed in perpendicular expression domains (Hoppe et al. 2006). Within the accessory or vomeronasal olfactory epithelium V1R and V2R-expressing receptor neurons segregate into an apical and a basal layer, respectively. Expression of individual VR genes appears to be restricted to apical-to-basal subdivisions within these layers (Ryba and Tirindelli 1997), although no systematic study has been performed to analyze segregation in this dimension.

Fish OR genes are expressed in spatial expression domains that are broader and overlap more than the mammalian expression zones, but otherwise have the same characteristics of several receptors intermingling in one domain (Weth et al. 1996).

4.2 Ligands for Teleost Olfactory Receptor Molecules

The olfactory receptor genes can be considered the first layer of olfactory information processing and in fact they define the nature of odorants, since any molecule becomes an odorant solely by virtue of its interaction with an olfactory receptor. Indeed, the olfactory receptors can be thought of as the dimensions of the multidimensional odor space, with each odorant or mixture of odorants defined by a unique set of coordinates in this odor space. Thus a thorough understanding of ligand/receptor interactions is an essential component to fully decipher the logic of olfactory coding.

Not many olfactory receptor genes are currently deorphanized, first due to the sheer complexity of the task, and second because heterologous expression is inefficient for many olfactory receptors. Consequently, detailed ligand response spectra so far only exist for a handful of mammalian olfactory receptors. The general conclusion is that of a relaxed specificity, i.e., several mostly related compounds can excite a particular receptor. With respect to the physicochemical nature of the ligands, one may expect drastic differences between teleost and tetrapod OR and TAAR receptors, since the latter ought to recognize hydrophobic, volatile substances. Teleost and tetrapod V1R and V2R could in principle have similar sets of ligands, because their ligands are transported in mucus towards the receptors, and thus are expected to be hydrophilic. However, the available data do not support this hypothesis.

The only fish olfactory receptor with identified ligands is a member of the OlfC family, OlfC a1 (Alioto and Ngai 2006). It recognizes several amino acids with graded affinity in an heterologous expression system. Interestingly, the optimal ligands for the goldfish receptor are basic amino acids, whereas the zebrafish receptor reacts most strongly to acidic amino acids. Mutation studies have identified the likely residues responsible for this shift in ligand binding characteristics (Luu et al. 2004). It is possible that most OlfC receptors will turn out to bind amino acids, since they share a predicted amino acid-binding motif (Alioto and Ngai 2006).

4.3 Odor Responses from the Cellular to the Behavioral Level

Known physiologically relevant odorants for teleost fish comprise amino acids, polyamines and nucleotides (food signals), bile acids, steroids and prostaglandins (pheromones) and alarm substances (of graded species-specificity, and so far unidentified molecular nature). These odor signals are detected and processed with high resolution; for example, fish can be trained to distinguish nearly all amino acids from one another (Valentincic et al. 2000, 2005). Fish possess three types of olfactory receptor neurons – ciliated and microvillous receptor neurons, and so-called crypt cells possessing both cilia and microvilli (Hansen and Zielinski 2005), which so far have not been observed in mammalian olfactory epithelia. In contrast to mammals with at least four segregated olfactory organs (Ma 2007), all fish olfactory neurons are intermingled in a single sensory surface (Hamdani and Doving 2007; Hansen et al. 2004), the olfactory epithelium. The projection areas of ciliated and microvillous receptor neurons have been established by dye tracing and genetic studies to be segregated, if somewhat intertwined (Hamdani and Doving 2002; Hansen et al. 2003; Morita and Finger 1998; Sato et al. 2005); for a detailed discussion see Yoshihara, this volume). This segregation seems to be carried over to the functionally different medial and lateral tracts of projection neurons innervating ciliated and microvillous receptor neurons, respectively (Hamdani and Doving 2007). Interestingly, in contrast to the mammalian system, ciliated neurons and microvillous receptor neurons terminate both in morphologically distinct glomeruli about 50–100 μm in diameter (cf. Baier and Korsching 1994; Sato et al. 2005), and smaller microglomeruli, sometimes referred to as aglomerular plexus (e.g., Baier and Korsching 1994). Anatomical studies have linked crypt cells to small groups of centrally lying ventral glomeruli (Hamdani and Doving 2006; Hansen et al. 2003), which may be innervated by a lateral subdivision of the medial olfactory tract (Hamdani and Doving 2007).

Electrophysiological studies of isolated receptor neurons (Nikonov and Caprio 2007; Restrepo et al. 1990) and projection neurons in the olfactory bulb (Lastein et al. 2006), as well as imaging of odor responses in the olfactory bulb (Friedrich and Korsching 1998) in conjunction with anatomical studies (Baier and Korsching 1994; Sato et al. 2005) have allowed some correlations of receptor neuron type with odor responses on the one hand and olfactory receptor classes on the other. The receptor families seem to be restricted to particular receptor neuron populations, ORs expressed in ciliated neurons, and OlfC in microvillous receptor neurons; see, e.g., Hansen et al. (2004) and Speca et al. (1999). It will be interesting to see whether ORAs are expressed in crypt cells, a sparse cell population, for which so far no receptors have been described (cf. Pfister and Rodriguez 2005; Saraiva and Korsching 2007). The orthologous mammalian V1R genes are expressed in microvillous cells. TAARs may be expressed in ciliated neurons like their mammalian counterparts. Fish microvillous receptor neurons appear to react to amino acids and nucleotides, whereas ciliated neurons may carry the response to bile acids, steroids and polyamines via ORs and TAARs, respectively. Crypt cells of a mackerel species have been shown to respond to amino acids (Schmachtenberg 2006; Vielma et al. 2008). However, this result appears to conflict with a combination of electrophysiological studies (Lastein et al. 2006) and backtracing experiments (Hamdani and Doving 2006) that show a response to steroids in the target region of crypt cells in the olfactory bulb of crucian carp.

Taken together, the fish olfactory bulb provides for the opportunity to study functionally segregated responses of all olfactory receptor neurons in a receptotopic map. Due to the small size and semi-transparent nature of the zebrafish olfactory bulb, it is to be expected that odor responses of all three receptor neuron populations could be measured simultaneously and possibly identified by spatial position. Indeed, in the zebrafish olfactory bulb it has been possible to measure odor responses in lateral, medial, and ventral glomeruli (Friedrich and Korsching 1997, 1998).

5 Outlook

The recent discovery of yet another olfactory receptor gene family (the TAARs) invites the speculation that the current repertoire of four different teleost olfactory receptor gene families may still not be complete. Indeed, an olfactory function has been shown for a mammalian member of the membrane-bound guanylate cyclase family (Hu et al. 2007; Leinders-Zufall et al. 2007). The corresponding gene family in teleost fish is known to be larger than in mammals (Yamagami and Suzuki 2005), but a systematic genome-wide study still needs to be done. For the four known families, the repertoires published for several teleost fish species appear reasonably complete. However, for none of these families is evolution in early vertebrates clearly understood. Future access to more and higher quality genome sequences of jawless and cartilaginous fish will enable such studies to be performed more thoroughly than currently possible.

The evolutionary path followed in teleost fish development has been distinctly different from that pursued in tetrapods, with massive diversification in the TAAR family but with a total lack of any expansion in the ORA family. It is expected that the respective ligand repertoires and corresponding biological function will turn out to be distinctly different as well. However, extensive progress in deorphanizing teleost olfactory receptors will be necessary to understand the evolution of ligand repertoires for the olfactory receptor gene families.

Acknowledgement I would like to thank Kim Robin Korsching for helping with the figures.

References

Alioto TS, Ngai J (2005) The odorant receptor repertoire of teleost fish. BMC Genomics 6:173

Alioto TS, Ngai J (2006) The repertoire of olfactory C family G protein-coupled receptors in zebrafish: candidate chemosensory receptors for amino acids. BMC Genomics 7:309

Baier H, Korsching S (1994) Olfactory glomeruli in the zebrafish form an invariant pattern and are identifiable across animals. J Neurosci 14:219–230

Barth AL, Dugas JC, Ngai J (1997) Noncoordinate expression of odorant receptor genes tightly linked in the zebrafish genome. Neuron 19:359–369

Berghard A, Dryer L (1998) A novel family of ancient vertebrate odorant receptors. J Neurobiol 37:383–392

Bingham J, Sudarsanam S (2000) Visualizing large hierarchical clusters in hyperbolic space. Bioinformatics 16:660–661

Borowsky B, Adham N, Jones KA, Raddatz R, Artymyshyn R, Ogozalek KL, Durkin MM, Lakhlani PP, Bonini JA, Pathirana S et al. (2001) Trace amines: identification of a family of mammalian G protein-coupled receptors. Proc Natl Acad Sci USA 98:8966–8971

Buck LB (2000) The molecular architecture of odor and pheromone sensing in mammals. Cell 100:611–618

Buck L, Axel R (1991) A novel multigene family may encode odorant receptors: a molecular basis for odor recognition. Cell 65:175–187

Bustamante CD, Fledel-Alon A, Williamson S, Nielsen R, Hubisz MT, Glanowski S, Tanenbaum DM, White TJ, Sninsky JJ, Hernandez RD et-al.. (2005) Natural selection on protein-coding genes in the human genome. Nature 437:1153–1157

Coulombe-Huntington J, Majewski J (2007) Characterization of intron loss events in mammals. Genome Res 17:23–32

Dulac C, Axel R (1995) A novel family of genes encoding putative pheromone receptors in mammals. Cell 83:195–206

Elgar G (1996) Quality not quantity: the pufferfish genome. Hum Mol Genet 5(Spec No):1437–1442

Freitag J, Beck A, Ludwig G, von Buchholtz L, Breer H (1999) On the origin of the olfactory receptor family: receptor genes of the jawless fish (*Lampetra fluviatilis*). Gene 226:165–174

Fried HU, Fuss SH, Korsching SI (2002) Selective imaging of presynaptic activity in the mouse olfactory bulb shows concentration and structure dependence of odor responses in identified glomeruli. Proc Natl Acad Sci USA 99:3222–3227

Friedrich RW, Korsching SI (1997) Combinatorial and chemotopic odorant coding in the zebrafish olfactory bulb visualized by optical imaging. Neuron 18:737–752

Friedrich RW, Korsching SI (1998) Chemotopic, combinatorial, and noncombinatorial odorant representations in the olfactory bulb revealed using a voltage-sensitive axon tracer. J Neurosci 18:9977–9988

Fukuda N, Yomogida K, Okabe M, Touhara K (2004) Functional characterization of a mouse testicular olfactory receptor and its role in chemosensing and in regulation of sperm motility. J Cell Sci 117:5835–5845

Fuss SH, Korsching SI (2001) Odorant feature detection: activity mapping of structure response relationships in the zebrafish olfactory bulb. J Neurosci 21:8396–8407

Gloriam DE, Bjarnadottir TK, Yan YL, Postlethwait JH, Schioth HB, Fredriksson R (2005) The repertoire of trace amine G-protein-coupled receptors: large expansion in zebrafish. Mol Phylogenet Evol 35:470–482

Grus WE, Shi P, Zhang YP, Zhang J (2005) Dramatic variation of the vomeronasal pheromone receptor gene repertoire among five orders of placental and marsupial mammals. Proc Natl Acad Sci USA 102:5767–5772

Hamdani El H, Doving KB (2002) The alarm reaction in crucian carp is mediated by olfactory neurons with long dendrites. Chem Senses 27:395–398

Hamdani El H, Doving KB (2006) Specific projection of the sensory crypt cells in the olfactory system in crucian carp, *Carassius carassius*. Chem Senses 31:63–67

Hamdani El H, Doving KB (2007) The functional organization of the fish olfactory system. Prog Neurobiol 82:80–86

Hansen A, Zielinski BS (2005) Diversity in the olfactory epithelium of bony fishes: development, lamellar arrangement, sensory neuron cell types and transduction components. J Neurocytol 34:183–208

Hansen A, Rolen SH, Anderson K, Morita Y, Caprio J, Finger TE (2003) Correlation between olfactory receptor cell type and function in the channel catfish. J Neurosci 23:9328–9339

Hansen A, Anderson KT, Finger TE (2004) Differential distribution of olfactory receptor neurons in goldfish: structural and molecular correlates. J Comp Neurol 477:347–359

Hashiguchi Y, Nishida M (2006) Evolution and origin of vomeronasal-type odorant receptor gene repertoire in fishes. BMC Evol Biol 6:76

Hashiguchi Y, Nishida M (2007) Evolution of trace amine associated receptor (TAAR) gene family in vertebrates: lineage-specific expansions and degradations of a second class of vertebrate chemosensory receptors expressed in the olfactory epithelium. Mol Biol Evol 24:2099–2107

Hoppe R, Lambert TD, Samollow PB, Breer H, Strotmann J (2006) Evolution of the "OR37" subfamily of olfactory receptors: a cross-species comparison. J Mol Evol 62:460–472

Hu J, Zhong C, Ding C, Chi Q, Walz A, Mombaerts P, Matsunami H, Luo M (2007) Detection of near-atmospheric concentrations of CO2 by an olfactory subsystem in the mouse. Science 317:953–957

Irie-Kushiyama S, Asano-Miyoshi M, Suda T, Abe K, Emori Y (2004) Identification of 24 genes and two pseudogenes coding for olfactory receptors in Japanese loach, classified into four subfamilies: a putative evolutionary process for fish olfactory receptor genes by comprehensive phylogenetic analysis. Gene 325:123–135

Kajiya K, Inaki K, Tanaka M, Haga T, Kataoka H, Touhara K (2001) Molecular bases of odor discrimination: Reconstitution of olfactory receptors that recognize overlapping sets of odorants. J Neurosci 21:6018–6025

Katoh K, Toh H (2008) Recent developments in the MAFFT multiple sequence alignment program. Brief Bioinform 9:286–298

Lane RP, Young J, Newman T, Trask BJ (2004) Species specificity in rodent pheromone receptor repertoires. Genome Res 14:603–608

Larsson MC, Domingos AI, Jones WD, Chiappe ME, Amrein H, Vosshall LB (2004) Or83b encodes a broadly expressed odorant receptor essential for Drosophila olfaction. Neuron 43:703–714

Lastein S, Hamdani El H, Doving KB (2006) Gender distinction in neural discrimination of sex pheromones in the olfactory bulb of crucian carp, *Carassius carassius*. Chem Senses 31:69–77

Leinders-Zufall T, Cockerham RE, Michalakis S, Biel M, Garbers DL, Reed RR, Zufall F, Munger SD (2007) Contribution of the receptor guanylyl cyclase GC-D to chemosensory function in the olfactory epithelium. Proc Natl Acad Sci USA 104:14507–14512

Liberles SD, Buck LB (2006) A second class of chemosensory receptors in the olfactory epithelium. Nature 442:645–650

Loh YH, Brenner S, Venkatesh B (2008) Investigation of loss and gain of introns in the compact genomes of pufferfishes (Fugu and Tetraodon). Mol Biol Evol 25:526–535

Luu P, Acher F, Bertrand HO, Fan J, Ngai J (2004) Molecular determinants of ligand selectivity in a vertebrate odorant receptor. J Neurosci 24:10128–10137

Ma M (2007) Encoding olfactory signals via multiple chemosensory systems. Crit Rev Biochem Mol Biol 42:463–480

Matsunami H, Buck LB (1997) A multigene family encoding a diverse array of putative pheromone receptors in mammals. Cell 90:775–784

Michel WC, Sanderson MJ, Olson JK, Lipschitz DL (2003) Evidence of a novel transduction pathway mediating detection of polyamines by the zebrafish olfactory system. J Exp Biol 206:1697–1706

Mombaerts P (2004) Genes and ligands for odorant, vomeronasal and taste receptors. Nat Rev Neurosci 5:263–278

Morita Y, Finger TE (1998) Differential projections of ciliated and microvillous olfactory receptor cells in the catfish, *Ictalurus punctatus*. J Comp Neurol 398:539–550

Nakatani Y, Takeda H, Kohara Y, Morishita S (2007) Reconstruction of the vertebrate ancestral genome reveals dynamic genome reorganization in early vertebrates. Genome Res 17:1254–1265

Ngai J, Dowling MM, Buck L, Axel R, Chess A (1993) The family of genes encoding odorant receptors in the channel catfish. Cell 72:657–666

Niimura Y, Nei M (2005) Evolutionary dynamics of olfactory receptor genes in fishes and tetrapods. Proc Natl Acad Sci USA 102:6039–6044

Nikonov AA, Caprio J (2007) Highly specific olfactory receptor neurons for types of amino acids in the channel catfish. J Neurophysiol 98:1909–1918

Pfister P, Rodriguez I (2005) Olfactory expression of a single and highly variable V1r pheromone receptor-like gene in fish species. Proc Natl Acad Sci USA 102:5489–5494

Pfister P, Randall J, Montoya-Burgos JI, Rodriguez I (2007) Divergent evolution among teleost V1r receptor genes. PLoS ONE 2:e379

Restrepo D, Miyamoto T, Bryant BP, Teeter JH (1990) Odor stimuli trigger influx of calcium into olfactory neurons of the channel catfish. Science 249:1166–1168

Rolen SH, Sorensen PW, Mattson D, Caprio J (2003) Polyamines as olfactory stimuli in the goldfish *Carassius auratus*. J Exp Biol 206:1683–1696

Ryba NJ, Tirindelli R (1997) A new multigene family of putative pheromone receptors. Neuron 19:371–379

Saraiva LR, Korsching SI (2007) A novel olfactory receptor gene family in teleost fish. Genome Res 17:1448–1457

Sato Y, Miyasaka N, Yoshihara Y (2005) Mutually exclusive glomerular innervation by two distinct types of olfactory sensory neurons revealed in transgenic zebrafish. J Neurosci 25:4889–4897

Sato Y, Miyasaka N, Yoshihara Y (2007) Hierarchical regulation of odorant receptor gene choice and subsequent axonal projection of olfactory sensory neurons in zebrafish. J Neurosci 27:1606–1615

Sato K, Pellegrino M, Nakagawa T, Nakagawa T, Vosshall LB, Touhara K (2008) Insect olfactory receptors are heteromeric ligand-gated ion channels. Nature 452:1002–1006

Schmachtenberg O (2006) Histological and electrophysiological properties of crypt cells from the olfactory epithelium of the marine teleost *Trachurus symmetricus*. J Comp Neurol 495:113–121

Shi P, Zhang J (2007) Comparative genomic analysis identifies an evolutionary shift of vomeronasal receptor gene repertoires in the vertebrate transition from water to land. Genome Res 17:166–174

Silvotti L, Giannini G, Tirindelli R (2005) The vomeronasal receptor V2R2 does not require escort molecules for expression in heterologous systems. Chem Senses 30:1–8

Speca DJ, Lin DM, Sorensen PW, Isacoff EY, Ngai J, Dittman AH (1999) Functional identification of a goldfish odorant receptor. Neuron 23:487–498

Spehr M, Gisselmann G, Poplawski A, Riffell JA, Wetzel CH, Zimmer RK, Hatt H (2003) Identification of a testicular odorant receptor mediating human sperm chemotaxis. Science 299:2054–2058

Sprague J, Bayraktaroglu L, Bradford Y, Conlin T, Dunn N, Fashena D, Frazer K, Haendel M, Howe DG, Knight J et al. (2008) The zebrafish information network: the zebrafish model organism database provides expanded support for genotypes and phenotypes. Nucleic Acids Res 36:D768D772

Studer RA, Penel S, Duret L, Robinson-Rechavi M (2008) Pervasive positive selection on duplicated and nonduplicated vertebrate protein coding genes. Genome Res 11:11

Thompson JD, Gibson TJ, Plewniak F, Jeanmougin F, Higgins DG (1997) The CLUSTAL_X windows interface: flexible strategies for multiple sequence alignment aided by quality analysis tools. Nucleic Acids Res 25:4876–4882

Valentincic T, Kralj J, Stenovec M, Koce A, Caprio J. (2000) The behavioral detection of binary mixtures of amino acids and their individual components by catfish. J Exp Biol 203:3307–3317

Valentincic T, Miklavc P, Dolenek J, Pliberek K (2005) Correlations between olfactory discrimination, olfactory receptor neuron responses and chemotopy of amino acids in fishes. Chem Senses 30(Suppl 1):i312–i314

Vielma A, Ardiles A, Delgado L, Schmachtenberg O (2008) The elusive crypt olfactory receptor neuron: evidence for its stimulation by amino acids and cAMP pathway agonists. J Exp Biol 211:2417–2422

Weth F, Nadler W, Korsching S (1996) Nested expression domains for odorant receptors in zebrafish olfactory epithelium. Proc Natl Acad Sci USA 93:13321–13326

Yamagami S, Suzuki N (2005) Diverse forms of guanylyl cyclases in medaka fish – their genomic structure and phylogenetic relationships to those in vertebrates and invertebrates. Zoolog Sci 22:819–835

Yasuoka A, Endo K, Asano-Miyoshi M, Abe K, Emori Y (1999) Two subfamilies of olfactory receptor genes in medaka fish, *Oryzias latipes*: genomic organization and differential expression in olfactory epithelium. J Biochem 126:866–873

Young JM, Trask BJ (2007) V2R gene families degenerated in primates, dog and cow, but expanded in opossum. Trends Genet 23:212–215

Zhang X, Rodriguez I, Mombaerts P, Firestein S (2004) Odorant and vomeronasal receptor genes in two mouse genome assemblies. Genomics 83:802–811

Odorant Receptor Gene Choice and Axonal Projection in the Mouse Olfactory System

T. Imai and H. Sakano

Abstract In the mouse olfactory system, each olfactory sensory neuron (OSN) expresses a single type of odorant receptor (OR) out of approximately 1,000 in a monoallelic manner. Furthermore, OSNs expressing the same OR converge their axons to a specific set of glomeruli on the olfactory bulb. These two basic principles are fundamental to the peripheral olfactory system, and are regulated by the expressed OR protein itself. Singular OR gene choice is ensured by the combination of stochastic enhancer–promoter interaction and negative-feedback regulation by OR proteins. In the axonal projection, OR-derived cyclic AMP signals and neuronal activity determine the expression levels of axon guidance/sorting molecules, and thereby direct glomerular positioning and axon sorting.

1 Introduction

In the mouse, odorants are detected by approximately 1,000 odorant receptors (ORs) expressed by olfactory sensory neurons (OSNs) (Buck and Axel 1991; Zhang and Firestein 2002; Zhao et al. 1998). ORs are G-protein-coupled receptors with seven-transmembrane domains and transduce odorant binding signals to neuronal activity via cyclic AMP (cAMP). The depolarization signals are transmitted to glomeruli in the olfactory bulb (OB) of the brain. There are two basic principles for organizing the peripheral olfactory system. One is the one neuron–one receptor rule: each OSN expresses only one functional OR gene in a monoallelic manner, thereby establishing the neuronal identity of OSNs (Chess et al. 1994; Malnic et al. 1999). The other is the OR-instructed axonal projection: OSNs expressing the same OR converge their axons to a specific set of glomeruli in the OB (Vassar et al. 1994; Ressler et al. 1994; Mombaerts et al. 1996). Thus, the odor information detected in the olfactory

H. Sakano (✉)
Department of Biophysics and Biochemistry, Graduate School of Science,
The University of Tokyo, 2-11-16 Yayoi, Bunkyo-ku, Tokyo 113-0032, Japan
e-mail: sakano@mail.ecc.u-tokyo.ac.jp

epithelium (OE) by approximately 1,000 different ORs are topographically represented by approximately 1,000 pairs of glomeruli in each OB. This conversion of odor information to an odor map, consisting of a set of activated glomeruli that is specific for each odor, enables the mammalian brain to detect and discriminate various odorants (for a review, see Mori et al. 2006). Remarkably, both basic rules mentioned above are maintained and regulated by the expressed OR protein itself. In this chapter, we review recent studies that have focused on the mechanisms of OR gene choice and OR-instructed axonal projection in the mouse.

2 Axonal Projection

2.1 Organization and Structure of the OR Genes

The mammalian olfactory system is capable of detecting and discriminating a wide variety of airborne chemicals. To cover such a diverse array of odorants, more than 1,000 functional OR genes in the mouse, comprising approximately 4% of all protein-coding genes in the genome, are dedicated to olfaction (Zhang and Firestein 2002). The discovery of OR genes by Buck and Axel (1991) was recognized with the Nobel prize in physiology or medicine in 2004. The mouse genome contains 1,375 OR genes, among which 1,194 have intact open reading frames (Zhang et al. 2007). Most OR genes are clustered at 43 different loci, spread out among nearly all chromosomes, while 21 OR genes are isolated. Each cluster can contain up to 262 different OR genes, with an average intergenic distance of approximately 25 kb. Most OR genes are composed of two to five exons, of which only one exon encodes the entire protein. Vertebrate OR genes are divided phylogenetically into two different classes, I and II. The class I genes resemble the fish OR genes, and are responsive to water-soluble odorants such as aliphatic acids, aldehydes, and alcohols (Malnic et al. 1999; Kobayakawa et al. 2007), while the class II genes are unique to terrestrial vertebrates. In the mouse, there are 158 class I OR genes, clustered at a single locus on chromosome 7. Most class I OR genes are expressed in the dorsomedial zone of the OE (Zhang et al. 2004; Tsuboi et al. 2006).

2.2 One Neuron–One Receptor Rule in the Mouse Olfactory System

To discriminate a variety of odorous information, the mammalian olfactory system uses a remarkable decoding strategy: the OE contains approximately 1,000 types of OSNs, each expressing only one type of OR. Earlier analyses of in situ hybridization led to the conclusion that each OSN expresses a limited member of OR genes, possibly only one (Strotmann et al. 1992; Ngai et al. 1993; Ressler et al. 1993; Vassar et al.1993). Subsequent complementary DNA analyses of single OSNs revealed that only one OR gene is expressed in each cell (Malnic et al. 1999). Furthermore, allelic exclusion was observed in the OR gene system when OR transcripts were assayed by PCR for polymorphisms that differ between parental alleles; the OR gene is expressed from either the maternal or the paternal allele (Chess et al. 1994). Gene-targeted mice, in which both alleles were differentially tagged, demonstrated the monoallelic expression of the OR gene (Strotmann et al. 2000): no OSN expressed both markers simultaneously. RNA fluorescence in situ hybridization (FISH) and DNA-FISH experiments also confirmed the monoallelic activation of OR genes within OSN nuclei (Ishii et al. 2001). In this last experiment, DNA-FISH detected two genomic loci in the nuclei, one for each parental allele, while RNA-FISH detected the primary transcript at only one of the two genomic sites. The allelic exclusion provides even more diversity for odor recognition by allowing the distinction of polymorphic allele differences.

Exclusion has also been found between endogenous and transgenic OR gene alleles, and even among tandemly linked transgenic alleles with identical coding and promoter sequences (Serizawa et al. 2000). These results demonstrate that under any circumstance, no more than one functional OR gene allele can be expressed in each neuron. The OR gene choice in mouse appears to be stochastic, in contrast to the genetically determined OR gene choice in the fly with only approximately 50 OR genes (Ray et al. 2007). The monogenic and monoallelic mode of OR gene expression is known as the "one neuron–one receptor rule." Such an unusual mode of gene expression has previously been known only for antigen receptor genes of the immune system.

2.3 cis-DNA Elements and Positive Regulation

How is it, then, that a single OR gene is chosen and activated from a repertoire of 2,000? On the basis of previous studies on other multigene families, three activation mechanisms have been considered for the choice and activation of OR genes (Kratz et al. 2002): (1) DNA recombination, which brings a promoter and an enhancer region into close proximity; (2) gene conversion, which transfers a copy of the gene into the expression cassette; and (3) locus control region (LCR), which physically interacts with only one particular OR gene promoter. Irreversible DNA changes, i.e., recombination and gene conversion, have been attractive explanations for ensuring single OR gene expression because of the many parallels between the immune and the olfactory system. However, two groups have independently cloned mice from postmitotic OSN nuclei, and found no irreversible DNA changes in the OR genes (Eggan et al. 2004;

Li et al. 2004). Furthermore, these mice did not exhibit monoclonal OR gene expression, suggesting that OR gene choice is regulated by epigenetic mechanisms.

The third possibility, the LCR model, was tested on the *MOR28* gene cluster. Sequence comparison of the mouse and human genomes revealed a 2-kb homology region (the H region) far upstream of the *MOR28* cluster (Nagawa et al. 2002). Deletion of H region abolished the expression of all OR gene members in the *MOR28* cluster on a yeast artificial chromosome transgenic construct (Serizawa et al. 2003). Interestingly, when H region was relocated closer to the cluster, the number of OSNs expressing the proximal OR gene was greatly increased. These results indicate that H region is a cis-acting LCR that activates the *MOR28* cluster (Fig. 1b). A recent biochemical study suggested that H region may even activate in trans other OR gene loci on different chromosomes (Lomvardas et al. 2006). However, more recent, targeted deletions of H region in knockout mice contradict the *trans*-acting enhancer model: H region acts locally only in cis, but not in *trans* (Fuss et al. 2007; Nishizumi et al. 2007). Indeed, multiple LCR-like elements are found for zebrafish OR genes (Nishizumi et al. 2007).

Mutational analyses of H region and promoter regions both revealed the existence of at least two essential DNA elements: an O/E-like site and a homeodomain site (Vassalli et al. 2002; Rothman et al. 2005; Nishizumi et al. 2007). Mutations of these elements abolish the expression of OR genes in the transgenic systems. O/E proteins that bind to O/E-like sites and Lhx2 that binds to a homeodomain site are both required for OSN development (Wang et al. 2004; Hirota and Mombaerts 2004; Kolterud et al. 2004; Hirota et al. 2007).

2.4 Negative-Feedback Regulation by OR Proteins

It appears that the activation complex formed in the LCR interacts with only one promoter site in the OR gene cluster. However, this would not preclude the activation of a second OR from another OR gene cluster. To achieve mutually exclusive expression, negative-feedback regulation may be needed to prohibit the further activation of other OR genes or OR gene clusters. Possible involvement of negative-feedback regulation in the maintenance of the one neuron–one receptor rule was demonstrated by the null or frame-shift mutation of OR genes (Serizawa et al. 2003; Lewcock and Reed 2004). When the coding region of an OR gene was deleted, OSNs expressing the deletion allele were found to coexpress other members of OR genes (Fig. 1a). Similar results were obtained with naturally occurring frame-shift mutants of OR genes (Fig. 1a). Forced expression of transgenic OR genes by artificial promoters also suppresses the expression of endogenous OR genes (Nguyen et al. 2007). These observations suggest that the OR gene product – not messenger RNA but protein – has a regulatory role in preventing the secondary activation of other OR genes (Fig. 1b). However, the exact nature of the negative-feedback signal is still to be explored. Given that the genetic manipulation that uncouples ORs from

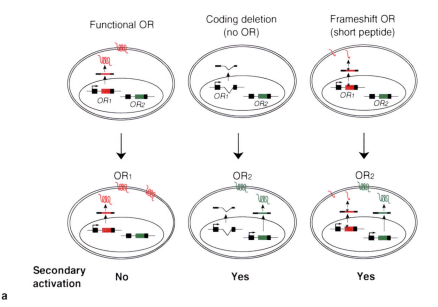

Fig. 1 Odorant receptor (*OR*) gene choice. **a** Negative-feedback regulation by the OR proteins ensures the one neuron–one receptor rule (Serizawa et al. 2003; Lewcock and Reed 2004). When the activated OR gene (*OR1*) expresses functional OR proteins, activation of other OR
(continued)

G-protein activation does not cause coexpression of multiple OR genes (Imai et al. 2006; Nguyen et al. 2007), negative feedback may not require G-protein-mediated signaling pathways. Targets of the feedback signals are also issues for future studies. Promoters of OR genes, enhancers of OR gene clusters, and/or protein factors binding them could be silenced by the OR-derived negative-feedback signals.

3 Axonal Projection

3.1 Axonal Convergence

How is odor information represented in the OB? In situ hybridization studies indicated that OSNs expressing the same OR converge their axons to specific sites on the OB; typically a pair of glomeruli per bulb (Vassar et al. 1994; Ressler et al. 1994). These results were later confirmed with a genetic labeling method (Mombaerts et al. 1996), which introduces an internal ribosome entry site and an axonal marker gene, *tau-lacZ*, into the 3′ untranslated region of the *P2 OR* gene. This technique has widely been used to visualize axonal projection for various OR genes (Fig. 2a).

Because OSNs expressing the same OR are scattered in the OE, topographic reorganization must occur during the process of axonal projection to the OB. This contrasts with the axonal projection seen in other sensory systems, e.g., visual, somatosensory, and auditory systems, where relative positional information is preserved between the periphery and the brain (for a review, see Flanagan 2006). Since such reorganization could be the basis for the complex informational processing in the brain, axonal projection of OSNs may serve as an excellent model system to study the molecular basis of neuronal wiring in the brain.

Fig. 1 (continued) genes (*OR2*) is inhibited (*left*). When the coding region is deleted, the activated mutant gene (OR1) cannot suppress the secondary activation of other OR genes (OR2) (*middle*). From the frame-shifted pseudogenes, the full-length messenger RNA is synthesized. However, such pseudogenes express only short abnormal peptides because of premature stop codons created by frame-shift mutations, and thus permit activation of other OR genes (OR2) (*right*). b A model for the OR gene expression. The activation complex formed in the locus control region (LCR) stochastically chooses one promoter (*P*) site by random collision, activating one particular OR gene member within the cluster (positive regulation). Once the activated gene has been expressed, the functional OR molecules transmit inhibitory signals to block further activation of additional OR genes or clusters (negative regulation). Stochastic activation of an OR gene and negative-feedback regulation by the OR gene product, together, can ensure the maintenance of the one receptor–one neuron rule in the mammalian olfactory system

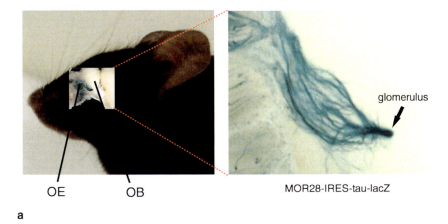

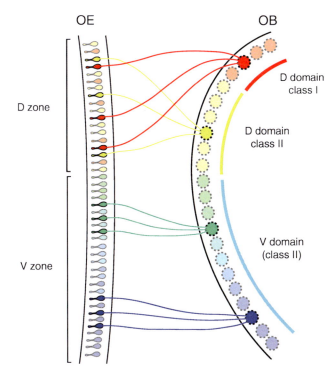

Fig. 2 Axonal projection in the mouse olfactory system. **a** The transgenic mouse, *MOR28-IRES-tau-lacZ* (Serizawa et al. 2000), stained with X-gal (lateral view). Olfactory sensory neurons (OSNs) expressing the *tau-lacZ*-tagged *MOR28* gene project their axons to a specific site forming a glomerulus in the olfactory bulb (*OB*). **b** Spatial correlation between the olfactory epithelium (*OE*) and the OB in the mouse olfactory system. OSNs in the dorsomedial zone (*D zone*) in the OE project their axons to the dorsal domain (*D domain*) of the OB. Class I ORs are mostly expressed by OSNs in the D zone in the OE, which target the anterodorsal cluster of the D domain in the OB. In the ventrolateral zone (*V zone*), each class II OR possesses its own unique expression

(continued)

3.2 Projection Along the Dorsal-Ventral Axis

What mechanisms could account for the striking axonal convergence of OSNs? One parameter used to define the axonal projection site is the spatial information in the OE. In situ hybridization demonstrated that each OR gene is expressed in a restricted area, or "zone," in the OE (Ressler et al. 1993; Vassar et al. 1993). Using zone-specific marker genes, one can divide the OE into the dorsomedial and ventrolateral zones. For example, the O-MACS and NQO-1 genes are specifically expressed in the dorsomedial zone of the OE (Oka et al. 2003; Gussing and Bohm 2004), whereas the OCAM gene is expressed in the ventrolateral zone (Yoshihara et al. 1997). Interestingly, phylogenetically distinct class I ORs are mostly expressed within the dorsomedial zone (Zhang et al. 2004; Tsuboi et al. 2006), although class II OR genes are expressed in both dorsomedial and ventrolateral zones (Zhang et al. 2004; Miyamichi et al. 2005). Within the dorsomedial zone, class I and class II expressing cells are intermingled and there is no restricted distribution for both classes (Tsuboi et al. 2006). In the ventrolateral zone, however, each class II OR gene possesses its unique expression area, which is distributed in a continuous and overlapping manner along the dorsomedial–ventrolateral axis in the OE (Iwema et al. 2004; Miyamichi et al. 2005).

It has been proposed that there is a spatial correlation between the OE and the OB in the OSN projection. Antibody staining of zone-specific molecules demonstrated the "zone-to-zone" correlation: OCAM-positive axons project to the ventral part of the OB (ventral domain), whereas NQO-1-positive axons project to the dorsal part of the OB (dorsal domain) (Yoshihara et al. 1997; Gussing and Bohm 2004). Systematic in situ hybridization and genetic labeling experiments indicated that OSNs expressing class I ORs target their axons to the most anterodorsal areas within the dorsal domain (Tsuboi et al. 2005; Kobayakawa et al. 2007). Zone-to-zone projection along the dorsal–ventral axis in the OB was also confirmed by dye tracing experiments (Astic and Saucier 1986; Saucier and Astic 1986; Miyamichi et al. 2005). Good correlation was demonstrated between dorsal–ventral positioning of glomeruli in the OB and dorsomedial–ventrolateral locations of OSNs in the OE, although no discernible correlation was found for anterior–posterior positioning. These observations suggest that spatial information in the OE contributes to the dorsal–ventral glomerular positioning in the OB (Fig. 2b).

Fig. 2 (continued) area, which is distributed in a continuous and overlapping manner along the dorsomedial–ventrolateral axis in the OE. The dorsomedial–ventrolateral expression area in the OE corresponds to the glomerular positioning along the dorsal–ventral axis in the OB; thus, the dorsal–ventral arrangement of glomeruli is roughly determined by the locations of OSNs in the OE. This positional information may be represented by the expression levels of guidance molecules, e.g., Robo2 and neuropilin-2, that form gradients along the dorsal–ventral axis. *V domain* domain ventral

This notion was also supported by the analyses of some transgenic mice, in which the expression areas of particular ORs were genetically altered. When the expression areas of ORs were shifted or broadened, the projection sites were accordingly changed along the dorsal–ventral axis in the OB (Vassalli et al. 2002; Nakatani et al. 2003; Miyamichi et al. 2005). Robo2/Slit1 and neuropilin-2 (Nrp2) are good candidate guidance molecules along the dorsal–ventral axis, because they demonstrate gradients of expression in the OE and OSN axon termini (Norlin et al. 2001; Walz et al. 2002; Cho et al. 2007); Robo2 is expressed in a dorsomedial-high and ventrolateral-low gradient, and Nrp-2 demonstrates the opposite gradient in the OE. Some dorsomedial-zone OSN axons mistarget to the ventral domain of the OB in mice deficient for Robo2 or its ligand Slit1. However, the molecular logics that coordinate the zone-specific OR gene expression and axonal targeting are still unclear.

3.3 *Projection Along the Anterior–Posterior Axis*

In contrast to the dorsal–ventral arrangement of glomeruli, anterior–posterior positioning appears to be independent of the OE zone and to be more dependent on the expressed OR species. Possible involvement of OR protein in OSN projection was indicated by coding-swap experiments of OR genes (Mombaerts et al. 1996; Wang et al. 1998; Feinstein and Mombaerts 2004). While it has been thought that OR molecules play an instructive role in forming the glomerular map, it has remained entirely unclear how this occurs at the molecular level (for reviews, see Mombaerts 2006; Imai and Sakano 2007). It was also unclear if OR-derived signaling was involved during the process of OSN projection. In odor detection, binding of an odorant to an OR converts the olfactory-specific G protein, G_{olf}, from a GDP-bound state to a GTP-bound state. G_{olf} in turn activates adenylyl cyclase type III (ACIII), generating cAMP, which opens cyclic-nucleotide-gated (CNG) channels. The CNG channel, together with the chloride channels, induces the depolarization of membrane potentials. Targeted deletions of the G_{olf} and CNGA2 genes cause severe anosmia. However, these knockouts do not demonstrate major defects in the initial process of glomerular map formation (Belluscio et al. 1998; Lin et al. 2000; Zheng et al. 2000). It was, therefore, assumed that OR-derived cAMP signals are not required for OSN projection.

Despite these observations, it was possible to assume that an alternative G protein mediates OR-instructed axonal projection. To examine this possibility, axonal projection was analysed for a mutant OR, whose conserved Asp-Arg-Tyr (DRY) motif, which is essential for G-protein coupling, was changed to RDY (Imai et al. 2006). Although null OR gene alleles, e.g., the coding deletion, allow secondary activation of other OR genes, the RDY mutant did not permit coexpression of other OR genes. Interestingly, axons of OSNs expressing the RDY mutant stayed in the anterior region of the OB, did not converge to a specific site, and failed to penetrate the glomerular layer in the OB (Fig. 3a). However, coexpression of a constitutively active G_s rescued the defective wiring of the RDY mutant. Partial rescue of the RDY phenotype was also achieved either by the constitutively active protein kinase A (PKA) or cAMP response element binding protein mutant. Thus, cAMP-dependent transcriptional regulation appears to play major roles in establishing the OR-instructed axonal projection (Imai et al. 2006). Although the functions of ORs at axon termini are yet to be clarified (Barnea et al. 2004; Strotmann et al. 2004), it is possible that axonal cAMP/PKA may modulate growth cone navigation (for a review, see Song and Poo 1999). Our results are consistent with the previous observation that G_s-coupled β2 adrenergic receptor can instruct OSN projection, whereas a G_{i2}-coupled vomeronasal receptor V1rb2 cannot (Feinstein et al. 2004).

The ability to rescue the RDY mutant by constitutively active G_s indicates that the receptor function of the OR is not required for the wiring specificity of OSN axons. How is it, then, that OR-derived cAMP signaling defines the wiring specificity? It was found that an increase in cAMP level by constitutively active G_s causes a posterior shift of glomeruli, whereas suppression of cAMP signals by the dominant-negative PKA mutant causes an anterior shift (Imai et al. 2006) (Fig. 3b). To screen for genes with expression levels correlated with cAMP signals, Imai et al. (2006) performed single-cell microarray analysis of different transgenic OSNs, and identified a gene coding for neuropilin-1 (Nrp1). The Nrp1 gene is expressed at elevated levels in the cAMP-high OSNs, and at low levels, if any, in the OSNs expressing the RDY mutant OR or the dominant-negative PKA mutant. Furthermore, the level of Nrp1, as a readout of cAMP signals, demonstrated an anterior-low/posterior-high gradient in the glomerular layer of the OB (Imai et al. 2006). Previous OR swapping experiments indicated that OR proteins may determine the projection sites along the anterior–posterior axis in the OB (Mombaerts et al. 1996; Wang et al. 1998). It was also found that expression levels of OR protein can affect OSN projection (Feinstein et al. 2004). It is conceivable that different ORs generate different levels of cAMP, which in turn define the expression levels of axon guidance molecules (e.g., Nrp1), to determine the OR-specific projection sites (Fig. 3c). Knockout of Sema3A, a repulsive ligand for Nrp1, alters the glomerular arrangement along the anterior–posterior axis (Schwarting et al. 2000; Taniguchi et al. 2003), suggesting that Nrp1 is involved in the establishment of the anterior–posterior topography. Gain-of-function experiments with transgenic mice indicated that changes in the level of Nrp1 indeed affect the axonal projection of OSNs along the anterior–posterior axis (our unpublished data).

Knockout studies of ACIII also support the involvement of cAMP signals in axonal projection of OSNs. A previous study demonstrated that the glomerular structure is severely disorganized in mice deficient for ACIII, a major adenylyl cyclase in OSNs (Trinh and Storm 2003). Recent analyses of ACIII-deficient mice demonstrated that Nrp1 expression is abolished in OSNs by the ACIII knockout and OR-specific glomerular map formation is perturbed along the anterior–posterior axis (Chesler et al. 2007; Col et al. 2007; Zou et al. 2007). Remarkably, two neighboring glomeruli for M71 and M72 no longer segregate into distinct glomeruli in ACIII mutant mice (Chesler et al. 2007).

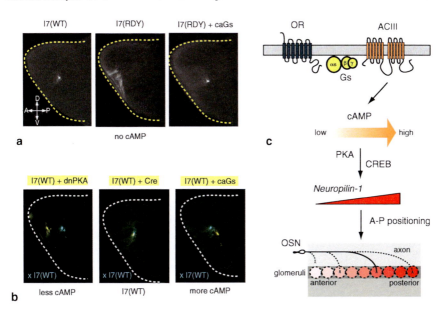

Fig. 3 OR-instructed glomerular positioning (Imai et al. 2006). **a** OR-derived cyclic AMP (*cAMP*) signals direct axonal targeting. A wild-type odorant receptor I7 navigated axons to a specific glomerulus in the OB. When the conserved Asp-Arg-Tyr (DRY) motif at the cytoplasmic end of transmembrane domain III was changed to RDY, G-protein coupling was disrupted, and accordingly OSN axons failed to innervate the glomerular layer. Coexpression of a constitutively active G_s mutant (*caG$_s$*) with the I7(RDY) mutant rescued the defective phenotype of I7(RDY) projection, although the glomerular position was not exactly the same as that for I7(WT). Axons were visualized with ECFP or EYFP fluorescence by whole mount of medial OBs. **b** Changes in the cAMP signals affect glomerular positioning along the anterior–posterior (A–P) axis. Coexpression of caG$_s$ with I7(WT) caused a posterior shift of the target glomeruli. In contrast, coexpression of the dominant-negative mutant of protein kinase A (*dnPKA*), which blocks cAMP signals, caused an anterior shift of glomeruli. **c** A model for the OR/cAMP-directed glomerular positioning along the A–P axis. Each OR generates a unique level of cAMP signals, driven by its intrinsic activity at an earlier stage of OSN projection. The level of cAMP signals is converted to a relative expression level of axon guidance molecules (e.g., neuropilin-1) via cAMP-dependent protein kinase A (*PKA*) and cAMP response element binding protein (*CREB*). Neuropilin-1, together with other guidance molecules, navigates OSN axons along the A–P axis according to its expression level. *ACIII* adenylyl cyclase type III, *WT* wild type

3.4 Axon Sorting To Form Discrete Glomeruli

It appears that a combination of dorsal–ventral patterning, based on anatomical locations of OSNs, and anterior–posterior patterning, based on OR-derived cAMP signals, establishes a rough OB topography. After OSN axons reach their approximate destinations in the OB, further refinement of the glomerular map may occur through fasciculation and segregation of axon termini in an activity-dependent manner. The CNGA2-null mouse, which almost entirely lacks odor-evoked neuronal activity, demonstrates defects in axonal convergence for some ORs (Zheng et al. 2000). Segregation of glomeruli for the CNGA2-positive and CNGA2-negative OSNs has been reported in mosaic knockout mice (Zheng et al. 2000). Genetic block of neuronal activity by the overexpression of the inward rectifying potassium channel, *Kir2.1*, severely affects axonal convergence (Yu et al. 2004). Developmental studies have shown that neighboring glomeruli are not well separated before birth, and discrete glomeruli emerge only after the early neonatal period (Sengoku et al. 2001; Conzelmann et al. 2001; Potter et al. 2001).

To study how OR-instructed axonal fasciculation is controlled, Serizawa et al. (2006) group searched for genes whose expression profiles are correlated with the expressed ORs. Using the transgenic mouse in which the majority of OSNs express a particular OR, such genes were identified. Examples include those that code for homophilic adhesive molecules Kirrel2/Kirrel3 and repulsive molecules ephrin-A5/EphA5. In the *CNGA2* knockout mouse, *Kirrel2* and *EphA5* were downregulated, while *Kirrel3* and *ephrin-A5* were upregulated, indicating that these genes are transcribed in an activity-dependent manner. Heterozygous females of X-linked *CNGA2* mutant mice generate separate CNGA2-positive and CNGA2-negative glomeruli for the same OR (Fig. 4a). Mosaic analysis demonstrated that gain of function of *Kirrel2/Kirrel3* genes also generates duplicated glomeruli (Fig. 4b). It is possible that specific sets of adhesive/repulsive molecules, whose expression levels are determined by OR molecules, regulate the axonal fasciculation of OSNs during glomerular map formation (Fig. 4c).

In *Caenorhabditis elegans*, the Kirrel2/Kirrel3 homolog SYG-1 plays a role in determining the location of specific HSNL synapses (Shen and Bargmann 2003). In our study, experiments using affinity probes in situ and on COS cells confirmed the homophilic, adhesive properties of Kirrel2 and Kirrel3 proteins (Serizawa et al. 2006). Homophilic interactions of these molecules at axon termini were also confirmed with *H-Kirrel* transgenic mice, in which Kirrel2 or Kirrel3 was overexpressed in a mosaic manner. These observations indicate potential roles for Kirrel2 and Kirrel3 in segregating like axons via homophilic, adhesive interactions. In addition to *Kirrel2* and *Kirrel3* genes, several other genes are also transcribed in OSNs at various levels that correlate with the expressed OR species. Among them, the *ephrin-A/EphA* family genes are particularly interesting. In other tissues, ephrin-As and EphAs are known to interact with each other, causing repulsion of the interacting cells (Flanagan 2006). In the mouse olfactory system, expression of ephrin and Eph proteins has been analyzed with various antibodies (St John et al. 2002). It has been reported that OSNs expressing different ORs express different levels of ephrin-A proteins on their axons (Cutforth et al. 2003). EphAs are also differentially expressed in different subsets of OSNs (Serizawa et al. 2006). Since ephrin-As and EphAs are expressed in a complementary manner in each subset of OSNs, the repulsive interaction between two different sets of axons, one that is ephrin-A^{high}/EphA^{low} and the other that is ephrin-A^{low}/EphA^{high}, may be important in the segregation of OSN axons (Serizawa et al. 2006).

It was demonstrated that the expression levels of OR-correlated cell-recognition molecules are affected by the *CNGA2* mutation in OSNs. Since the CNG channel

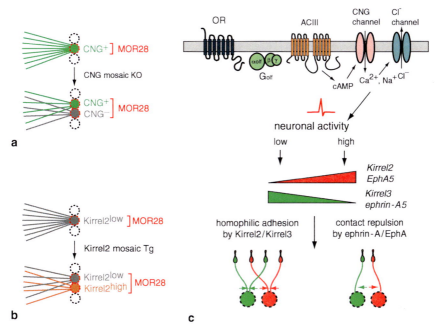

Fig. 4 OR-specific and activity-dependent axon sorting (Serizawa et al. 2006). **a** A mosaic analysis demonstrating the activity dependency of axon sorting. Heterozygous females of X-linked *CNGA2* mutant mice were used for the mosaic analysis, taking advantage of the random X chromosome inactivation. In these mice CNGA2-positive and CNGA2-negative axons formed distinct, neighboring glomeruli even for the same OR (e.g., MOR28), suggesting that CNG channel mediated neuronal activity has an instructive role in sorting axons. **b** A mosaic analysis demonstrating the Kirrel2-mediated axon sorting. With use of the H enhancer and negative feedback by OR molecules, a transgenic system was devised, which generates an additional repertoire of OSNs expressing a particular gene of interest. In Kirrel2 mosaic mice, a subset of MOR28-positive OSNs expressed Kirrel2 at an elevated level. In these mice, Kirrel2-low and Kirrel2-high MOR28 axons were segregated, forming separate glomeruli. **c** A model for the OR-specific and activity-dependent axon sorting to form a discrete map. Different ORs generate different neuronal activities through the CNG channel, at a later stage of OSN projection. It appears that CNG channel mediated neuronal activity upregulates *Kirrel2* and *EphA5* genes, and downregulates *Kirrel3* and *ephrin-A5*. Homophilic cell adhesion molecules, e.g., Kirrel2 and Kirrel3, induce homotypic axonal fasciculation, whereas ephrin-A5 and EphA5 may facilitate heterotypic axonal segregation through their contact-induced repulsive activities

converts OR activity to a change in membrane potential and calcium entry, an intriguing possibility is that the OR-mediated neuronal activity regulates the expression of cell-recognition molecules (Fig. 4c). OSNs may set the rate of neuronal activity, depending upon the expressed OR species. Neuronal activity and calcium influx can regulate the expression of a particular set of genes in other systems (West et al. 2002; Hanson and Landmesser 2004). Similarly, in OSNs, neuronal activity, most likely set by the particular OR expressed, may also determine the expression pattern of cell-recognition molecules.

3.5 Refinement and Maintenance of the Map

Activity-dependent refinement, which follows the initial targeting processes, plays an important role in many sensory systems during development (Goodman and Shatz 1993). Activity inputs are also essential to maintain the neuronal map. In the mouse olfactory system, satellite glomeruli are ectopically formed in young animals. Such minor glomeruli are gradually and eventually eliminated with age, refining the glomerular map. In mice deficient for *CNGA2* or whose naris is surgically occluded, ectopic glomeruli persist (Nakatani et al. 2003; Zheng et al. 2000; Zou et al. 2004).

Taking advantage of the X-linked *CNGA2* mutant, Zhao and Reed (2001) demonstrated that neuronal activity is required for the maintenance of the glomerular map. In mosaic female mice, CNGA2-negative cells are eliminated in a competitive condition with odor exposure, but survive in a noncompetitive condition without odors. The refinement and maintenance appears to be regulated at the level of cell survival, rather than axonal retraction/rewiring (Zhao and Reed 2001; Nakatani et al. 2003; Yu et al. 2004; Zou et al. 2004). However, the exact molecular mechanisms for the selective elimination remain elusive.

Although higher-order olfactory circuits are beginning to be elucidated by recent studies (Lodovichi et al. 2003; Yan et al. 2008; Zou and Buck 2006), the progress in wiring mechanisms is hampered by the lack of powerful genetic labeling/manipulation techniques. In *Drosophila*, wiring patterns of projection neurons, as well as OSNs, are genetically determined (Komiyama and Luo 2006; Komiyama et al. 2007; Luo and Flanagan 2007); however, those of mice appear to be distinct and rely more on OR-derived inputs to instruct the OB circuitry (Belluscio et al. 2002). Recently, it was reported that the odor-evoked neuronal activity is required for the maintenance of the precise intrabulber neuronal connections (Marks et al. 2006). Unlike in the visual system (Hensch 2005), no discernible critical period was found for plasticity. How the higher-order circuits are organized and how information is extracted from the bulbar map in the olfactory cortex are important issues for future studies.

4 Concluding Remarks

Since the discovery of OR genes, it has remained entirely elusive how each OSN expresses only one OR gene, and how OSNs expressing the same OR converge their axons to a specific set of glomeruli (Buck and Axel 1991). In recent years, clear answers were provided to these problems. Singular OR gene choice appears to be ensured by the combination of a rate-limiting enhancer–promoter interaction and negative-feedback regulation by OR proteins. For the OR-instructed axonal projection, it was assumed that OR molecules at axon termini recognize guidance cues on the OB and also mediate the homophilic interactions of like axons.

However, recent studies demonstrated that OR-instructed axonal projection is established by OR-derived cAMP signals, and not by the direct action of OR molecules. The levels of cAMP establish the anterior–posterior topography of axonal projection via cAMP-dependent PKA at an early stage of development (Imai et al. 2006). The neuronal activity generated by cAMP via CNG channels promotes the fasciculation of OSN axons at a later stage (Serizawa et al. 2006).

For the dorsal–ventral arrangement of glomeruli, the locations of OSNs in the OE determine the target sites of OSN axons. This positional information may be represented by the expression levels of guidance molecules, e.g., Robo2 and Nrp2, forming gradients along the dorsal–ventral axis. Along the anterior–posterior axis, a different set of guidance molecules, e.g., Nrp1, is involved, whose expression levels are correlated with the OR species via cAMP. After axons have been guided to approximate destinations in the OB, axon termini are further sorted on the basis of the expressed OR species. It is conceivable that a unique combination of axon guidance/sorting molecules, whose expression levels are determined by OR molecules and neuronal activity, constitutes the "neuronal identity code," and contributes to the discrete glomerular map formation during the process of olfactory development in the mouse. Although much remains to be investigated, the mammalian olfactory system will continue to provide insightful information to increase our understanding of the logics underlying the neuronal circuit formation in the brain.

Acknowledgements This work was supported by the CREST program of the Japan Science and Technology Agency, Japan Society for the Promotion of Science, Mitsubishi Foundation, and the Special Promotion Research Grants from the Ministry of Education, Culture and Science of Japan.

References and Recommended Reading

Astic L, Saucier D (1986) Anatomical mapping of the neuroepithelial projection to the olfactory bulb in the rat. Brain Res Bull 16(4):445–454

Barnea G, O'Donnell S, Mancia F, Sun X, Nemes A, Mendelsohn M, Axel R (2004) Odorant receptors on axon termini in the brain. Science 304(5676):1468

Belluscio L, Gold GH, Nemes A, Axel R (1998) Mice deficient in G(olf) are anosmic.Neuron 20(1):69–81

Belluscio L, Lodovichi C, Feinstein P, Mombaerts P, Katz LC (2002) Odorant receptors instruct functional circuitry in the mouse olfactory bulb.Nature 419(6904):296–300

Buck L, Axel R (1991) A novel multigene family may ncode odorant receptors: a molecular basis for odor recognition.Cell 65:175–187

Chesler AT, Zou DJ, Le Pichon CE, Peterlin ZA, Matthews GA, Pei X, Miller MC, Firestein S (2007) A G protein/cAMP signal cascade is required for axonal convergence into olfactory glomeruli.Proc Natl Acad Sci USA 104(3):1039–1044

Chess A, Simon I, Cedar H, Axel R (1994) Allelic inactivation regulates olfactory receptor gene expression.Cell 78(5):823–834

Cho JH, Lépine M, Andrews W, Parnavelas J, Cloutier JF (2007) Requirement for Slit-1 and Robo-2 in zonal segregation of olfactory sensory neuron axons in the main olfactory bulb.J Neurosci 27(34):9094–9104

Col JA, Matsuo T, Storm DR, Rodriguez I (2007) Adenylyl cyclase-dependent axonal targeting in the olfactory system.Development 134(13):2481–2489

Conzelmann S, Malun D, Breer H, Strotmann J (2001) Brain targeting and glomerulus formation of two olfactory neuron populations expressing related receptor types.Eur J Neurosci 14(10):1623–1632

Cutforth T, Moring L, Mendelsohn M, Nemes A, Shah NM, Kim MM, Frisen J, Axel R (2003) Axonal ephrin-As and odorant receptors: coordinate determination of the olfactory sensory map.Cell 114(3):311–322

Eggan K, Baldwin K, Tackett M, Osborne J, Gogos J, Chess A, Axel R, Jaenisch R (2004) Mice cloned from olfactory sensory neurons. Nature 428(6978):44–49

Feinstein P, Mombaerts P (2004) A contextual model for axonal sorting into glomeruli in the mouse olfactory system.Cell 117(6):817–831

Feinstein P, Bozza T, Rodriguez I, Vassalli A, Mombaerts P (2004) Axon guidance of mouse olfactory sensory neurons by odorant receptors and the beta2 adrenergic receptor.Cell 117(6):833–846

Flanagan JG (2006) Neural map specification by gradients.Curr Opin Neurobiol 16(1):59–66

Fuss SH, Omura M, Mombaerts P (2007) Local and cis effects of the H element on expression of odorant receptor genes in mouse.Cell 130(2):373–384

Goodman CS, Shatz CJ (1993) Developmental mechanisms that generate precise patterns of neuronal connectivity.Cell 72(Suppl):77–98

Gussing F, Bohm S (2004) NQO1 activity in the main and the accessory olfactory systems correlates with the zonal topography of projection maps.Eur J Neurosci 19(9):2511–2518

Hanson MG, Landmesser LT (2004) Normal patterns of spontaneous activity are required for correct motor axon guidance and the expression of specific guidance molecules.Neuron 43(5):687–701

Hensch TK (2005) Critical period plasticity in local cortical circuits.Nat Rev Neurosci 6(11):877–888

Hippenmeyer S, Kramer I, Arber S (2004) Control of neuronal phenotype: what targets tell the cell bodies.Trends Neurosci 27(8):482–488

Hirota J, Mombaerts P (2004) The LIM-homeodomain protein Lhx2 is required for complete development of mouse olfactory sensory neurons.Proc Natl Acad Sci USA 101(23):8751–8755

Hirota J, Omura M, Mombaerts P (2007) Differential impact of Lhx2 deficiency on expression of class I and class II odorant receptor genes in mouse.Mol Cell Neurosci 34(4):679–688

Imai T, Sakano H (2007) Roles of odorant receptors in projecting axons in the mouse olfactory system.Curr Opin Neurobiol17:507–515

Imai T, Suzuki M, Sakano H (2006) Odorant receptor-derived cAMP signals direct axonal targeting.Science 314(5799):657–661

Ishii T, Serizawa S, Kohda A, Nakatani H, Shiroishi T, Okumura K, Iwakura Y, Nagawa F, Tsuboi A, Sakano H (2001) Monoallelic expression of the odourant receptor gene and axonal projection of olfactory sensory neurones.Genes Cells 6(1):71–78. Erratum (2001) Genes Cells 6(6):573

Iwema CL, Fang H, Kurtz DB, Youngentob SL, Schwob JE (2004) Odorant receptor expression patterns are restored in lesion-recovered rat olfactory epithelium.J Neurosci 24(2):356–369

Jacobs GH, Williams GA, Cahill H, Nathans J (2007) Emergence of novel color vision in mice engineered to express a human cone photopigment.Science 315(5819):1723–1725

Kobayakawa K, Kobayakawa R, Matsumoto H, Oka Y, Imai T, Ikawa M, Okabe M, Ikeda T, Itohara S, Kikusui T, Mori K, Sakano H (2007) Innate versus learned odour processing in the mouse olfactory bulb.Nature 450(7169):503–508

Kolterud A, Alenius M, Carlsson L, Bohm S (2004) The Lim homeobox gene Lhx2 is required for olfactory sensory neuron identity.Development 131(21):5319–5326

Komiyama T, Luo L (2006) Development of wiring specificity in the olfactory system.Curr Opin Neurobiol 16(1):67–73

Komiyama T, Sweeney LB, Schuldiner O, Garcia KC, Luo L (2007) Graded expression of semaphorin-1a cell-autonomously directs dendritic targeting of olfactory projection neurons.Cell 128(2):399–410

Kratz E, Dugas JC, Ngai J (2002) Odorant receptor gene regulation: implications from genomic organization.Trends Genet 18(1):29–34

Lewcock JW, Reed RR (2004) A feedback mechanism regulates monoallelic odorant receptor expression.Proc Natl Acad Sci USA 101(4):1069–1074

Li J, Ishii T, Feinstein P, Mombaerts P (2004) Odorant receptor gene choice is reset by nuclear transfer from mouse olfactory sensory neurons. Nature 428(6981):393–399

Lin DM, Wang F, Lowe G, Gold GH, Axel R, Ngai J, Brunet L (2000) Formation of precise connections in the olfactory bulb occurs in the absence of odorant-evoked neuronal activity. Neuron 26(1):69–80

Lodovichi C, Belluscio L, Katz LC (2003) Functional topography of connections linking mirror-symmetric maps in the mouse olfactory bulb. Neuron 38(2):265–276

Lomvardas S, Barnea G, Pisapia DJ, Mendelsohn M, Kirkland J, Axel R (2006) Interchromosomal interactions and olfactory receptor choice.Cell 126(2):403–413

Luo L, Flanagan JG (2007) Development of continuous and discrete neural maps.Neuron 56(2):284–300

Malnic B, Hirono J, Sato T, Buck LB (1999) Combinatorial receptor codes for odors.Cell 96(5):713–723

Marks CA, Cheng K, Cummings DM, Belluscio L (2006) Activity-dependent plasticity in the olfactory intrabulbar map.J Neurosci 26(44):11257–11266

Miyamichi K, Serizawa S, Kimura HM, Sakano H (2005) Continuous and overlapping expression domains of odorant receptor genes in the olfactory epithelium determine the dorsal/ventral positioning of glomeruli in the olfactory bulb J Neurosci 25(14):3586–3592

Mombaerts P (2006) Axonal wiring in the mouse olfactory system.Annu Rev Cell Dev Biol22:713–737

Mombaerts P, Wang F, Dulac C, Chao SK, Nemes A, Mendelsohn M, Edmondson J, Axel R (1996) Visualizing an olfactory sensory map.Cell 87(4):675–686

Mori K, Takahashi YK, Igarashi KM, Yamaguchi M (2006) Maps of odorant molecular features in the Mammalian olfactory bulb.Physiol Rev 86(2):409–433

Nagawa F, Yoshihara S, Tsuboi A, Serizawa S, Itoh K, Sakano H (2002) Genomic analysis of the murine odorant receptor MOR28 cluster: a possible role of gene conversion in maintaining the olfactory map. Gene 292(1–2):73–80

Nakatani H, Serizawa S, Nakajima M, Imai T, Sakano H (2003) Developmental elimination of ectopic projection sites for the transgenic OR gene that has lost zone specificity in the olfactory epithelium.Eur J Neurosci 18(9):2425–2432

Ngai J, Chess A, Dowling MM, Necles N, Macagno ER, Axel R (1993) Coding of olfactory information: topography of odorant receptor expression in the catfish olfactory epithelium.Cell 72(5):667–680

Nguyen MQ, Zhou Z, Marks CA, Ryba NJ, Belluscio L (2007) Prominent roles for odorant receptor coding sequences in allelic exclusion.Cell 131(5):1009–1017

Nishizumi H, Kumasaka K, Inoue N, Nakashima A, Sakano H (2007) Deletion of the core-H region in mice abolishes the expression of three proximal odorant receptor genes in cis.Proc Natl Acad Sci USA 104(50):20067–20072

Norlin EM, Alenius M, Gussing F, Fagglund M, Vedin V, Bohm S (2001) Evidence for gradients of gene expression correlating with zonal topography of the olfactory sensory map. Mol Cell Neurosci 18(3):283-295

Oka Y, Kobayakawa K, Nishizumi H, Miyamichi K, Hirose S, Tsuboi A, Sakano H (2003) O-MACS, a novel member of the medium-chain acyl-CoA synthetase family, specifically expressed in the olfactory epithelium in a zone-specific manner. Eur J Biochem 270(9):1995–2004

Potter SM, Zheng C, Koos DS, Feinstein P, Fraser SE, Mombaerts P (2001) Structure and emergence of specific olfactory glomeruli in the mouse.J Neurosci 21(24):9713–9723

Ray A, van Naters WG, Shiraiwa T, Carlson JR (2007) Mechanisms of odor receptor gene choice in Drosophila. Neuron 53(3):353–369

Ressler KJ, Sullivan SL, Buck LB (1993) A zonal organization of odorant receptor gene expression in the olfactory epithelium.Cell 73(3):597–609

Ressler KJ, Sullivan SL, Buck LB (1994) Information coding in the olfactory system: evidence for a stereotyped and highly organized epitope map in the olfactory bulb.Cell 79(7):12451255

Rothman A, Feinstein P, Hirota J, Mombaerts P (2005) The promoter of the mouse odorant receptor gene M71.Mol Cell Neurosci 28(3):535–546

Saucier D, Astic L (1986) Analysis of the topographical organization of olfactory epithelium projections in the rat. Brain Res Bull 16(4):455–462

Schwarting GA, Kostek C, Ahmad N, Dibble C, Pays L, Püschel AW (2000) Semaphorin 3A is required for guidance of olfactory axons in mice.J Neurosci 20(20):7691–7697

Sengoku S, Ishii T, Serizawa S, Nakatani H, Nagawa F, Tsuboi A, Sakano H (2001) Axonal projection of olfactory sensory neurons during the developmental and regeneration processes. Neuroreport 12(5):1061–1066

Serizawa S, Ishii T, Nakatani H, Tsuboi A, Nagawa F, Asano M, Sudo K, Sagakami J, Sakano H, Ijiri T, Matsuda Y, Suzuki M, Yamamori T, Iwakura Y, Sakano H (2000) Mutually exclusive expression of odorant receptor transgenes. Nat Neurosci 3(7):687–693

Serizawa S, Miyamichi K, Nakatani H, Suzuki M, Saito M, Yoshihara Y, Sakano H (2003) Negative feedback regulation ensures the one receptor-one olfactory neuron rule in mouse. Science 302(5653):2088–2094

Serizawa S, Miyamichi K, Sakano H (2004) One neuron-one receptor rule in the mouse olfactory system. Trends Genet 20(12):648–653

Serizawa S, Miyamichi K, Takeuchi H, Yamagishi Y, Suzuki M, Sakano H (2006) A neuronal identity code for the odorant receptor-specific and activity-dependent axon sorting. Cell 127(5):1057–1069

Shen K, Bargmann CI (2003) The immunoglobulin superfamily protein SYG-1 determines the location of specific synapses in C. elegans. Cell 112(5):619–630

Song HJ, Poo MM (1999) Signal transduction underlying growth cone guidance by diffusible factors. Curr Opin Neurobiol 9(3):355–363

St John JA, Pasquale EB, Key B (2002) EphA receptors and ephrin-A ligands exhibit highly regulated spatial and temporal expression patterns in the developing olfactory system. Brain Res Dev Brain Res 138(1):1–14

Strotmann J, Wanner I, Krieger J, Raming K, Breer H (1992) Expression of odorant receptors in spatially restricted subsets of chemosensory neurones. Neuroreport 3(12):1053–1056

Strotmann J, Conzelmann S, Beck A, Feinstein P, Breer H, Mombaerts P (2000) Local permutations in the glomerular array of the mouse olfactory bulb. J Neurosci 20(18):6927–6938

Strotmann J, Levai O, Fleischer J, Schwarzenbacher K, Breer H (2004) Olfactory receptor proteins in axonal processes of chemosensory neurons. J Neurosci 24(35):7754–7761

Taniguchi M, Nagao H, Takahashi YK, Yamaguchi M, Mitsui S, Yagi T, Mori K, Shimizu T (2003) Distorted odor maps in the olfactory bulb of semaphorin 3A-deficient mice. J Neurosci 23(4):1390–1397

Trinh K, Storm DR (2003) Vomeronasal organ detects odorants in absence of signaling through main olfactory epithelium. Nat Neurosci 6(5):519–525

Tsuboi A, Miyazaki T, Imai T, Sakano H (2006) Olfactory sensory neurons expressing class I odorant receptors converge their axons on an antero-dorsal domain of the olfactory bulb in the mouse. Eur J Neurosci 23(6):1436–1444

Vassalli A, Rothman A, Feinstein P, Zapotocky M, Mombaerts P (2002) Minigenes impart odorant receptor-specific axon guidance in the olfactory bulb. Neuron 35(4):681–696

Vassar R, Ngai J, Axel R (1993) Spatial segregation of odorant receptor expression in the mammalian olfactory epithelium. Cell 74(2):309–318

Vassar R, Chao SK, Sitcheran R, Nuñez JM, Vosshall LB, Axel R (1994) Topographic organization of sensory projections to the olfactory bulb. Cell 79(6):981–991

Vrieseling E, Arber S (2006) Target-induced transcriptional control of dendritic patterning and connectivity in motor neurons by the ETS gene Pea3. Cell 127(7):1439–1452

Walz A, Rodriguez I, Mombaerts P (2002) Aberrant sensory innervation of the olfactory bulb in neuropilin-2 mutant mice. J Neurosci 22(10):4025–4035

Wang F, Nemes A, Mendelsohn M, Axel R (1998) Odorant receptors govern the formation of a precise topographic map. Cell 93(1):47–60

Wang SS, Lewcock JW, Feinstein P, Mombaerts P, Reed RR (2004) Genetic disruptions of O/E2 and O/E3 genes reveal involvement in olfactory receptor neuron projection. Development 131(6):1377–1388

West AE, Griffith EC, Greenberg ME (2002) Regulation of transcription factors by neuronal activity. Nat Rev Neurosci 3(12):921–931

Yan Z, Tan J, Qin C, Lu Y, Ding C, Luo M (2008) Precise circuitry links bilaterally symmetric olfactory maps. Neuron 58(4):613–624

Yoshihara Y, Kawasaki M, Tamada A, Fujita H, Hayashi H, Kagamiyama H, Mori K (1997) OCAM: A new member of the neural cell adhesion molecule family related to zone-to-zone projection of olfactory and vomeronasal axons. J Neurosci 17(15):5830–5842

Yu CR, Power J, Barnea G, O'Donnell S, Brown HE, Osborne J, Axel R, Gogos JA (2004) Spontaneous neural activity is required for the establishment and maintenance of the olfactory sensory map. Neuron 42(4):553–566

Zhang X, Firestein S (2002) The olfactory receptor gene superfamily of the mouse. Nat Neurosci 5(2):124–133

Zhang X, Rogers M, Tian H, Zhang X, Zou DJ, Liu J, Ma M, Shepherd GM, Firestein SJ (2004) High-throughput microarray detection of olfactory receptor gene expression in the mouse. Proc Natl Acad Sci USA 101(39):14168–14173

Zhang X, Zhang X, Firestein S (2007) Comparative genomics of odorant and pheromone receptor genes in rodents. Genomics 89(4):441–450

Zhao H, Reed RR (2001) X inactivation of the OCNC1 channel gene reveals a role for activity-dependent competition in the olfactory system. Cell 104(5):651–660

Zhao H, Ivic L, Otaki JM, Hashimoto M, Mikoshiba K, Firestein S (1998) Functional expression of a mammalian odorant receptor. Science 279(5348):237–242

Zheng C, Feinstein P, Bozza T, Rodriguez I, Mombaerts P (2000) Peripheral olfactory projections are differentially affected in mice deficient in a cyclic nucleotide-gated channel subunit. Neuron 26(1):81–91

Zou Z, Buck LB (2006) Combinatorial effects of odorant mixes in olfactory cortex. Science 311(5766):1477–1481

Zou DJ, Feinstein P, Rivers AL, Mathews GA, Kim A, Greer CA, Mombaerts P, Firestein S (2004) Postnatal refinement of peripheral olfactory projections. Science 304(5679):1976–1979

Zou DJ, Chesler AT, Le Pichon CE, Kuznetsov A, Pei X, Hwang EL, Firestein S (2007) Absence of adenylyl cyclase 3 perturbs peripheral olfactory projections in mice. J Neurosci 27(25):6675–6683

Pheromone Sensing in Mice

I. Rodriguez and U. Boehm

Abstract Beginning with the neuroepithelium of the vomeronasal organ, the accessory olfactory system in rodents runs parallel to the main olfactory system and is specialized in the detection of pheromones. Only a small number of vomeronasal agonists carrying pheromonal information have been identified this far. These structurally diverse classes of chemicals include peptides secreted by exocrine glands and range from small volatile molecules to proteins and fragments thereof present in urine. Most pheromones activate both vomeronasal and main olfactory sensory neurons, making the identification of functionally relevant populations of sensory neurons difficult. Analyses of gene-targeted mice selectively affecting either vomeronasal or main olfactory signaling have attempted to elucidate the functional contribution of the different chemosensory epithelia to pheromone sensing in mice. These mouse models suggest that both the main and the accessory olfactory systems can converge and synergize to express the complex array of stereotyped behaviors and hormonal changes triggered by pheromones.

1 Pheromone Detection in the Rodent Nose

Pheromones are chemical cues, which are released by animals and act on members of the same species to regulate their social interactions and the size of their populations (Halpern and Martinez-Marcos 2003; Brennan and Zufall 2006). Pheromone effects in rodents range from intermale aggression to sexual behaviors and long-term neuroendocrine alterations. Although the original description of a pheromone

---

I. Rodriguez (✉)
Department of Zoology and Animal Biology, University of Geneva, 30 Quai Ernest Ansermet, Geneva, 1211 Switzerland
e-mail: Ivan.Rodriguez@zoo.unige.ch

U. Boehm (✉)
Center for Molecular Neurobiology, Institute for Neural Signal Transduction, Falkenried 94, 20251 Hamburg, Germany
e-mail: ulrich.boehm@zmnh.uni-hamburg.de

is almost 50 years old (Karlson and Luscher 1959), our list of chemical structures with pheromone-like effects is still quite short, in great contrast to a repertoire of more than 1,300 chemosensory receptors found in the rodent nose (Mombaerts 2004). Chemicals carrying pheromonal information are structurally diverse and include small volatile molecules found in urine, proteins, and fragments thereof, and specific peptides secreted by exocrine glands. The rodent nose houses at least four distinct chemosensory epithelia, the main olfactory epithelium (MOE) lining the nasal cavity, the neuroepithelium of the vomeronasal organ (VNO), which is part of the accessory olfactory system, the Grüneberg ganglion, and the septal organ (reviewed in Breer et al. 2006). Classical studies in which the VNO has been removed or its connection with the brain severed (VNX) have clearly established its pivotal role in pheromone perception. However vomeronasal signaling is not functionally equivalent to pheromone signaling and experiments encompassing several species have shown that pheromone signals are not exclusively perceived by the VNO, but can also be processed by the main olfactory system. New mouse models have started to address the functional contributions of the different olfactory epithelia to pheromone responses in mice.

2 Vomeronasal Agonists

The nature of the pheromone cues regulating mouse behavior is largely unknown (this observation is in fact true for all mammals). A few compounds with pheromonal effects have, however, been identified. Mice employ urine investigation as a tool to discriminate between individuals. Most research has thus naturally focused on components of this bodily fluid, which has in addition the useful characteristic of being available in relatively large quantities. Several small and volatile molecules present in urine were identified to be vomeronasal sensory neuron (VSN) agonists, including 2-*sec*-butyl-4,5-dihydrothiazole, 3,4-dehydro-*exo*-brevicomin, farnesene, *n*-pentylacetate, 6-hydroxy-6-methyl-3-heptanone, isobutylamine, 2-heptanone, and 2,5-dimethylpyrazine (Fig. 1; Novotny 2003). These molecules are detected by nonoverlapping small sets of sensory neurons located in the apical part of the vomeronasal neuroepithelium (Fig. 2), suggesting narrow tuning properties in these neurons (Leinders-Zufall et al. 2000). In addition, nonvolatile major histocompatibility complex (MHC) class I peptides (Fig. 1) have been shown to selectively activate a limited number of VSNs located in the basal zone of the VNO (Fig. 2; Leinders-Zufall et al. 2004). These observations, although limited, suggest that the VSN populations in the apical and basal zones of the vomeronasal neuroepithelium are specialized in detecting different types of chemosensory signals, possibly volatile molecules and peptides, respectively. Neither the volatile urine compounds listed above nor the MHC peptides are exclusively detected by the VNO however; they are also detected by olfactory sensory neurons (OSNs) in the MOE (Spehr et al. 2006; Wang et al. 2006; reviewed in Wang et al. 2007; Zufall and Leinders-Zufall 2007).

Pheromone Sensing in Mice

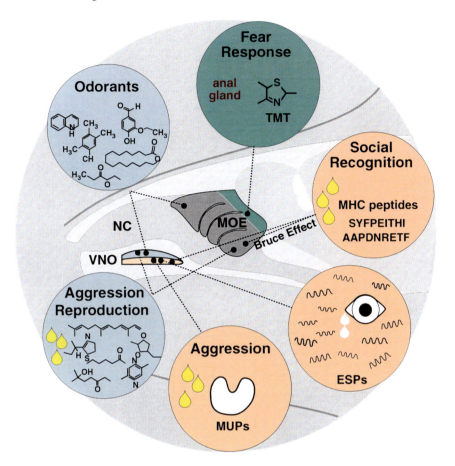

Fig. 1 Pheromones: from aggression to social recognition. Several classes of agonists activate vomeronasal sensory neurons (VSNs) and/or olfactory sensory neurons (OSNs). Fear response: trimethylthiazoline (*TMT*) secreted from the anal gland of fox (although stricto senso not a pheromone but a kairomone) triggers fear responses in mice via OSNs in the main olfactory epithelium (*MOE*). Social recognition: amino acid sequence of two major histocompatibility complex (*MHC*) class I presented peptides (specific to H-2b and H-2d haplotypes of C57BL/6 and BALB/C mice, respectively) that activate V1R VSNs and mediate pregnancy block. Extraorbital lacrimal gland specific peptides (*ESPs*) present in tear fluid activate VSNs located at the base of the vomeronasal neuroepithelium. Mouse urinary proteins (*MUPs*) activate V2R VSNs and trigger aggressive behavior. Aggression and reproduction: small volatile molecules isolated from urine (structures of 2-*sec*-butyl-4,5-dihydrothiazole, 3,4-dehydro-*exo*-brevicomin, farnesene, *n*-pentylacetate, 6-hydroxy-6-methyl-3-heptanone, isobutylamine, 2-heptanone, and 2,5-dimethyl pyrazine are shown) activate V1R VSNs and OSNs. Odorants: structures for odorants that activate V1Rs (and OSNs) such as indole, 16-hexadecanolide, durene, ethyl propionate (acetate), and ethyl vanillin are shown. The *orange area* and the *blue area* of the vomeronasal organ (*VNO*) correspond to basal (V2R-expressing) and apical (V1R-expressing) zones, respectively. *Yellow drops* indicate the presence of the compounds in urine. *NC* nasal cavity

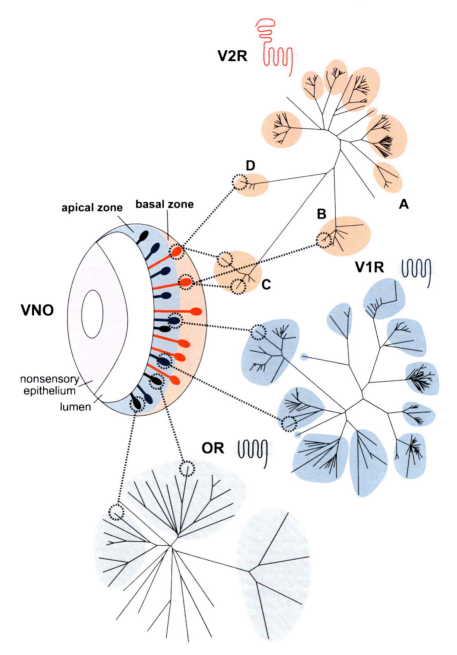

Fig. 2 Vomeronasal receptors. Schematic coronal section through a mouse VNO. The apical sensory epithelium (containing V1R- and OR-expressing VSNs) is in *blue*, while the basal (V2R-expressing VSNs) zone is in *orange*. All members of the V1R, V2R, and OR mouse repertoires expressed in the VNO are shown by unrooted phylogenetic trees. *Dotted lines* indicate the combinatorial expression of V2Rs and the monogenic expression of V1Rs in a given VSN. Nonclassical class I MHC receptors were not included in this scheme because evidence for a direct role in chemodetection is lacking. *OR* odorant receptor

In addition to urine, mice also investigate other secretions, among which are those produced by glands on the face and perceived by physical contact with the head. Specific peptides, called extraorbital lacrimal gland specific peptides (ESPs), are secreted by the extraorbital lacrimal glands. In mice, there are 38 ESP genes and two of them, ESP1 and ESP36, are expressed in a sexually dimorphic fashion (Kimoto et al. 2005, 2007). All ESPs elicit electrical responses in VSNs (ESP1 activates VSNs in the basal zone) but do not trigger responses in OSNs (Kimoto et al. 2007). The functional role played by these peptides is still unknown, but it appears that they are only found in rodents, suggesting that they may play a role specific to this phylogenetic group.

Multiple approaches have been taken to understand the nature of pheromones in rodents. Bottom-up approaches start with assessing the effect of candidate molecules on physiological function. Top-down approaches start with a behavior and search for the chemical signals that trigger it.

2.1 From Molecules to Physiological Effects

Olfaction-mediated pregnancy block (or Bruce effect) takes place when a recently inseminated female is exposed to a strange male rather than to the stud male (Bruce 1960). This striking effect is mediated by the vomeronasal system (reviewed in Halpern and Martinez-Marcos 2003; Boehm and Zufall 2006). What chemical signals inform the female that the new individual's olfactory imprint is not that of the stud male? The answer may lie in molecules involved in the immune response. Class I MHC molecules comprise a highly polymorphic family of proteins, which present small peptides on the surface of nucleated cells. These peptides can be shed from the cell surface and are therefore present in bodily fluids (Singh et al. 1987). As the peptides vary between individuals depending on MHC haplotype, they generate a signature specific to each individual. The potential effect of MHC class I peptide ligands on VSNs was tested electrophysiologically (Leinders-Zufall et al. 2004). Two prototypal representatives of two H-2 haplotypes (the H-2b and H-2d haplotypes of C57BL/6 and BALB/C mice, respectively) proved to be excellent agonists, at picomolar concentrations, for subsets of VSNs in the basal zone. The same peptides, used alone or in combination with urine as individuality signals after mating, were able to recapitulate pregnancy block, suggesting that they may represent, at least in part, the individuality signals underlying mate recognition (Leinders-Zufall et al. 2004; Thompson et al. 2007).

2.2 From Behavior to Chemical

A few very robust and innate behaviors are triggered by pheromonal stimuli in mice. These include protection of pups by lactating mothers, mounting of females in estrus by males, or intermale aggression (Halpern and Martinez-Marcos 2003; Novotny 2003). Many, taking advantage of this lattermost behavior, have tried to identify the molecules triggering fight, by analyzing the composition of male urine. The olfactory perception of dehydro-*exo*-brevicomin, *sec*-butyl dihydrothiazole, 3-amino-*s*-triazole, 4-ethyl phenol, and 2,7-dimethyl octane were shown to facilitate aggression in male mice (Achiraman and Archunan 2002; Novotny 2003). Very recently, Chamero et al. (2007) also identified molecules triggering intermale aggression. Fractionation, purification, and characterization of components of male urine allowed them to show that aggression-promoting activity was dependent on the presence of mouse urinary proteins. Only basal zone VSNs responded to the stimuli and the aggressive behavior induced was apparently exclusively mediated by the VNO (Chamero et al. 2007).

3 Vomeronasal Receptors

After the discovery of odorant receptors (ORs) by Buck and Axel in 1991 the identification of the receptors responsible for vomeronasal chemodetection was thought to represent a relatively easy task (Buck and Axel 1991). This was not the case, however, as although vomeronasal receptors (VRs) (like ORs) are putative seven-transmembrane receptors, they do not share any sequence similarity with ORs.

Two types of VRs have been identified (reviewed in Mombaerts 2004; Rodriguez 2004). Roughly half of the VSN population expresses members of the V1R superfamily (Dulac and Axel 1995), while the remaining half expresses V2Rs (Fig. 2; Herrada and Dulac 1997; Matsunami and Buck 1997; Ryba and Tirindelli 1997). This leads to two physically separated neuronal groups in the vomeronasal epithelium, the V1R-expressing neurons lying on top of the V2R-transcribing ones, thus being closer to the outside world (Fig. 2). These vomeronasal subpopulations are not homogenous since each individual neuron transcribes a very limited number of VR genes, chosen from remarkably large gene repertoires (Dulac and Axel 1995; Herrada and Dulac 1997; Matsunami and Buck 1997; Ryba and Tirindelli 1997). This results in a significant number of molecularly and possibly functionally distinct sensory populations in the VNO, each composed of a few hundred neurons.

3.1 V1Rs

V1R genes are found in all vertebrate species, including humans (Rodriguez et al. 2000; Rodriguez and Mombaerts 2002). The size of the V1R repertoire is highly variable among species: it ranges from a few genes in fish species (Pfister et al. 2007; Saraiva and Korsching 2007) to 180 in mice (Fig. 2, Rodriguez et al. 2002; Zhang et al. 2004), and reaches over 300 in platypus (Grus et al. 2007). The intraspecies and interspecies content of these repertoires is also surprisingly variable, such that orthologous V1Rs are often difficult or impossible to identify even between closely related species such as rat and mouse. This diversity appears to be the result of rapid birth and death of V1R genes and of positive Darwinian selection, an evolutionary pace which likely reflects the production of molecular tools allowing the detection of species-specific chemicals (Mundy and Cook 2003; Grus and Zhang 2004; Lane et al. 2004; Grus et al. 2005; Shi et al. 2005).

The molecular mechanisms resulting in the choice of a given VR gene, its maintenance, and the nontranscription of other VR genes is not understood, but monogenic and monoallelic V1R transcription appears to be the rule (reviewed in Rodriguez 2007). This expression characteristic is dependent on the transcription of a functional V1R since the expression of a nonfunctional *V1rb2* allele in a given VSN leads to the coexpression of another, functional V1R gene (Roppolo et al. 2007). This obviously suggests that the expressed receptor itself plays a role in the nonexpression of other V1R genes, through some kind of negative-feedback mechanism. A second level of regulation, termed "gene cluster lock," appears to be at work. V1R genes are organized in clusters (Del Punta et al. 2000, 2002a, b; Lane et al. 2002); after the choice of a nonfunctional V1R gene, any member of the cluster, if located in cis, becomes incompetent for coexpression, while other V1R genes, either located in trans and from the same cluster or belonging to other V1R families, are available for coexpression (Roppolo et al. 2007). This observation strongly suggests that the organization of V1R genes in clusters does not simply reflect evolutionary proximity, but also underlies mechanisms that regulate V1R expression.

The striking size of the VR gene repertoire relative to the number of known mouse pheromones has been and still is difficult to explain if VR genes are to represent the molecular tools responsible for the recognition of these chemical signals. Does this reflect a terribly incomplete picture of the pheromones used by mice or is it that VRs play roles independent of their supposed chemosensory function? This latter suggestion is at least partly justified, as the first function experimentally shown for a V1R turned out to be a role in axon guidance: sensory neurons expressing null *V1rb2* or *V1ra1* alleles are unable to recapitulate the wild-type axonal projection patterns in the accessory olfactory bulb (AOB) (Belluscio et al. 1999; Rodriguez et al. 1999).

Today V1R receptors are considered to be primarily pheromone detectors. The functional role played by these membrane proteins (or at least some of them) in chemodetection was shown by two independent genetic approaches.

First, the deletion via chromosome engineering of a V1R gene cluster containing all members of the V1Ra and V1Rb subfamilies V1Rab–/–mice, representing 10% of the mouse V1R gene repertoire, led to a drastic alteration of the electrophysiological characteristics of VSNs, resulting in loss of their ability to respond to *n*-pentylacetate, 6-hydroxy-6-methyl-3-heptanone, and isobutylamine (Del Punta et al. 2002a). The most surprising result was not that V1Rs allowed pheromone detection, but rather that a very limited alteration of the V1R gene repertoire abolished responses to about one third of the pheromones tested, suggesting that VR agonist specialization may not only functionally distinguish V1R- and V2R-expressing neurons, but that it also characterizes subfamilies of V1R receptors.

Second, a knock-in/knock-out approach allowing visualization of VSNs expressing the *V1rb2* gene showed that VSNs expressing a functional V1RB2 receptor did respond at picomolar concentrations to the pheromone 2-heptanone, a compound present in mouse urine and able to extend estrus in females (Boschat et al. 2002). The functional correlation was confirmed by a loss-of-function experiment in which the expression of a null V1rb2 allele under the same conditions resulted in a lack of response to 2-heptanone. To date, this represents the only identified rodent V1R-pheromone pair.

3.2 V2Rs

V2Rs are, like V1Rs, seven-transmembrane G-coupled receptors (Fig. 2). They, however, possess long N-termini, unlike V1Rs, and share significant sequence similarities with taste T1R, calcium-sensing, and metabotropic glutamate receptors. The origin of V2R genes is ancient, since genomic evidence of their presence is found in all vertebrate species analyzed so far; but the use of this receptor superfamily appears restricted to a few species. Opossums, rats, and mice possess respectively 86, 79, and 121 intact V2R genes, while cows, dogs, and humans only carry V2R pseudogenes (between nine and 20) (Fig. 2; Yang et al. 2005; Young and Trask 2007). The V2R superfamily is divided into four families (A–D), with family C V2Rs being very divergent from the other three families (Yang et al. 2005; Shi and Zhang 2007; Young and Trask 2007). Each basal VSN expresses a specific combination of V2Rs, consisting of a single member of the A, B, or D family, together with a member of the C family (Fig. 2; Silvotti et al. 2007).

No ligand–V2R pair has been identified in mammals. However, the rapidly increasing number of identified molecules activating V2R-expressing VSNs (for example, mouse urinary proteins; Chamero et al. 2007), together with the ability to express V2Rs in vitro will surely lead to the identification of V2R–ligand pairs.

3.3 Odorant Receptors

Recently, members of another G-protein-coupled receptor superfamily were added to the vomeronasal chemosensory receptor repertoire. Levai et al. (2006) showed that a total of 44 different OR genes are expressed in the apical zone of the vomeronasal epithelium. On the basis of their morphology, axonal projections, and molecular profile, OR-expressing cells in the apical zone were similar to V1R-expressing VSNs. It is, however, not known whether these apical zone OR VSNs coexpress V1R receptors. Likewise it is unclear which transduction cascade is activated by these ORs, since most known OR-specific signaling elements are absent in VSNs (see Sect. 4).

3.4 Nonclassical Class I MHC Receptors

It is no longer blasphemy to discuss the potential role played by MHC molecules outside the immune system. In addition to presenting intracellular peptides to T cells, these molecules function in other systems, including neuronal networks, where they are involved in plasticity and development (Huh et al. 2000). The unexpected finding of an association between nonclassical class I MHC molecules from the M10 family and V2Rs in VSNs (Ishii et al. 2003; Loconto et al. 2003) led to the hypothesis that the MHC molecules themselves could participate in the recognition of pheromones. The crystallographic structure of one M10 member, M10.5, showed a groove potentially able to accommodate a peptide (Olson et al. 2005), a feature usually lost in non-antigen-presenting MHC molecules. However, this peptide is unknown, and its existence is still putative. On the basis of our limited understanding of these nonclassical MHC molecules in the VNO we can only suggest potential functions. These include a role as V2R molecular chaperones, in plasticity, in the modulation of V2R specificity, and, of course, as direct peptide receptors.

4 Vomeronasal Signal Transduction

Chemosensory signal transduction in VSNs is distinct from that in OSNs, but is poorly understood. Major components of the signal transduction cascade in OSNs (like the G-protein α subunit $G_{\alpha olf}$ or one subunit of the olfactory cyclic-nucleotide-gated, CNG, cation channel) are not expressed by VSNs (Berghard and Buck 1996). Instead, V1R- and V2R-positive VSNs express different G-protein α subunits, $G_{\alpha i2}$ and $G_{\alpha o}$, respectively, although it still needs to be demonstrated whether these play any role in VSN sensory signal transduction. Signal transduction in VSNs involves diacylglycerol (Spehr et al. 2002) and a diacylglycerol-activated cation channel (Lucas et al. 2003), which partially depends on TRPC2, a VNO-specific member of the transient receptor potential family of calcium channels (Zufall 2005). Interestingly, studies performed in mice deficient for TRPC2 suggest differential signal transduction cascades in the apical and the basal zone of the epithelium, respectively. Whereas signal transduction in apical zone V1R VSNs is strongly reduced in TRPC2–/– mice (Leypold et al. 2002; Stowers et al. 2002), MHC peptide sensing by basal zone V2R VSNs is unimpaired in this mouse model, implying an as yet undefined signal transduction mechanism in this VSN subpopulation (Kelliher et al. 2006).

5 Vomeronasal Projections

5.1 From VNO to Accessory Olfactory Bulb

VSNs in the apical and basal zones of the VNO synapse in separate parts of the AOB (Jia and Halpern 1996; Yoshihara et al. 1997), maintaining the strict zonal segregation observed in the VNO (Fig. 3). V1R-expressing VSNs (as well as OR-expressing VSNs (Levai et al. 2006) project to the anterior part of the AOB, whereas the V2R VSNs project to the posterior part (Fig. 3). Consistent with this segregation, the innervation of the numerous dendrites of a given mitral cell in the AOB is restricted to either anterior or posterior glomeruli (Jia and Halpern 1997; von Campenhausen et al. 1997).

Each VSN projects a single axon to the AOB, and axons from about 500–1,000 VSNs expressing a given V1R converge on ten to 20 glomeruli (Belluscio et al. 1999;

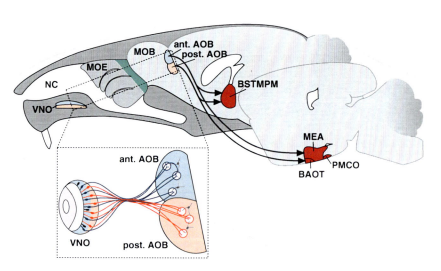

Fig. 3 Vomeronasal system. Schematic representation of a rodent nasal cavity and brain (lateral view). Accessory olfactory bulb (*AOB*) mitral cells project to vomeronasal and extended amygdala. Inset: The VNO is a bilateral tubular structure located at the base of the nasal septum. VSNs that express the same V1R or V2R converge on a small number of glomeruli in the AOB. Sensory neurons located in the apical layer of the epithelium project to the anterior part of the AOB, whereas those present in the basal layer project to the posterior part. *MOE* main olfactory epithelium, *MOB* main olfactory bulb, *BSTMPM* posteromedial bed nucleus of the stria terminalis, *MEA* medial amygdaloid nucleus, *BAOT* bed nucleus of the accessory olfactory tract, *PMCO* posteromedial cortical amygdaloid area

Rodriguez et al. 1999), whereas axons from VSNs expressing the same V2R receptor project six to ten glomeruli (Del Punta et al. 2002b; Fig. 3 inset). AOB mitral cells can have from one to six dendrites contacting multiple glomeruli (Takami and Graziadei 1990, 1991). These dendrites may integrate signals from glomeruli innervated by neurons expressing the same V1R or V2R (Del Punta et al. 2002b), but can also receive input from glomeruli associated with different although closely related V1Rs (i.e., members of the same V1R subfamily; Wagner et al. 2006). This selective heterotypic connectivity suggests that the AOB may play a role not only in the simple convergence of information via dendritic convergence, but also in V1R information integration processes (reviewed in Dulac and Wagner 2006). Several studies of biochemical aspects of VNO activation or neural activation of the AOB raise the possibility that signals generated by the two receptor families are eventually targeted to brain regions that mediate different behavioral and physiological effects (Dudley and Moss 1999; Inamura et al. 1999; Krieger et al. 1999; Kumar et al. 1999; Matsuoka et al. 1999; Halem et al. 2001; Kimoto et al. 2005).

5.2 Vomeronasal System Signaling Beyond the Bulb

Sensory signals generated in the VNO follow neural pathways separate from those carrying odor signals from the MOE (reviewed in Halpern and Martinez-Marcos 2003; Boehm 2006). Whereas MOE signals are relayed through the primary olfactory cortex to higher cortical areas as well as limbic areas controlling basic drives and emotions (Shipley et al. 2004), signals initiating in the neuroepithelium of the VNO are – via relay through the amygdala – transmitted to the hypothalamus, which is implicated in mediating the behavioral effects and neuroendocrine alterations triggered by pheromones. AOB mitral cells project to the vomeronasal amygdala (consisting of the medial amygdaloid nucleus and the posteromedial cortical amygdaloid area), as well as to the bed nucleus of the accessory olfactory tract and the posteromedial bed nucleus of the stria terminalis (also called "extended amygdala") (Fig. 3) (Scalia and Winans 1975; Alheid et al. 1995; Newman 1999; Halpern and Martinez-Marcos 2003).

Despite recent progress in identifying neurons relaying farnesene signals to hypothalamic centers controlling reproduction (Boehm et al. 2005; see Sect. 5.3), the individual neurons and the neural circuits representing a given vomeronasal agonist in the brain are by and large unknown. It is, for example, not known if signals representing two different VRs converge onto the same neuron(s) in the vomeronasal amygdala or in the hypothalamus. Likewise it is unclear whether the anatomical segregation of signals generated by V1R (and OR) or V2R VSNs is maintained beyond the bulb. Although segregated projections from the anterior and posterior AOB to the vomeronasal amygdala were observed in opposum (Martinez-Marcos and Halpern 1999) and rat (Mohedano-Moriano et al. 2007), complementary studies tracing the anterograde pathway from the different zones in the AOB or the retrograde pathway from the medial amygdaloid nucleus or the posteromedial cortical amygdaloid area did not find evidence for segregation in the mouse (von Campenhausen and Mori 2000; Salazar and Brennan 2001). However, the dorsal and ventral parts of the posterior medial amygdaloid nucleus are differentially activated by reproductive or defensive stimuli, respectively (Fernandez-Fewell and Meredith 1994; Bressler and Baum 1996; Heeb and Yahr 1996; Kollack-Walker and Newman 1997; Dielenberg et al. 2001; McGregor et al. 2004; Choi et al. 2005), and projections from these two medial amygdala subnuclei to the hypothalamus are anatomically segregated (Choi et al. 2005; reviewed in Swanson 2000; Canteras 2002).

5.3 Neural Circuits Linking Pheromone Sensing and Reproduction

Ultimately, vomeronasal signals are relayed to specific neurons in the hypothalamus, which initiate and control the behavioral and hormonal responses triggered by pheromones. At the center of hypothalamic control of reproduction are gonadotropin releasing hormone (GnRH) neurons, which regulate the reproductive endocrine status in mammals by secreting GnRH (Gore 2002). In the mouse, approximately 800 GnRH neurons form a loose continuum extending through the basal forebrain, with the majority concentrated in the preoptic area of the hypothalamus (reviewed in Herbison 2006). GnRH triggers the release of gonadotropins from the pituitary, which regulate puberty onset, gametogenesis, and estrus cycling. Recent studies using transneuronal tracers have shown that GnRH neurons appear to integrate both vomeronasal signals (Boehm et al. 2005) and main olfactory signals (Boehm et al. 2005; Yoon et al. 2005). These studies indicate direct synaptic connections between GnRH neurons and neurons relaying vomeronasal signals in both parts of the vomeronasal amygdala as well as the posteromedial bed nucleus of the stria terminalis (Boehm et al. 2005). Consistent with this, some neurons in the vomeronasal amygdala identified by the transneuronal tracer are activated in female mice exposed to α-farnesene (Boehm et al. 2005), a pheromone present in male urine inducing estrus in group-housed females (Novotny 2003).

Surprisingly, these studies also indicate direct synaptic connections between GnRH neurons and neurons relaying information from the main olfactory system such as the anterior cortical nucleus of the amygdala and the piriform cortex (Boehm et al. 2005; Yoon et al. 2005), both of which receive their major input from the main olfactory bulb (MOB) (Dulac and Wagner 2006). Some neurons in the anterior cortical nucleus of the amygdala identified by the transneuronal tracer were activated by α-farnesene, suggesting an involvement of the main olfactory system in response to pheromones and raising the possibility that signals representing the same chemical, but originating in the main and accessory olfactory systems, respectively, may converge onto the same neuron(s) at a deeper level in the brain (Boehm et al. 2005; Boehm 2006).

In addition, these studies have revealed feedback loops between the neuroendocrine hypothalamus and both the main and the accessory olfactory systems (Boehm et al. 2005), suggesting that the animal's neuroendocrine status might modulate its susceptibility to chemosensory cues (Boehm 2006).

6. Vomeronasal Effects

6.1 Mouse Models Addressing the Role of the VNO in Pheromone-Mediated Responses

Different knockout mouse models have addressed the functional role of the VNO in response to pheromones. Gene-targeted mice with functional ablations of the G-protein α subunits $G_{\alpha i2}$ or $G_{\alpha o}$ showed some behavioral and anatomical differences compared with wild-type mice (Table 1; Tanaka et al. 1999; Luo et al. 2002; Norlin et al. 2003). However, there is no formal proof that these G-protein α subunits, which are also expressed in other parts of the central nervous system (as well as in the MOE), are indeed indispensable for the VR signal transduction cascades. Unfortunately these loci have not yet been targeted using tissue-specific knockout approaches, leaving the possibility that the observed phenotypes are the result of alteration of nonvomeronasal structures. Consistent with this, behavioral analysis of mice lacking the α subunit of G_o, which is also expressed in all OSNs (Wekesa and Anholt 1999), revealed impaired olfactory exploratory behavior (Luo et al. 2002).

Particularly interesting is a mouse deficient for the transient receptor channel TRPC2 (TRPC2–/–). It was expected that the TRPC2–/– mice would essentially be VNO-deficient and thus would genetically replicate earlier studies with VNX mice. Surprisingly, TRPC2–/– mice exhibited behavioral abnormalities differing from those described in VNX mice (Leypold et al. 2002; Stowers et al. 2002; Pankevich et al. 2004; Keller et al. 2006). Strikingly, male TRPC2–/– mice mount male and female mice indiscriminately (Leypold et al. 2002; Stowers et al. 2002), while TRPC2–/– females mount both males and females (Kimchi et al. 2007), suggesting a major role played by the VNO in gender discrimination. Although the findings were apparently difficult to reconcile with those of VNX studies, the authors showed that the VNX procedure often affects the MOE in a dramatic and random way, possibly explaining the nonconcordant observations (Kimchi et al. 2007). Alternatively, the contradicting experimental observations might be explained by the fact that – owing to the conventional knockout strategy – the TRPC2–/– animals never possess a fully functional VNO, whereas the VNX animals are unimpaired until adulthood, suggesting an important VNO function during development. As mentioned, TRPC2–/– females show some characteristics of male reproductive behavior, raising the possibility that pheromone input via the VNO may repress male-specific behavioral circuits in the brain (Kimchi et al. 2007).

Table 1 Genetic dissection of pheromone sensing

	Aggressive behavior		Reproductive behavior		Neuroendocrine phenotype		Remarks
	Male	Female	Male	Female	Male	Female	
Affecting VNO							
Both zones							
TRPC2–/–	Eliminated[a,b]	Eliminated[c]	Altered: loss of gender preference[a,b]	Altered: absence of maternal aggression[a]; display male-specific behaviors[c]	Normal testosterone and estradiol levels[b,c]	Normal estradiol levels[c]; Bruce effect unimpaired[d]	Not confirmed by all VNX experiments[e,f]
Apical zone							
V1Rab–/–	Normal[g]	Reduced[g]	Reduced[g]	Not reported	Not reported	Normal estrus cycling[g]	Normal maternal behavior[g]
Gαi2–/–	Reduced[h]	Eliminated[h]	Normal[h]	Not reported	Not reported	Not reported	Also expressed in MOE as well as CNS[i]
Basal zone							
Gαo–/–	Not reported	Not reported	Not reported	Not reported	Not reported	Not reported	Also expressed in MOE[i]; impaired olfactory behavior[j]
β2m–/–	Reduced[k]	Not reported	Reduced[k]	Not reported	Not reported	Not reported	
Affecting MOE							
CNG2a–/–	Eliminated[l]	Not reported	Impaired[l,m]	Not reported	Not reported	Not reported	Confirmed by MOE ablation[m]
AC3–/–	Eliminated[n]	Not reported	Eliminated[n] or unaffected[o]	Not reported	Normal testosterone levels[n]	Not reported	Confirmed by MOE ablation[n]
Gαolf–/–	Not reported	Not reported	Fertile[p]	Fertile, poor nursing[p]	Not reported	Not reported	Also expressed in CNS[p]

VNO vomeronasal organ, *VNX* connection of the vomeronasal organ with the brain has been severed, *MOE* main olfactory epithelium, *CNS* central nervous system. [a]Leypold et al. (2002). [b]Stowers et al. (2002). [c]Kimchi et al. (2007). [d]Kelliher et al. (2006). [e]Pankevich et al. (2004). [f]Keller et al. (2006). [g]Del Punta et al. (2002a, b). [h]Norlin et al. (2003). [i]Wekesa and Anholt (1999). [j]Luo et al. (2002). [k]Loconto et al. (2003). [l]Mandiyan et al. (2005). [m]Yoon et al. (2005). [n]Wang et al. (2006). [o]Wong et al. (2000). [p]Belluscio et al. (1998)

Behavioral deficits of V1Rab–/– mice bearing a deletion of a gene cluster comprising 16 intact V1R genes include impaired male sexual behavior towards females and reduced maternal aggression to intruder males (Del Punta et al. 2002a). As discussed above, this mouse model provided functional evidence that at least some V1Rs are pheromone receptors, and clearly demonstrates the pivotal role of the VNO in pheromone-triggered behaviors.

Although a complementary mouse model deleting parts of the V2R repertoire has not yet been reported, behavioral analysis of β2-microglobulin-deficient mice, which do not properly traffic some V2R receptors, revealed impaired male–male aggression in this mouse model (Loconto et al. 2003).

6.2 Mouse Models Addressing the Contribution of the MOE to Pheromone-Mediated Responses

The contribution of both the main and the accessory olfactory systems to stereotyped behaviors and long-term neuroendocrine alterations triggered by chemosensory cues is evident in experiments across rodent species. Whereas complete removal of the olfactory bulbs (bulbectomy) eliminates mating and aggression, VNX or $ZnSO_4$-induced MOE ablation alone have subtler effects. To analyze the contribution of the main olfactory system to behaviors and hormonal changes thought to be triggered by pheromones, selective inactivation of MOE signaling (leaving VNO signaling unimpaired) has also been attempted genetically in mice. Deletion of a subunit of the CNG channel which is expressed in most OSNs but not of VSNs of CNG2a–/– mice leads to striking behavioral deficits. Male CNG2a–/– mice display impaired reproductive behaviors and have reduced but not completely abolished reproductive success (Mandiyan et al. 2005; Yoon et al. 2005). In addition, male CNG2a–/– mice fail to display aggressive behavior to intruder males (Mandiyan et al. 2005). Consistent with this, male mice lacking type III adenylyl cyclase (AC3–/–), which exhibit no sensory signaling through the MOE, display impaired sexual behaviors as well as aggressive behaviors towards intruders (Wang et al. 2006; however, also see Wong et al. 2000 and Table 1). These data demonstrate the contribution of the main olfactory system to both aggressive and reproductive behaviors.

In summary, data obtained from mouse models selectively affecting either MOE or VNO signaling strongly suggest that both the main and the accessory olfactory systems can converge and synergize to express stereotyped behaviors and hormonal changes triggered by chemosensory cues in rodents.

Most recently, Sakano's group reported zone-specific as well as class-specific OSN ablation in transgenic mice with striking behavioral deficits (Kobayakawa et al. 2007). Mice lacking OSNs expressing class II ORs fail to display freezing behavior in response to predator odorants such as trimethylthiazoline, which is secreted by the anal gland of fox (Fig. 1), suggesting hardwired neural circuits triggering stereotyped behavioral responses initiating in the main olfactory system.

Acknowledgements We are indebted to Dörte Clausen for artistic drawings and Libby Guethlein for critical comments on the manuscript. We thank Patrick Pfister for the OR alignment and the generation of the OR sunshine tree.

References

Achiraman S, Archunan G (2002) Characterization of urinary volatiles in Swiss male mice (Mus musculus): bioassay of identified compounds. J Biosci 27:679–686

Alheid GF, de Olmos JS, Betramino CA (1995) Amygdala and extended amygdala. In: G Paxinos (ed) The rat nervous system. Academic, San Diego, pp 495–578

Belluscio L, Gold GH, Nemes A, Axel R (1998) Mice deficient in G_{olf} are anosmic. Neuron 20:69–81

Belluscio L, Koentges G, Axel R, Dulac C (1999) A map of pheromone receptor activation in the mammalian brain. Cell 97:209–220

Berghard A, Buck LB (1996) Sensory transduction in vomeronasal neurons: evidence for $G_{\alpha o}$, $G_{\alpha i2}$, and adenylyl cyclase II as major components of a pheromone signaling cascade. J Neurosci 16:909–918

Boehm U (2006) The vomeronasal system in mice: from the nose to the hypothalamus- and back! Semin Cell Dev Biol 17:471–479

Boehm T, Zufall F (2006) MHC peptides and the sensory evaluation of genotype. Trends Neurosci 29:100–107

Boehm U, Zou Z, Buck LB (2005) Feedback loops link odor and pheromone signaling with reproduction. Cell 123:683–695

Boschat C, Pelofi C, Randin O, Roppolo D, Luscher C, Broillet MC, Rodriguez I (2002) Pheromone detection mediated by a V1r vomeronasal receptor. Nat Neurosci 5:1261–1262

Breer H, Fleischer J, Strotmann J (2006) The sense of smell: multiple olfactory subsystems. Cell Mol Life Sci 63:1465–1475

Brennan PA, Zufall F (2006) Pheromonal communication in vertebrates. Nature 444:308–315

Bressler SC, Baum MJ (1996) Sex comparison of neuronal Fos immunoreactivity in the rat vomeronasal projection circuit after chemosensory stimulation. Neuroscience 71:1063–1072

Bruce HM (1960) A block to pregnancy in the mouse caused by proximity of strange males. J Reprod Fertil 1:96–103

Buck LB, Axel R (1991) A novel multigene family may encode odorant receptors: a molecular basis for odor recognition. Cell 65:175–187

Canteras NS (2002) The medial hypothalamic defensive system: hodological organization and functional implications. Pharmacol Biochem Behav 71:481–491

Chamero P, Marton TF, Logan DW, Flanagan K, Cruz JR, Saghatelian A, Cravatt BF, Stowers L (2007) Identification of protein pheromones that promote aggressive behaviour. Nature 450:899–902

Choi GB, Dong HW, Murphy AJ, Valenzuela DM, Yancopoulos GD, Swanson LW, Anderson DJ (2005) Lhx6 delineates a pathway mediating innate reproductive behaviors from the amygdala to the hypothalamus. Neuron 46:647–660

Del Punta K, Leinders-Zufall T, Rodriguez I, Jukam D, Wysocki CJ, Ogawa S, Zufall F, Mombaerts P (2002a) Deficient pheromone responses in mice lacking a cluster of vomeronasal receptor genes. Nature 419:70–74

Del Punta K, Puche A, Adams NC, Rodriguez I, Mombaerts P (2002b) A divergent pattern of sensory axonal projections is rendered convergent by second-order neurons in the accessory olfactory bulb. Neuron 35:1057–1066

Del Punta K, Rothman A, Rodriguez I, Mombaerts P (2000) Sequence diversity and genomic organization of vomeronasal receptor genes in the mouse. Genome Res. 10:1958–1567

Dielenberg RA, Hunt GE, McGregor IS (2001) 'When a rat smells a cat': the distribution of Fos immunoreactivity in rat brain following exposure to a predatory odor. Neuroscience 104:1085–1097

Dudley CA, Moss RL (1999) Activation of an anatomically distinct subpopulation of accessory olfactory bulb neurons by chemosensory stimulation. Neuroscience 91:1549–1556

Dulac C, Axel R (1995) A novel family of genes encoding putative pheromone receptors in mammals. Cell 83:195–206

Dulac C, Wagner S (2006) Genetic analysis of brain circuits underlying pheromone signaling. Annu Rev Genet 40:449–467

Fernandez-Fewell GD, Meredith M (1994) c-fos expression in vomeronasal pathways of mated or pheromone-stimulated male golden hamsters: contributions from vomeronasal sensory input and expression related to mating performance. J Neurosci 14:3643–3654

Gore AC (2002) GnRH: the master molecule of reproduction. Kluwer, Dordecht

Grus WE, Zhang J (2004) Rapid turnover and species-specificity of vomeronasal pheromone receptor genes in mice and rats. Gene 340:303–312

Grus WE, Shi P, Zhang YP, Zhang J (2005) Dramatic variation of the vomeronasal pheromone receptor gene repertoire among five orders of placental and marsupial mammals. Proc Natl Acad Sci USA 102:5767–5772

Grus WE, Shi P, Zhang J (2007) Largest vertebrate vomeronasal type 1 receptor gene repertoire in the semiaquatic platypus. Mol Biol Evol 24:2153–2157

Halem HA, Baum MJ, Cherry JA (2001) Sex difference and steroid modulation of pheromone-induced immediate early genes in the two zones of the mouse accessory olfactory system. J Neurosci 21:2474–2480

Halpern M, Martinez-Marcos A (2003) Structure and function of the vomeronasal system: an update. Prog Neurobiol 70:245–318

Heeb MM, Yahr P (1996) c-Fos immunoreactivity in the sexually dimorphic area of the hypothalamus and related brain regions of male gerbils after exposure to sex-related stimuli or performance of specific sexual behaviors. Neuroscience 72:1049–1071

Herbison AE (2006) Physiology of the gonadotropin-releasing hormone neuronal network. In: Neill JD, Plant TM, Pfaff DW (eds) Knobil and Neill's physiology of reproduction. Elsevier, St Louis, pp 1415–1482

Herrada G, Dulac C (1997) A novel family of putative pheromone receptors in mammals with a topographically organized and sexually dimorphic distribution. Cell 90:763–773

Huh GS, Boulanger LM, Du H, Riquelme PA, Brotz TM, Shatz CJ (2000) Functional requirement for class I MHC in CNS development and plasticity. Science 290:2155–2159

Inamura K, Kashiwayanagi M, Kurihara K (1999) Regionalization of Fos immunostaining in rat accessory olfactory bulb when the vomeronasal organ was exposed to urine. Eur J Neurosci 11:2254–2260

Ishii T, Hirota J, Mombaerts P (2003) Combinatorial coexpression of neural and immune multigene families in mouse vomeronasal sensory neurons. Curr Biol 13:394–400

Jia C, Halpern M (1996) Subclasses of vomeronasal receptor neurons: differential expression of G proteins (G_{ia2} and G_{oa}) and segregated projections to the accessory olfactory bulb. Brain Res 719:117–128

Jia C, Halpern M (1997) Segregated populations of mitral/tufted cells in the accessory olfactory bulb. Neuroreport 8:1887–1890

Karlson P, Luscher M (1959) Pheromones': a new term for a class of biologically active substances. Nature 183:55–56

Keller M, Pierman S, Douhard Q, Baum MJ, Bakker J (2006) The vomeronasal organ is required for the expression of lordosis behaviour, but not sex discrimination in female mice. Eur J Neurosci 23:521–530

Kelliher KR, Spehr M, Li XH, Zufall F, Leinders-Zufall T (2006) Pheromonal recognition memory induced by TRPC2-independent vomeronasal sensing. Eur J Neurosci 23:3385–3390

Kimchi T, Xu J, Dulac C (2007) A functional circuit underlying male sexual behaviour in the female mouse brain. Nature 448:1009–1014

Kimoto H, Haga S, Sato K, Touhara K (2005) Sex-specific peptides from exocrine glands stimulate mouse vomeronasal sensory neurons. Nature 437:898–901

Kimoto H, Sato K, Nodari F, Haga S, Holy TE, Touhara K (2007) Sex- and strain-specific expression and vomeronasal activity of mouse ESP family peptides. Curr Biol 17:1879–1884

Kobayakawa K, Kobayakawa R, Matsumoto H, Oka Y, Imai T, Ikawa M, Okabe M, Ikeda T, Itohara S, Kikusui T, Mori K, Sakano H (2007) Innate versus learned odour processing in the mouse olfactory bulb. Nature 450:503–508

Kollack-Walker S, Newman SW (1997) Mating-induced expression of c-fos in the male Syrian hamster brain: role of experience, pheromones, and ejaculations. J Neurobiol 32:481–501

Krieger J, Schmitt A, Lobel D, Guderman T, Schultz G, Breer H, Boekhoff I (1999) Selective activation of G protein subtypes in the vomeronasal organ upon stimulation with urine-derived compounds. J Biol Chem 274:4655–4662

Kumar A, Dudley CA, Moss RL (1999) Functional dichotomy within the vomeronasal system: distinct zones of neuronal activity in the accessory olfactory bulb correlate with sex-specific behaviors. J Neurosci 19:RC32

Lane RP, Cutforth T, Axel R, Hood L, Trask BJ (2002) Sequence analysis of mouse vomeronasal receptor gene clusters reveals common promoter motifs and a history of recent expansion. Proc Natl Acad Sci USA 99:291–296

Lane RP, Young J, Newman T, Trask BJ (2004) Species specificity in rodent pheromone receptor repertoires. Genome Res 14:603–608

Leinders-Zufall T, Lane AP, Puche AC, Ma W, Novotny MV, Shipley MT, Zufall F (2000) Ultrasensitive pheromone detection by mammalian vomeronasal neurons. Nature 405:792–796

Leinders-Zufall T, Brennan P, Widmayer P,S PC, Maul-Pavicic A, Jager M, Li XH, Breer H, Zufall F, Boehm T (2004) MHC class I peptides as chemosensory signals in the vomeronasal organ. Science 306:1033–1037

Levai O, Feistel T, Breer H, Strotmann J (2006) Cells in the vomeronasal organ express odorant receptors but project to the accessory olfactory bulb. J Comp Neurol 498:476–490

Leypold BG, Yu CR, Leinders-Zufall T, Kim MM, Zufall F, Axel R (2002) Altered sexual and social behaviors in trp2 mutant mice. Proc Natl Acad Sci USA 99:6376–6381

Loconto J, Papes F, Chang E, Stowers L, Jones EP, Takada T, Kumanovics A, Fischer Lindahl K, Dulac C (2003) Functional expression of murine V2R pheromone receptors involves selective association with the M10 and M1 families of MHC class Ib molecules. Cell 112:607–618

Lucas P, Ukhanov K, Leinders-Zufall T, Zufall F (2003) A diacylglycerol-gated cation channel in vomeronasal neuron dendrites is impaired in TRPC2 mutant mice: mechanism of pheromone transduction. Neuron 40:551–561

Luo AH, Cannon EH, Wekesa KS, Lyman RF, Vandenbergh JG, Anholt RR (2002) Impaired olfactory behavior in mice deficient in the alpha subunit of G_o. Brain Res 941:62–71

Mandiyan VS, Coats JK, Shah NM (2005) Deficits in sexual and aggressive behaviors in Cnga2 mutant mice. Nat Neurosci 8:1660–1662

Martinez-Marcos A, Halpern M (1999) Differential projections from the anterior and posterior divisions of the accessory olfactory bulb to the medial amygdala in the opossum, *Monodelphis domestica*. Eur J Neurosci 11:3789–3799

Matsunami H, Buck LB (1997) A multigene family encoding a diverse array of putative pheromone receptors in mammals. Cell 90:775–784

Matsuoka M, Yokosuka M, Mori Y, Ichikawa M (1999) Specific expression pattern of Fos in the accessory olfactory bulb of male mice after exposure to soiled bedding of females. Neurosci Res 35:189–195

McGregor IS, Hargreaves GA, Apfelbach R, Hunt GE (2004) Neural correlates of cat odor-induced anxiety in rats: region-specific effects of the benzodiazepine midazolam. J Neurosci 24:4134–4144

Mohedano-Moriano A, Pro-Sistiaga P, Ubeda-Banon I, Crespo C, Insausti R, Martinez-Marcos A (2007) Segregated pathways to the vomeronasal amygdala: differential projections from the anterior and posterior divisions of the accessory olfactory bulb. Eur J Neurosci 25:2065–2080

Mombaerts P (2004) Genes and ligands for odorant, vomeronasal and taste receptors. Nat Rev Neurosci 5:263–278

Mundy NI, Cook S (2003) Positive selection during the diversification of class I vomeronasal receptor-like (V1RL) genes, putative pheromone receptor genes, in human and primate evolution. Mol Biol Evol 20:1805–1810

Newman SW (1999) The medial extended amygdala in male reproductive behavior. A node in the mammalian social behavior network. Ann N Y Acad Sci 877:242–257

Norlin EM, Gussing F, Berghard A (2003) Vomeronasal phenotype and behavioral alterations in $G_{\alpha i2}$ mutant mice. Curr Biol 13:1214–1219

Novotny MV (2003) Pheromones, binding proteins and receptor responses in rodents. Biochem Soc Trans 31:117–122

Olson R, Huey-Tubman KE, Dulac C, Bjorkman PJ (2005) Structure of a pheromone receptor-associated MHC molecule with an open and empty groove. PLoS Biol 3:e257

Pankevich DE, Baum MJ, Cherry JA (2004) Olfactory sex discrimination persists, whereas the preference for urinary odorants from estrous females disappears in male mice after vomeronasal organ removal. J Neurosci 24:9451–9457

Pfister P, Randall J, Montoya-Burgos JI, Rodriguez I (2007) Divergent evolution among teleost V1r receptor genes. PLoS ONE 2:e379

Rodriguez I (2004) Pheromone receptors in mammals. Horm Behav 46:219–230

Rodriguez I (2007) Odorant and pheromone receptor gene regulation in vertebrates. Curr Opin Genet Dev 17:465–470

Rodriguez I, Mombaerts P (2002) Novel human vomeronasal receptor-like genes reveal species-specific families. Curr Biol 12:R409–411

Rodriguez I, Feinstein P, Mombaerts P (1999) Variable patterns of axonal projections of sensory neurons in the mouse vomeronasal system. Cell 97:199–208

Rodriguez I, Greer CA, Mok MY, Mombaerts P (2000) A putative pheromone receptor gene expressed in human olfactory mucosa. Nat Genet 26:18–19

Rodriguez I, Del Punta K, Rothman A, Ishii T, Mombaerts P (2002) Multiple new and isolated families within the mouse superfamily of V1r vomeronasal receptors. Nat Neurosci 5:134–140

Roppolo D, Vollery S, Kan CD, Luscher C, Broillet MC, Rodriguez I (2007) Gene cluster lock after pheromone receptor gene choice. EMBO J 26:3423–3430

Ryba NJ, Tirindelli R (1997) A new multigene family of putative pheromone receptors. Neuron 19:371–379

Salazar I, Brennan PA (2001) Retrograde labelling of mitral/tufted cells in the mouse accessory olfactory bulb following local injections of the lipophilic tracer DiI into the vomeronasal amygdala. Brain Res 896:198–203

Saraiva LR, Korsching SI (2007) A novel olfactory receptor gene family in teleost fish. Genome Res 17:1448–1457

Scalia F, Winans SS (1975) The differential projections of the olfactory bulb and accessory olfactory bulb in mammals. J Comp Neurol 161:31–55

Shi P, Zhang J (2007) Comparative genomic analysis identifies an evolutionary shift of vomeronasal receptor gene repertoires in the vertebrate transition from water to land. Genome Res 17:166–174

Shi P, Bielawski JP, Yang H, Zhang YP (2005) Adaptive diversification of vomeronasal receptor 1 genes in rodents. J Mol Evol 60:566–576

Shipley MT, Ennis M, Puche A (2004) Olfactory system. In: Paxinos G (ed) The rat nervous system. Elsevier, San Diego, pp 923–964

Silvotti L, Moiani A, Gatti R, Tirindelli R (2007) Combinatorial co-expression of pheromone receptors, V2Rs. J Neurochem 103:1753–1763

Singh PB, Brown RE, Roser B (1987) MHC antigens in urine as olfactory recognition cues. Nature 327:161–164

Spehr M, Hatt H, Wetzel CH (2002) Arachidonic acid plays a role in rat vomeronasal signal transduction. J Neurosci 22:8429–8437

Spehr M, Kelliher KR, Li XH, Boehm T, Leinders-Zufall T, Zufall F (2006) Essential role of the main olfactory system in social recognition of major histocompatibility complex peptide ligands. J Neurosci 26:1961–1970

Stowers L, Holy TE, Meister M, Dulac C, Koentges G (2002) Loss of sex discrimination and male–male aggression in mice deficient for TRP2. Science 295:1493–1500

Swanson LW (2000) Cerebral hemisphere regulation of motivated behavior. Brain Res 886:113–164

Takami S, Graziadei PP (1990) Morphological complexity of the glomerulus in the rat accessory olfactory bulb – a Golgi study. Brain Res 510:339–342

Takami S, Graziadei PP (1991) Light microscopic Golgi study of mitral/tufted cells in the accessory olfactory bulb of the adult rat. J Comp Neurol 311:65–83

Tanaka M, Treloar H, Kalb RG, Greer CA, Strittmatter SM (1999) $G_{\alpha 0}$ protein-dependent survival of primary accessory olfactory neurons. Proc Natl Acad Sci USA 96:14106–14111

Thompson RN, McMillon R, Napier A, Wekesa KS (2007) Pregnancy block by MHC class I peptides is mediated via the production of inositol 1,4,5-trisphosphate in the mouse vomeronasal organ. J Exp Biol 210:1406–1412

von Campenhausen H, Mori K (2000) Convergence of segregated pheromonal pathways from the accessory olfactory bulb to the cortex in the mouse. Eur J Neurosci 12:33–46

von Campenhausen H, Yoshihara Y, Mori K (1997) OCAM reveals segregated mitral/tufted cell pathways in developing accessory olfactory bulb. Neuroreport 8:2607–2612

Wagner S, Gresser AL, Torello AT, Dulac C (2006) A multireceptor genetic approach uncovers an ordered integration of VNO sensory inputs in the accessory olfactory bulb. Neuron 50:697–709

Wang Z, Balet Sindreu C, Li V, Nudelman A, Chan GC, Storm DR (2006) Pheromone detection in male mice depends on signaling through the type 3 adenylyl cyclase in the main olfactory epithelium. J Neurosci 26:7375–7379

Wang Z, Nudelman A, Storm DR (2007) Are pheromones detected through the main olfactory epithelium? Mol Neurobiol 35:317–323

Wekesa KS, Anholt RR (1999) Differential expression of G proteins in the mouse olfactory system. Brain Res 837:117–126

Wong ST, Trinh K, Hacker B, Chan GC, Lowe G, Gaggar A, Xia Z, Gold GH, Storm DR (2000) Disruption of the type III adenylyl cyclase gene leads to peripheral and behavioral anosmia in transgenic mice. Neuron 27:487–497

Yang H, Shi P, Zhang YP, Zhang J (2005) Composition and evolution of the V2r vomeronasal receptor gene repertoire in mice and rats. Genomics 86:306–315

Yoon H, Enquist LW, Dulac C (2005) Olfactory inputs to hypothalamic neurons controlling reproduction and fertility. Cell 123:669–682

Yoshihara Y, Kawasaki M, Tamada A, Fujita H, Hayashi H, Kagamiyama H, Mori K (1997) OCAM: A new member of the neural cell adhesion molecule family related to zone-to-zone projection of olfactory and vomeronasal axons. J Neurosci 17:5830–5842

Young JM, Trask BJ (2007) V2R gene families degenerated in primates, dog and cow, but expanded in opossum. Trends Genet 23:212–215

Zhang X, Rodriguez I, Mombaerts P, Firestein S (2004) Odorant and vomeronasal receptor genes in two mouse genome assemblies. Genomics 83:802–811

Zufall F (2005) The TRPC2 ion channel and pheromone sensing in the accessory olfactory system. Naunyn Schmiedebergs Arch Pharmacol 371:245–250

Zufall F, Leinders-Zufall T (2007) Mammalian pheromone sensing. Curr Opin Neurobiol 17:483–489

Molecular Genetic Dissection of the Zebrafish Olfactory System

Y. Yoshihara

Abstract Zebrafish is now becoming one of the most useful model organisms in neurobiology. In addition to its general advantageous properties (external fertilization, rapid development, transparency of embryos, etc.), the zebrafish is amenable to various genetic engineering technologies such as transgenesis, mutagenesis, gene knockdown, and transposon-mediated gene transfer. A transgenic approach unraveled two segregated neural circuits originating from ciliated and microvillous sensory neurons in the olfactory epithelium to distinct regions of the olfactory bulb, which likely convey different types of olfactory information (e.g., pheromones and odorants) to the higher olfactory centers. Furthermore, the two basic principles identified in mice, so-called one neuron–one receptor rule and convergence of like axons to target glomeruli, are basically preserved also in the zebrafish, rendering this organism a suitable model vertebrate for studies of the olfactory system. This review summarizes recent advances in our knowledge on genetic, molecular, and cellular mechanisms underlying the development and functional architecture of the olfactory neural circuitry in the zebrafish.

1 Introduction

Olfaction, the sense of smell, is an important neural system in various animal species, including fish, for their life. Fish can detect a variety of odorants emitted from objects and dissolved in the water, such as amino acids, bile salts, nucleotides, polyamines, prostaglandins, and steroids. The fish olfactory system is extensively developed to receive and discriminate these odorant molecules, to transmit their signals to the brain, and to mediate fundamental behaviors such as food finding, alarm response, predator avoidance, social communication, reproductive activity, and spawning migration (Sorensen and Caprio 1998; Zielinski and Hara 2007).

Y. Yoshihara (✉)
Laboratory for Neurobiology of Synapse, RIKEN Brain Science Institute, 2-1 Hirosawa, Wako, Saitama 351-0198, Japan
e-mail: yoshihara@brain.riken.jp

Since the discovery of the odorant receptor (OR) multigene family in rodents by Buck and Axel (1991), a rapid and remarkable advance has been made in our understanding of how the olfactory information is received in the olfactory epithelium (OE) and coded in the olfactory bulb (OB). Molecular biological and modern genetic techniques contributed a great deal to the elucidation of two basic principles underlying the establishment of functional architecture of the olfactory system. One is the "one neuron–one receptor" rule. In mouse, for example, each olfactory sensory neuron (OSN) expresses only one type of OR gene out of a repertoire of up to 1,000 genes equipped in the genome (Chess et al. 1994; Malnic et al. 1999; Serizawa et al. 2003; Lewcock and Reed 2004; Shykind et al. 2004). This principle enables individual OSNs to respond to a range of odorants that bind to the expressed ORs. In other words, OSNs expressing a given OR are tuned to a particular molecular receptive range. The other is the "convergence of like axons to target glomeruli." The OSNs expressing a given OR project and converge their axons to a pair of topographically fixed glomeruli in the OB: one on the medial side and the other on the lateral side (Vassar et al. 1994; Ressler et al. 1994; Mombaerts et al. 1996; Wang et al. 1998). Thus, the odor information received by a given OR is converged onto and represented by a particular pair of OB glomeruli. Through this elegantly wired neural circuit, an "OR map" or "odor map" is developed on the glomerular sheet of the OB (Mori et al. 1999; Mori et al. 2006).

In addition to mouse (*Mus musculus*), various model organisms amenable to genetic engineering have been proved useful in studies of the olfactory system, including zebrafish (*Danio rerio*), fruit fly (*Drosophila melanogaster*), and nematode worm (*Caenorhabditis elegans*). Among them, the zebrafish, phylogenetically situated between mammals and insects/worms, is becoming an attractive vertebrate model organism suitable for olfactory research. This review will highlight recent progress in neuroanatomical, developmental, and functional research on the zebrafish olfactory system with special reference to molecular biological and genetic correlates.

2 Zebrafish as an Excellent Vertebrate Model Organism

2.1 General Advantages

Zebrafish, a freshwater small teleost commonly available in pet shops, offers numerous advantages over other vertebrates for biological studies. Zebrafish are easy to grow and produce large clutches of eggs (100–200 per mating) through external fertilization (Westerfield 1995). The embryos develop quickly, hatching at as early as 3 days after fertilization and starting to swim at 5 days after fertilization. The zebrafish embryos are optically transparent throughout early development, enabling us to observe organogenesis and morphogenesis in vivo. In particular, transgenic expression of green fluorescent protein (GFP) and its derivatives in selective cell types greatly facilitates the live imaging of various dynamic developmental events such as cell division, cell migration, and neural circuit formation.

2.2 Forward and Reverse Genetics

One major advantage of using the zebrafish in biological studies is its amenability of various genetic engineering techniques as follows:

- *Mutagenesis*: As has been proved in *Drosophila* and *C. elegans*, forward genetics (from phenotypes to genes) is a powerful strategy of fishing out mutants with perturbation in various biological processes and subsequent identification of genes responsible for the mutant phenotypes. Zebrafish is the first vertebrate species on which large-scale mutant screens were carried out (Haffter et al. 1996; Driever et al. 1996). The publication of a series of 37 articles on approximately 2,000 mutations perturbing the zebrafish development in a special issue of the journal*Development* in 1996 triggered the identification of a number of mutagenized zebrafish which show defects in various aspects of development, morphogenesis, neural functions, and behaviors. Among them, there are several mutants displaying abnormalities in development and behavior of the olfactory system (Miyasaka et al. 2005, 2007; Vitebsky et al. 2005) (see Sect. 5).
- *Transgenesis*: Generation of transgenic animals is a fundamental technique for assessing gene function, identifying transcriptional promoter/enhancer elements, and labeling specific cell types with reporter molecules. The use of a heat-shock promoter enables us to switch on ubiquitous expression of transgenes at any developmental stage (Halloran et al. 2000; Miyasaka et al. 2005, 2007). Conditional transgene expression can be performed with the aid of the Cre/loxP system and the Gal4/UAS system that have been successfully used in mice and *Drosophila*, respectively. Extremely long DNA fragments encompassing approximately 100 kb can be stably introduced into the zebrafish genome by using bacterial artificial chromosome (BAC) transgenes (Yang et al. 2006; Sato et al. 2007b; Nishizumi et al. 2007). In addition, the method of making transgenic zebrafish has been greatly improved by the introduction of the Tol2 transposable element (Kawakami and Shima 1999; Kawakami 2004; Kwan et al. 2007) and *I-SceI* meganuclease (Thermes et al. 2002) for efficient integration of transgenes.
- *Gene knockdown and knockout*: Reverse genetics (from genes to phenotypes) is also applicable in zebrafish. Gene knockdown (reduced expression of gene products) is a very popular and powerful tool in studies of zebrafish development. It is easily achieved by the injection of morpholino antisense oligonucleotide into yolk of fertilized eggs (Nasevisius and Ekker 2000; Ekker 2000). Furthermore, a recently established method, targeting induced local lesions in genomes (TILLING), has made it possible to obtain knockout zebrafish of any genes by PCR-based screening of heteroduplex formation and resequencing of a DNA library prepared from mutagenized fish (Wienholds et al. 2003; Sood et al. 2006).

In addition to the abovementioned basic techniques, more-advanced genetic methods have been developed in the zebrafish, including a Tol2 transposon-meditated gene trap approach combined with the Gal4/UAS system (Asakawa et al. 2008), retrovirus-mediated large-scale enhancer detection (Ellingsen et al. 2005), and Cre/loxP/Gal4/UAS-mediated single-cell mosaic analysis (Sato et al. 2007a). Thus, the zebrafish is a remarkably useful vertebrate species with which we can perform both forward and reverse genetic analyses, similar to *Drosophila* and *C. elegans*.

2.3 Disadvantages

Despite numerous advantageous features over other model organisms, several disadvantages are apparent that should be taken into consideration when using zebrafish. First, the generation time of zebrafish is not so short (approximately 3 months), which is comparable to that of mouse (2–3 months) but much longer than that of *Drosophila* (approximately 7 days) and *C. elegans* (approximately 3 days). Second, the duplication of the fish genome after the phylogenetic divergence of fish and mammals resulted in gene redundancy in the zebrafish genome. In some cases, this redundancy makes it difficult and complicated to analyze the functions of zebrafish orthologs of a particular mammalian gene. In other cases, however, complex localizations and functions of a given mammalian gene are separated and allocated to different paralogous genes in zebrafish, rendering the redundancy into an advantageous property. Third, the activity of an internal ribosome entry site (IRES) from encephalomyocarditis virus for bicistronic expression of two transgene products is very low in zebrafish, although the IRES has been proved to be extremely useful in mouse for simultaneous expression of various transgenes (e.g., β-galactosidase, GFP derivatives, Cre recombinase, tetracycline transactivator, etc.) in OSNs together with a given OR (Mombaerts et al. 1996; Wang et al. 1998; Serizawa et al. 2003; Shykind et al. 2004; Yu et al. 2004).

3 Molecular Neuroanatomy of the Zebrafish Olfactory Epithelium

3.1 Three Types of Olfactory Sensory Neurons

In most mammalian species, two functionally distinct classes of chemicals (odorants and pheromones) are detected and processed through anatomically segregated neural pathways: the main olfactory system and the vomeronasal (accessory olfactory) system (Buck 2000; Mombaerts 2004). Volatile odorants are received by a large repertoire of ORs expressed on ciliated OSNs in the OE and the information is transmitted to the main OB. By contrast, pheromones are mostly received by two families of vomeronasal receptors (V1Rs and V2Rs) expressed on microvillous sensory neurons in the vomeronasal organ that project their axons to the accessory OB.

On the other hand, the anatomical situation is completely different in the olfactory system of fish. In the fish nose, there is only a single type of olfactory organ (olfactory rosette), containing three types of OSNs, ciliated, microvillous, and crypt cells, all of which innervate the same OB via a tightly fasciculated bundle of olfactory nerves (Fig. 1a–c). Two major types of OSNs are the ciliated and microvillous neurons that clearly differ from one another with respect to the morphology and relative positions in the OE. The ciliated OSNs are situated in the deep layer of the OE, project a long dendrite, and extend several long cilia into the lumen of the nasal cavity. The microvillous OSNs are located in the superficial layer, project a short dendrite, and emanate tens of short microvilli.

In fish, there is a third type of OSN called crypt cells that account for only a small population in the OE (Hansen and Zeiske 1998; Hansen and Finger 2000; Hansen et al. 2003, 2004; Hansen and Zielinski 2005). The crypt cells are located in the most superficial layer of the OE and have unique ovoid-shaped cell bodies bearing microvilli as well as submerged short cilia. Although it was reported that the crypt cells show S100 calcium-binding protein-like and nerve growth factor receptor TrkA-like immunoreactivities (Catania et al. 2003; Germana et al. 2004, 2007), a detailed molecular expression profile has not been clarified for this unique cell type. In particular, there is little information on what type(s) of chemosensory receptors are expressed in the zebrafish crypt cells.

3.2 Zebrafish Olfactory Receptors

In addition to the abovementioned morphological differences, a discrimination between the ciliated and microvillous OSNs can be clearly made from their distinct molecular expression profiles. There are approximately 140 OR-type, six V1R-type, and approximately 50 V2R-type olfactory receptor genes in the zebrafish genome (Alioto and Ngai 2005, 2006; Hashiguchi and Nishida 2006; Ngai and Alioto 2007; Saraiva and Korsching 2007). The expression of OR-type olfactory receptors is observed in the ciliated OSNs in teleost fishes, while V2R-type olfactory receptors are found in the microvillous OSNs (Fig. 1d) (Cao et al. 1998; Speca et al. 1999; Hansen et al. 2004; Sato et al. 2005). Thus, the ciliated and microvillous OSNs likely detect distinct types of chemosensory signals through different families of olfactory receptors. This notion is supported by several lines of evidence obtained in molecular biological, electrophysiological, and activity-dependent labeling experiments (Michel and Derbidge 1997; Speca et al. 1999; Michel 1999; Lipschitz and Michel 2002; Nikonov and Caprio 2007). By contrast, it has not been clearly demonstrated which type(s) of OSNs express V1R-type olfactory receptors, although the messenger RNA (mRNA) expression of a V1R-type receptor (zV1R1; ORA1) is observed in cells in the apical part of the OE (Pfister and Rodriguez 2005). In addition, another multigene family of receptors, called "trace amine-associated receptors" (TAARs), has been discovered as chemosensory receptors expressed in mouse OSNs (Liberles and Buch 2006). In silico database searches revealed that there are 15 TAAR genes in mouse and 109 TAAR genes in zebrafish (Gloriam et al. 2005; Hashiguchi and Nishida

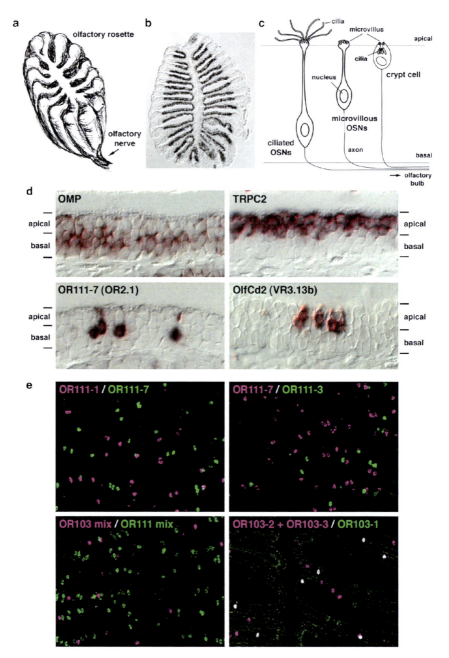

Fig. 1 Molecular neuroanatomy of the zebrafish olfactory epithelium. **a** The zebrafish olfactory rosette. **b** A section of the olfactory rosette hybridized with an olfactory marker protein (OMP) complementary RNA probe. *Dark signals* represent ciliated olfactory sensory neurons (OSNs) in the olfactory epithelium (OE). **c** Three types of olfactory sensory neurons in fish. **d** Distinct locations and molecular signatures of ciliated and microvillous OSNs. The ciliated OSNs locate in a

2007). It has not been examined in detail whether the zebrafish TAARs are expressed in OSNs as chemosensory receptors. Nevertheless, TAARs are likely candidate receptors for polyamines that activate fish OSNs through a unique signaling pathway (Michel et al. 2003; Rolen et al. 2003).

The ciliated OSNs express signal transduction machineries downstream of OR-type olfactory receptors that are shared with mammals, such as olfactory-specific GTP-binding protein subunit (Golf) and cyclic nucleotide-gated cation channel A2 subunit (CNGA2) (Hansen et al. 2003; Sato et al. 2005). Olfactory marker protein (OMP) is detected only in the ciliated OSNs, but not in the microvillous OSNs in zebrafish (Fig. 1d) (Sato et al. 2005). This situation is different from that in mammals, where OMP is expressed in all the chemosensory neurons in both the OE and the vomeronasal organ (Monti Graziade et al. 1980). The microvillous OSNs in zebrafish express transient receptor potential channel C2 (TRPC2) (Fig. 1d) (Sato et al. 2005), whose mouse ortholog plays a central role in the signal transduction cascade of vomeronasal sensory neurons for social and sexual behaviors (Stowers et al. 2002; Leypold et al. 2002; Kimchi et al. 2007).

3.3 One Neuron – How Many Receptors?

Is the one neuron–one receptor rule applicable also to the zebrafish olfactory system, as is the case in mice and *Drosophila*? Individual OR-type olfactory receptor genes are expressed in a small population of OSNs ranging from 0.5 to 2% (Barth et al. 1996). Double-label in situ hybridization experiments revealed that most combinations of different OR-type olfactory receptor probes label nonoverlapping populations of OSNs (Fig. 1e) (Barth et al. 1997; Sato et al. 2007b). These results support the notion that the zebrafish OSNs obey the one neuron–one receptor rule fundamentally. However, two exceptional cases have been reported for particular olfactory receptors, in which "one neuron–*multiple receptors*" is true. One is the case for a subpopulation of ciliated OSNs expressing the zOR103 family members. zOR103-1-positive OSNs simultaneously express zOR103-2 and/or zOR103-5 (Fig. 1e) (Sato et al. 2007b). Coexpression of multiple chemoreceptors has been reported in several types of chemosensory cells, including *C. elegans* OSNs, *Drosophila* OSNs, and mammalian "bitter" taste receptor cells (Troemel et al. 1995; Goldman et al. 2005; Adler et al. 2000). For instance, a single AWC neuron in *C. elegans* expresses multiple olfactory receptors, responds to various odorants such as aldehydes, ketones, alcohols, and thiazoles without discrimination,

---

Fig. 1 (continued) deep layer and express OMP and odorant receptor (OR) type olfactory receptors (*left*), whereas the microvillous OSNs locate in a superficial layer and express transient receptor potential channel C2 (*TRPC2*) and V2R-type olfactory receptors. **e** Double-label in situ hybridization analysis of OE sections with olfactory receptor probes. In most combinations (*upper-left*, *upper-right*, and *lower-left panels*), two signals are nonoverlapping, supporting the one neuron–one receptor rule also in zebrafish. Exceptionally, OR103-1-expressing OSNs are always positive for OR103-2 and/or OR103-3 (*lower-right panel*). (Adopted from Sato et al. 2005, 2007b)

and mediates attractive behavior to all these odorants (Bargmann et al. 1993; Troemel et al. 1995). By analogy, it is likely that zebrafish do not need to discriminate a range of odorants received by the individual zOR103 subfamily members. These OSNs expressing multiple zOR103 members thus may integrate odor information at the most peripheral level, leading to particular behavioral or hormonal responses. The other case is a broad expression of zV2R5.3 in almost all microvillous OSNs. zV2R5.3 is detected ubiquitously throughout the superficial layer of the OE similar to TRPC2 (Sato et al. 2005), as is reported for goldfish V2R-type receptors 5.3 and 5.24 (Speca et al. 1999). This situation is reminiscent of *Drosophila* Or83b and mouse V2R2 olfactory receptors (Larsson et al. 2004; Neuhaus et al. 2005; Martini et al. 2001). *Drosophila* Or83b is broadly expressed in almost all OSNs together with a selectively expressed OR and plays a general role as a heterodimerization partner for the selected regular OR. The coexpression increases the functional activity of the regular OR by improving the OR trafficking to plasma membrane, altering the binding sensitivity to odorants, or enhancing the signal transduction efficacy (Spehr and Leinders-Zufall 2005). In conclusion, both "one neuron–one receptor" and "one neuron–multiple receptors" rules are observed in zebrafish, depending on the different families of olfactory receptors and the divergence of relevant functions in distinct types of OSNs.

4 Olfactory Neural Circuitry in the Zebrafish

4.1 Segregated Neural Pathways from Distinct Types of Olfactory Sensory Neurons

A number of studies employed a classical neuroanatomical tracing method for analysis of neural circuits in the fish olfactory system. The lipophilic fluorescent tracer 1,1′-dioctadecyl-3,3,3′,3′-tetramethylindocarbocyanine perchlorate (DiI) was injected into a small area of the OB, taken up by olfactory axon terminals in glomeruli, and retrogradely transported to cell bodies of OSNs in the OE. Subsequently, the types of DiI-labeled OSNs were determined on the basis of cellular morphology and location in the OE. In catfish, for example, the medial and ventral regions of the OB are innervated mostly by ciliated OSNs, whereas the dorsal region appears to be innervated by microvillous OSNs (Morita and Finger 1998; Hansen et al. 2003). The crypt cells can be labeled only after DiI injection into two discrete areas in the ventral OB in catfish (Hansen et al. 2003). In carp, the medial, lateral, and ventral regions of the OB are likely to be innervated by ciliated, microvillous, and crypt OSNs, respectively (Hamdani et al. 2001a, Hamdani and Doving 2002, 2006). However, these results implied only a tendency of axonal segregation from the distinct types of OSNs to different regions of the OB. Thus, it was impossible with this conventional method to elucidate detailed patterns of innervation from the distinct types of OSNs to individual glomeruli.

Recently, the introduction of the transgenic technique opened a new window in zebrafish olfactory research and unambiguously solved the issue on axonal wiring from the OE to the OB (Sato et al. 2005, 2007b). The ciliated and microvillous OSNs can be differentially labeled with spectrally distinct fluorescent proteins (red fluorescent protein and Venus) under the control of zebrafish OMP and TRPC2 gene promoters, respectively (Fig. 2a). The transparency of zebrafish embryos makes it possible to visualize developing olfactory axons from the two types of OSNs by an in vivo time-lapse imaging analysis in the same individual (Fig. 2b). Fluorescence images of whole-mount OB in the adult transgenic zebrafish clearly show that the ciliated OSNs project axons mostly to the dorsal and medial regions of the OB, whereas the microvillous OSNs project axons to the lateral region of the OB (Fig. 2c). A histological analysis of horizontal OB sections indicates that the two distinct types of OSNs innervate different glomeruli in a mutually exclusive manner (Fig. 2d). According to the nomenclature of zebrafish OB glomeruli of Baier and Korsching (1994), the ciliated OSNs project their axons onto the dorsal cluster, dorsal-cluster-associated glomeruli, anterior plexus, medial glomeruli, medioventral posterior glomeruli, ventromedial glomeruli, and lateroposterior glomeruli, whereas the microvillous OSNs target the lateral chain and ventrolateral glomeruli. Importantly, there is no double-positive glomerulus that receives convergent inputs from both types of OSNs. In contrast, several glomeruli such as the mediodorsal cluster and a few glomeruli in the lateral chain are double-negative, raising a possibility that they may be innervated by the third type of OSNs, crypt cells. Together with the morphological, physiological, and molecular differences between the ciliated and microvillous OSNs, the two segregated neural pathways are responsible for coding and processing of different types of olfactory information, at least at the level of the OB, in the zebrafish olfactory system.

4.2 Olfactory Axon Convergence to Target Glomeruli

In rodents, the olfactory axons originating from OSNs expressing a given OR converge onto a specific pair of glomeruli in the OB. This elegantly wired neural circuitry underlies the basis for the "one glomerulus–one receptor" principle, leading to the establishment of the odor map in the OB. The olfactory axon convergence to target glomeruli was clearly demonstrated in rodents with three different methods. First, in situ hybridization analysis with specific OR probes revealed that mRNAs for individual ORs are detectable not only in the cell bodies but also in the axons of OSNs projecting onto a few topographically fixed glomeruli (Vassar et al. 1994; Ressler et al. 1994). Second, the generation of OR-IRES-reporter knock-in mice by gene targeting in embryonic stem cells succeeded in visualization of olfactory axon convergence to target glomeruli from OSNs expressing a particular OR gene together with chromogenic or fluorescent reporters (tau-LacZ or tau-GFP) (Mombaerts et al. 1996; Wang et al. 1998). Third, the immunohistochemical demonstration for the presence of OR proteins in olfactory axon terminals in

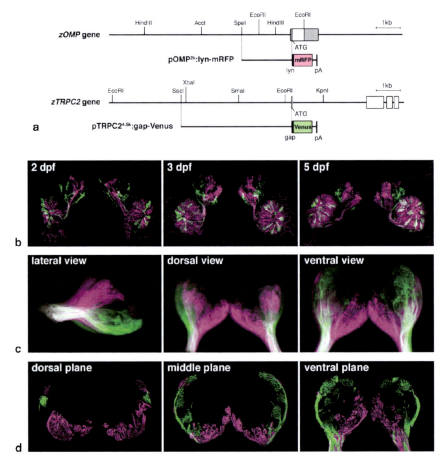

Fig. 2 Segregated neural pathways in the zebrafish olfactory system. **a** Transgene constructs for visualization of distinct olfactory pathways. The OMP promoter drives the expression of membrane-targeted red fluorescent protein (*mRFP*) (*lyn-mRFP*) specifically in ciliated OSNs, while the TRPC2 promoter drives the expression of membrane-targeted Venus (*gap-Venus*) in the microvillous OSNs. **b** Time-lapse imaging of axon projection from ciliated (*magenta*) and microvillous (*green*) OSNs in the OE to presumptive olfactory bulb (OB) in the double-transgenic zebrafish at 2, 3, and 5 days after fertilization (*dpf*). Segregated innervation is observed from these very early stages of development. **c** Whole-mount fluorescence images of the adult OB in the double-transgenic zebrafish viewed from lateral, dorsal, and ventral sides. **d** Horizontal sections of the adult OB of the double-transgenic zebrafish in dorsal, middle, and ventral planes. Two types of OSNs (*magenta* ciliated, *green* microvillous) innervate OB glomeruli in a mutually exclusive manner. (Adopted from Sato et al. 2005)

specific glomeruli provided another line of direct evidence for the one glomerulus–one receptor rule (Barnea et al. 2004; Strotmann et al. 2004). In *Drosophila*, the availability of transcriptional promoters upstream of individual olfactory receptor genes made it easy to selectively label the receptor-defined OSNs for demonstration of the convergent axon projection into glomeruli in the antennal lobe (Vosshall et al. 2000; Gao et al. 2000).

In zebrafish, by contrast, it was difficult to prove the olfactory axon convergence to target glomeruli because of the lower abundance of OR mRNA and protein in OSNs, the absence of embryonic stem cells for gene targeting, and the inefficiency of bicistronic IRES activity. However, the use of BAC transgenes made it possible to label a small population of zebrafish OSNs under the control of a OR gene promoter/enhancer. Because BAC harbors a genomic DNA fragment as long as 100 kb, a tandem array of clustered olfactory receptor genes can be contained in a BAC transgene together with putative *cis*-acting transcriptional regulatory elements. A stable line of BAC transgenic zebrafish was generated, which contains a transgene (95 kb) with 16 OR-type olfactory receptor genes on zebrafish chromosome 15 (Sato et al. 2007b). To visualize functional subsets of OSNs, the BAC transgene was modified so that two of the receptors (zOR103-1 and zOR111-7) were replaced with complementary DNAs encoding membrane-targeted, spectrally distinct fluorescent proteins (yellow fluorescent protein and cyan fluorescent protein) (Fig. 3a). As a result, the fluorescently labeled OSNs made the second choice for expression of olfactory receptor genes mostly within the same subfamily and their axons targeted a topographically fixed cluster of glomeruli in the medial OB (Fig. 3b,c). This finding, for the first time, provided suggestive evidence for the convergence of like axons to target glomeruli in zebrafish.

4.3 Odor Map in the Olfactory Bulb

The odor map is a central representation of various structural features in odorants that are systematically arranged on a glomerular sheet of the first relay station along the olfactory neural circuitry (Mori et al. 1999, 2006; Korsching 2002; Vosshall and Stocker 2007). In other words, individual glomeruli represent a single olfactory receptor and are thus tuned to specific molecular features of odorants that can activate the receptor. The concept of an odor map was clearly demonstrated first in rabbits by electrophysiological single-unit recordings of spike discharges from mitral/tufted cells to odor stimuli (Mori et al. 1992; Imamura et al. 1992; Katoh et al. 1993) and was subsequently confirmed in various mammalian and insect species after the emergence of neural activity imaging techniques (Rubin and Katz 1999; Uchida et al. 2000; Meister and Bonhoeffer 2001; Wang et al. 2003; Meijerink et al. 2003; Takahashi et al. 2004; Igarashi and Mori 2005).

Similarly, the odor map in the OB was analyzed in zebrafish by neural activity imaging techniques (Friedrich and Korsching 1997, 1998; Fuss and Korsching 2001; Li et al. 2005a), in channel catfish by electrophysiological recordings of mitral cell activities (Nikonov and Caprio 2001, 2004; Rolen and Caprio 2007), and in salmonid fishes by electroencephalogram recordings (Hara and Zhang 1998). These experiments with distinct methods in different teleosts yielded essentially similar results on spatial representations of various odorant structures on the OB (Fig. 4):

Adult zebrafish generally exhibit a clear attractive response to amino acids, recognizing them as potential feeding cues (Steele et al. 1990, 1991), although an

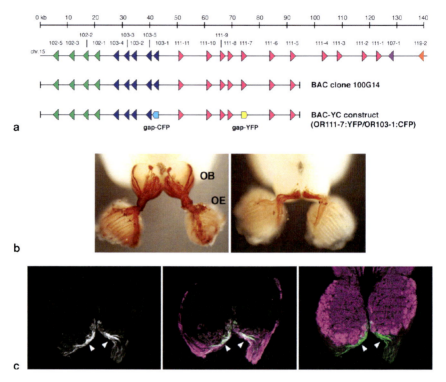

Fig. 3 Olfactory axon convergence to target glomeruli in the zebrafish. **a** Organization of a cluster of OR-type olfactory receptors located on zebrafish chromosome 15 (*top*), a bacterial artificial chromosome (*BAC*) clone containing 16 receptors (*middle*), and a BAC transgene construct for visualization of small populations of ciliated OSNs with membrane-targeted cyan fluorescent protein (*CFP*) and yellow fluorescent protein (*YFP*) (*bottom*). **b** Projection of CFP/YFP-expressing OSN axons to a small region in the medial OB of adult BAC transgenic fish (*right*), compared with OMP:YFP fish in which all axons from the ciliated OSNs are labeled (*left*). Whole-mount OBs together with the OE were stained with anti green fluorescent protein (anti-GFP) antibody. **c** Targeting of YFP/CFP-expressing OSNs to specific glomeruli. A horizontal section of the OB from adult BAC:YFP/CFP and OMP:RFP double-transgenic fish was triple-labeled with anti-GFP, anti-RFP (*magenta* in the *middle panel*), and anti-SV2 (synaptic marker; *magenta* in the *right panel*) antibodies. (Adopted from Sato et al. 2007b)

aversive response to particular amino acids is also reported (Vitebsky et al. 2005). Teleost fishes can also discriminate between different amino acids in behavioral studies (Zippel et al. 1993; Valentincic et al. 2000). Amino acids are detected mostly by microvillous OSNs through V2R-type olfactory receptors (Speca et al. 1999; Hansen et al. 2003; Luu et al. 2004) and activate a chain of glomeruli located on the lateral side of the OB (Friedrich and Korsching 1997). Defined structural features of amino acids (e.g., long or short side chains; hydrophilic or hydrophobic; acidic, neutral, or basic) are represented in a combinatorial fashion in spatially confined groups of the lateral chain glomeruli. These findings are corroborated by

Molecular Genetic Dissection of the Zebrafish Olfactory System 109

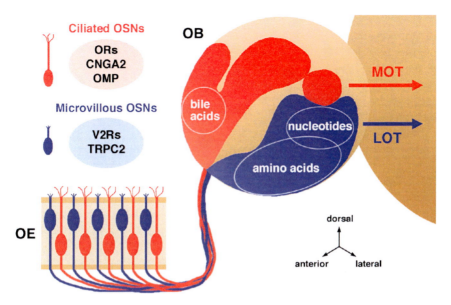

Fig. 4 Odor map in the zebrafish olfactory bulb. Amino acids and nucleotides are received mostly by V2R-type olfactory receptors on microvillous OSNs and are represented in the lateral region of the OB. In contrast, bile acids are received by OR-type olfactory receptors on ciliated OSNs and are represented in the medial region of the OB. *LOT* lateral olfactory tract, *MOT* medial olfactory tract

the transgenic experiment visualizing the innervation of lateral chain glomeruli by microvillous OSNs (Sato et al. 2005).

Bile acids are biliary steroids synthesized in the liver, stored in the gall bladder, secreted into the intestine, and reabsorbed by the enterohapatic system. In sea lamprey, some bile acids are released into environment that act as sex or migratory pheromones (Li et al. 2002; Sorensen et al. 2005). Although it remains unclear what types of behavioral responses are elicited by bile acids in teleost fishes, they are potent odorants that activate ciliated OSNs possibly via the OR-type odorant receptors/Golf/cyclic AMP signaling cascade (Hansen et al. 2003). In the zebrafish OB, bile acids elicit strong responses mainly in a cluster of glomeruli in the anteromedial part and a weaker response in the lateroposterior subregion (Friedrich and Korsching 1998) in perfect accordance with the target glomeruli of ciliated OSNs visualized in the transgenic zebrafish (Sato et al. 2005). The activity patterns induced by different bile acids show a similar but not identical distribution, indicating that distinct molecular features of bile acids are represented in a combinatorial manner.

Nucleotides such as ATP, IMP, and ITP induce excitatory responses in a sub-population of fish OSNs and bulbar neurons (Kang and Caprio 1995; Nikonov and Caprio 2001), possibly acting as feeding cues (Carr 1988). They activate glomeruli located in the lateral portion of the OB that partly overlaps with the amino acid sensitive region (Friedrich and Korsching 1998) and possibly innervated by microvillous OSNs (Sato et al. 2005). Again, different nucleotides are represented by combinatorial activity patterns within the responsive region.

In contrast to the three classes of the abovementioned ordinary odorants, unique single foci of glomerular activity are induced by two putative reproductive pheromones: prostaglandin $F_2\alpha$ (PGF) and $17\alpha,20\beta$-dihydroxy-4-pregnene-3-one-20-sulfate (17,20P-S). PGF specifically activates a single large glomerulus in the ventromedial region of the OB, which shows no response to other related prostaglandins (Friedrich and Korsching 1998). An electrophysiological experiment suggested a very restricted localization of PGF-responsive neurons in the ventromedial OB in lake whitefish (Laberge and Hara 2003). The steroid 17,20P-S selectively activates a small focus in the medial region of zebrafish OB (Friedrich and Korsching 1998). Both of these responsive glomerui are likely innervated by ciliated OSNs expressing OR-type odorant receptors (Sato et al. 2005). Thus, the two pheromones are represented by noncombinatorial patterns of glomerular activity, indicating the existence of straightforward and simple neural pathways from pheromonal inputs to stereotyped behavioral and endocrine outputs.

In addition to the spatial representations of odorant structural features on the surface of the OB, several electrophysiological studies proposed the temporal coding of odor quality and intensity by neuronal populations in the fish OB (Kang and Caprio 1995; Friedrich and Laurent 2001). For details, see the reviews by Laberge and Hara (2001) and Friedrich (2002).

4.4 Odor Coding in the Higher Olfactory Centers

The odor and pheromone information represented in the glomerular map of the OB is next transferred via mitral cell axons to several higher olfactory centers in the forebrain. In the higher olfactory centers, the information is decoded and processed to perceive, discriminate, and memorize odorants, to change hormonal secretion and reproductive activity, and to elicit various olfactory behaviors such as attraction to foods, aversion from predators, social communication with companies, and spawning migration to home rivers. Compared with the great progress in our knowledge on functional correlates in the OE and OB, however, the higher olfactory centers still remain a mystery. Little has been elucidated on the molecular, cellular, and circuit mechanisms underlying odor information coding in the higher olfactory centers in fish, although there have been some physiological and behavioral studies; these are described next.

Different from mammals, in which the mitral cells send axons from the OB to the olfactory cortex through a single tract (the lateral olfactory tract, LOT), there are two major bundles in the secondary olfactory pathways of fish: the LOT and the medial olfactory tract (MOT). From the results of neuroanatomical, electrophysiological, and behavioral experiments in cod, goldfish, and carp (Doving and Selset 1980; Stacey and Kyle 1983; Kyle et al. 1987; Sorensen et al. 1991; Hamdani et al. 2000, 2001a, b, 2002; Hamdani and Doving 2003, 2006; Weltzien et al. 2003), the

LOT and the MOT appear to mediate different types of olfactory behaviors that are induced by stimulation of distinct types of OSNs. Hamdani and Doving (2007) proposed a simplified hypothesis of "three labeled lines" in the fish olfactory pathways: (1) microvillous OSNs – LOT – feeding behavior, (2) ciliated OSNs – a medial bundle of the MOT – alarm reaction, and (3) crypt OSNs – a lateral bundle of the MOT – reproductive behavior. This hypothesis will be verified in zebrafish with the aid of various advantageous techniques of genetic manipulations (cell ablation, silencing, activation,*trans*-synaptic tracing, activity imaging, etc.) on specific types of neurons at different peripheral–central levels along the individual olfactory pathways.

How is the odor information encoded in the higher olfactory centers? Nikonov et al. (2005) reported electrophysiological evidence for the relative conservation of the OB odor map at the one-step-higher level in the catfish forebrain. As in the OB, amino acids and nucleotides are represented in a lateral portion of the forebrain, whereas bile acids are represented in a medial portion, supporting the "labeled line" hypothesis. However, the response properties of the forebrain neurons are not necessarily the same as those of the mitral cells. For example, some neurons in the lateral forebrain respond to several amino acids belonging to different categories and other neurons respond to both amino acids and nucleotides, showing complex tuning profiles that are never observed in the mitral cells. Thus, the convergence of different types of odor information emerges at the level of the forebrain, as reported also in the mouse olfactory cortex (Zou and Buck 2006).

5 Zebrafish Mutants with Abnormalities in the Olfactory System

Mutagenesis is a powerful technique of forward genetics successfully used in various animal species. In zebrafish, large-scale mutant screens were conducted originally focusing on embryogenesis and neural development (Haffter et al. 1996; Driever et al. 1996) and afterwards on various biological aspects, including sensory-system-mediated behaviors (Baier 2000; Neuhauss 2003; Vitebsky et al. 2005). Several zebrafish mutants have been identified that affect development and function of the olfactory system (Miyasaka et al. 2005; Vitebsky et al. 2005).

5.1 *odysseus*

Chemokines and their receptors were originally discovered as signaling partners regulating leukocyte trafficking in the immune system (Rossi and Zlotnik 2000) and also identified as molecules playing crucial roles in cell migration and axon projection in the developing nervous system (Tran and Miller 2003). *odysseus* is a zebrafish mutant with a loss of function of chemokine receptor Cxcr4b, which displays abnormality in the germ cell migration (Knaut et al. 2003), the lateral line primordia migration

(Gilmour et al. 2004), the trigeminal sensory ganglia assembly (Knaut et al. 2005), and the retinal axon projection (Li et al. 2005b). Recently, a dual role of Cxcr4b-mediated signaling has been clarified in the zebrafish olfactory system (Miyasaka et al. 2007). At an early stage of development, Cxcr4b and its cognate ligand Cxcl12a are expressed in the migrating olfactory placodal precursors and in the abutting anterior neural tube, respectively. The Cxcr4b expression persists in the immature OSNs at the initial phase of their axon pathfinding, while the Cxcl12a expression prefigures the route and target of OSN axons (Fig. 5a). In *odysseus* mutant, many olfactory placodal precursors abnormally migrate ventromedially and fail to form a proper olfactory placode. In addition, OSN axons frequently stall at the OE–telencephalon border and fail to enter the OB in *odysseus* mutant (Fig. 5c). Thus, Cxcl12a/Cxcr4b signaling is required for both the olfactory placode assembly and the OSN axon pathfinding in zebrafish. Interestingly, a unilateral, but not bilateral, defect of OSN axon projection is observed in approximately 30% of the *odysseus* mutant fish, thus serving as a good experimental model to study the influence of OSN innervation on the morphogenesis and functions of the OB and higher olfactory centers.

5.2 *astray*

Slits and Roundabouts (Robos) are chemorepellents and receptors, which play prominent roles in axon pathfinding in the nervous system (Nguyen-Ba-Charvet and Chedotal 2002; Dickson and Gilestro 2006). *astray* is a loss-of-function mutant of Robo2, originally identified in a large-scale screen for mutations affecting development of the visual system (Karlstrom et al. 1996; Fricke et al. 2001; Hutson and Chien 2002). A Robo/Slit-mediated chemorepulsion is important also in the neural circuit formation in the olfactory system (Miyasaka et al. 2005). Robo2 is only transiently expressed in the immature OSNs at an early stage of axon pathfinding, while four Slit homologs are present in regions adjacent to the olfactory axon trajectory. In *astray* mutant embryos, some OSN axons misroute ventromedially to cross the midline and/or posteriorly to invade into the diencephalon without

Fig. 5 *odysseus* and *astray*: zebrafish mutants affecting the olfactory axon pathfinding. **a** The expression patterns of a chemokine Cxcl12a and its receptor Cxcr4b in the developing olfactory system of zebrafish. Cxcr4b is expressed in immature OSNs (*green*). Cxcl12a is expressed along the olfactory placode–telencephalon border and at the anterior tip of the telencephalon (*blue*), prefiguring the route and target of olfactory axons. **b** The expression patterns of chemorepellents Slit1a/Slit1b/Slit2/Slit3 and their receptor Robo2 in the developing olfactory system of zebrafish Robo2 is expressed in immature OSNs (*green*). Four Slit homologs are expressed in regions adjacent to the olfactory axon trajectory (*red*), forming surround repulsive barriers against olfactory axons. **c** Abnormal axon projection in*odysseus* (Cxcr4b mutant, *middle panels*) and*astray* (Robo2 mutant, *right panels*), compared with wild-type fish (*left panels*). In *odysseus* mutant, OSN axons stall at the OE–telencephalon border and fail to enter the OB. In *astray* mutant, some olfactory axons misroute ventromedially to cross the midline and posteriorly to invade into the diencephalon without reaching the OB. *Upper panels* frontal views, *lower panels* dorsal views

Molecular Genetic Dissection of the Zebrafish Olfactory System

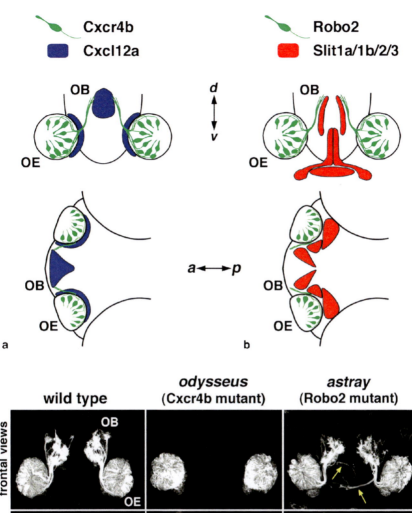

reaching the OB. Thus, Robo2/Slit signaling is essential for proper navigation of the early growing olfactory axons toward the OB. In addition, the spatial arrangement in OB glomeruli is perturbed in adult *astray* fish, suggesting that the precise formation of the initial axon scaffold mediated by Robo2 is crucial for establishment of a sound glomerular map in the adult OB.

5.3 *laure*

laure was isolated by a genetic screen for mutations affecting an olfactory behavior (Vitebsky et al. 2005). *laure* mutant fish are defective in their responses to amino acids, but not to bile acids, and display impaired olfactory axon projection. Identification of a gene responsible for the *laure* mutant phenotypes is awaited.

6 Conclusions

Recent decades have witnessed significant progress in our understandings of the molecular and cellular mechanisms underlying the development and functional organization of the primary olfactory system. Multigene families encoding odorant and pheromone receptors have been cloned in various animal species. The modes of neural wiring from the peripheral olfactory organs (OE and antenna) to the first relay station in the brain (OB and antennal lobe) have been clarified by visualization of OSN axons in genetically engineered animals. The concept of "odor map" has been established as the internal representation of odorant molecular features in the OB of vertebrates and in the antennal lobe of insects, demonstrating the importance of glomerular modules as functional units for odor coding and processing. These splendid discoveries have been made mostly in mouse and *Drosophila*, the two animal species amenable to the powerful technology of forward and reverse genetics. Towards understanding the vertebrate olfactory system as a whole, the zebrafish will undoubtedly become an ideal model organism as a *Drosophila*-like vertebrate in the next decade.

Acknowledgements The author thanks Nobuhiko Miyasaka and Yuki Sato for preparation of the figures and critical reading of the manuscript. The work was supported in part by a Grant-in-Aid for Scientific Research (B) and a Grant-in-Aid for Scientific Research on Priority Area (Cellular Sensor) from the Ministry of Education, Culture, Sports, Science, and Technology of Japan.

References

Adler E, Hoon MA, Mueller KL, Chandrashekar J, Ryba NJ, Zuker CS (2000) A novel family of mammalian taste receptors. Cell 100:693–702
Alioto TS, Ngai J (2005) The odorant receptor repertoire of teleost fish. BMC Genomics 6:173. doi:10.1186/1471-2164-6-173

Alioto TS, Ngai J (2006) The repertoire of olfactory C family G protein-coupled receptors in zebrafish: candidate chemosensory receptors for amino acids. BMC Genomics 7:309. doi:10.1186/1471-2164-7-309

Asakawa K, Suster ML, Mizusawa K, Nagayoshi S, Kotani T, Urasaki A, Kishimoto Y, Hibi M, Kawakami K (2008) Genetic dissection of neural circuits by Tol2 transposon-mediated Gal4 gene and enhancer trapping in zebrafish. Proc Natl Acad Sci USA 105:1255–1260

Baier H (2000) Zebrafish on the move: towards a behavior-genetic analysis of vertebrate vision. Curr Opin Neurobiol 10:451–455

Baier H, Korsching S (1994) Olfactory glomeruli in the zebrafish form an invariant pattern and are identifiable across animals. J Neurosci 14:219–230

Bargmann CI, Hartwieg E, Horvitz HR (1993) Odorant-selective genes and neurons mediate olfaction in *C. elegans*. Cell 74:515–527

Barnea G, O'Donnell S, Mancia F, Sun X, Nemes A, Mendelsohn M, Axel R (2004) Odorant receptors on axon termini in the brain. Science 304:1468

Barth AL, Nicholas JJ, Ngai J (1996) Asynchronous onset of odorant receptor expression in the developing zebrafish olfactory system. Neuron 16:23–34

Barth AL, Dugas JC, Ngai J (1997) Noncoordinate expression of odorant receptor genes tightly linked in the zebrafish genome. Neuron 19:359–369

Buck LB (2000) The molecular architecture of odor and pheromone sensing in mammals. Cell 100:611–618

Buck L, Axel R (1991) A novel multigene family may encode odorant receptors: a molecular basis for odor recognition. Cell 65:175–187

Cao Y, Oh BC, Stryer L (1998) Cloning and localization of two multigene receptor families in goldfish olfactory epithelium. Proc Natl Acad Sci USA 95:11987–11992

Carr WES (1988) The molecular nature of chemical stimuli in the aquatic environment. In: Atema J, Fay RR, Popper AN, Tavolga WN (eds) Sensory biology of aquatic animals. Springer, New York, pp 3–27

Catania S, Germana A, Laura R, Gonzalez-Martinez T, Ciriaco E, Vega JA (2003) The crypt neurons in the olfactory epithelium of the adult zebrafish express TrkA-like immunoreactivity. Neurosci Lett 350:5–8

Chess A, Simon I, Cedar H, Axel R (1994) Allelic inactivation regulates olfactory receptor gene expression. Cell 78:823–834

Dickson BJ, Gilestro GF (2006) Regulation of commissural axon pathfinding by slit and its Robo receptors. Annu Rev Cell Dev Biol 22:651–675

Doving KB, Selset R (1980) Behavior patterns in cod released by electrical stimulation of olfactory tract bundles. Science 207:559–560

Driever W, Solnica-Krezel L, Schier AF, Neuhauss SC, Malicki J, Stemple DL, Stainier DY, Zwartkruis F, Abdelilah S, Rangini Z, Belak J, Boggs C (1996) A genetic screen for mutations affecting embryogenesis in zebrafish 123:37–46

Ekker SC (2000) Morphants: a new systematic vertebrate functional genomics approach. Yeast 17:302–306

Ellingsen S, Laplante MA, Konig M, Kikuta H, Furmanek T, Hoivik EA, Becker TS (2005) Large-scale enhancer detection in the zebrafish genome. Development 132:3799–3811

Fricke C, Lee JS, Geiger-Rudolph S, Bonhoeffer F, Chien CB (2001)*Astray*, a zebrafish roundabout homolog required for retinal axon guidance. Science 292:507–510

Friedrich RW (2002) Real time odor representations. Trends Neurosci 25:487–489

Friedrich RW, Korsching SI (1997) Combinatorial and chemotopic odorant coding in the zebrafish olfactory bulb visualized by optical imaging. Neuron 18:737–;752

Friedrich RW, Korsching SI (1998) Chemotopic, combinatorial, and noncombinatorial odorant representations in the olfactory bulb revealed using a voltage-sensitive axon tracer. J Neurosci 18:9977–9988

Friedrich RW, Laurent G (2001) Dynamic optimization of odor representations by slow temporal patterning of mitral cell activity. Science 291:889–894

Fuss SH, Korsching SI (2001) Odorant feature detection: activity mapping of structure response relationships in the zebrafish olfactory bulb. J Neurosci 21:8396–8407

Gao Q, Yuan B, Chess A (2000) Convergent projections of Drosophila olfactory neurons to specific glomeruli in the antennal lobe. Nat Neurosci 3:780–785

Germana A, Montalbano G, Laura R, Ciriaco E, dell Calle ME, Vega JA (2004) S100 protein-like immunoreactivity in the crypt olfactory neurons of the adult zebrafish. Neurosci Lett 371:196–198

Germana A, Paruta S, Germana GP, Ochoa-Erena FJ, Montalbano G, Cobo J, Vega JA (2007) Differential distribution of S100 protetin and calretinin in mechanosensory and chemosensory cells of adult zebrafish (Danio rerio). Brain Res 1162:48–55

Gilmour D, Knaut H, Maischein HM, Nusslein-Volhard C (2004) Towing of sensory axons by their migrating target cells *in vivo*. Nat Neurosci 7:491–492

Gloriam DEI, Bjarnadottir TK, Yan YL, Postlethwait JH, Schioth HB, Fredriksson R (2005) The repertoire of trace amine G-protein-coupled receptors: large expansion in zebrafish. Mol Phylogenet Evol 35:470–482

Goldman AL, Van der Goes van Naters W, Lessing D, Warr CG, Carlson JR (2005) Coexpression of two functional odor receptors in one neuron. Neuron 45:661–666

Haffter P, Granato M, Brand M, Mullins MC, Hammerschmidt M, Kane DA, Odenthal J, van Eeden FJ, Jiang YJ, Heisenberg CP, Kelsh RN, Furutani-Seiki M, Vogelsang E, Beuchle D, Schach U, Fabian C, Nusslein-Volhard C (1996) The identification of genes with unique and essential functions in the development of the zebrafish, *Danio rerio*. Development 123:1–36

Halloran MC, Sato-Maeda M, Warren JT, Su F, Lele Z, Krone PH, Kuwada JY, Shoji W (2000) Laser-induced gene expression in specific cells of transgenic zebrafish. Development 127:1953–1960

Hamdani EH, Doving KB (2002) The alarm reaction in crucian carp is mediated by olfactory neurons with long dendrites. Chem Senses 27:395–398

Hamdani EH, Doving KB (2003) Sensitivity and selectivity of neurons in the medial region of the olfactory bulb to skin extract from conspecifics in crucian carp, *Carassius carassius*. Chem Senses 28:181–191

Hamdani EH, Doving KB (2006) Specific projection of the sensory crypt cells in the olfactory system in crucian carp, *Carassius carassius*. Chem Senses 31:63–67

Hamdani EH, Doving KB (2007) The functional organization of the fish olfactory system. Prog Neurobiol 82:80–86

Hamdani EH, Stabell OB, Alexander G, Doving KB (2000) Alarm reaction in the crucian carp is mediated by the medial bundle of the medial olfactory tract. Chem Senses 25:103–109

Hamdani EH, Alexander G, Doving KB (2001a) Projection of sensory neurons with microvilli to the lateral olfactory tract indicates their participation in feeding behaviour in crucian carp. Chem Senses 26:1139–1144

Hamdani EH, Kasumyan A, Doving KB (2001b) Is feeding behaviour in crucian carp mediated by the lateral olfactory tract? Chem Senses 26:1133–1138

Hansen A, Finger TE (2000) Phyletic distribution of crypt-type olfactory receptor neurons in fishes. Brain Behav Evol 55:100–110

Hansen A, Zeiske E (1998) The peripheral olfactory organ of the zebrafish, Danio rerio: an ultrastructural study. Chem Senses 23:39–48

Hansen A, Rolen SH, Anderson K, Morita Y, Caprio J, Finger TE (2003) Correlation between olfactory receptor cell type and function in the channel catfish. J Neurosci 23:347–359

Hansen A, Anderson KT, Finger TE (2004) Differential distribution of olfactory receptor neurons in goldfish: structural and molecular correlates. J Comp Neurol 477:347–359

Hansen A, Zielinski BS (2005) Diversity in the olfactory epithelium of bony fishes: development, lamellar arrangement, sensory neuron cell types and transduction components. J Neurocytol 34:183–208

Hara TH, Zhang C (1998) Topographic bulbar projections and dual neural pathways of the primary olfactory neurons in salmonid fishes. Neuroscience 82:301–313

Hashiguchi Y, Nishida M (2006) Evolution and origin of vomeronasal-type odorant receptor gene repertoire in fishes. BMC Evol Biol 6:76. doi:10.1186/1471-2148-6-76

Hashiguchi Y, Nishida M (2007) Evolution of trace amine-associated receptor (TAAR) gene family in vertebrates: lineage-specific expansions and degradations of a second class of vertebrate chemosensory receptors expressed in the olfactory epithelium. Mol Biol Evol 24:2099–2107

Hutson LD, Chien CB (2002) Pathfinding and error correction by retinal axons: the role of *astray/robo2*. Neuron 33:205–217

Igarashi KM, Mori K (2005) Spatial representation of hydrocarbon odorants in the ventrolateral zones of the rat olfactory bulb. J Neurophysiol 93:1007–1019

Imamura K, Mataga N, Mori K (1992) Coding of odor molecules by mitral/tufted cells in rabbit olfactory bulb. I. Aliphatic compounds. J Neurophysiol 68:1986–2002

Kang J, Caprio J (1995) Electrophysiological response of single olfactory bulb neurons to amino acids in the channel catfish, *Ictalurus punctatus*. J Neurophysiol 74:1421–1434

Karlstrom RO, Trowe T, Klostermann S, Baier H, Brand M, Crawford AD, Grunewald B, Haffter P, Hoffmann H, Meyer SU, Muller BK, Richter S, van Eeden FJ, Nusslein-Volhard C, Bonhoeffer F (1996) Zebrafish mutations affecting retinotectal axon pathfinding. Development 123:427–438

Katoh K, Koshimoto H, Tani A, Mori K (1993) Coding of odor molecules by mitral/tufted cells in rabbit olfactory bulb. II. Aromatic compounds. J Neurophysiol 70:2161–2175

Kawakami K (2004) Transgenesis and gene trap methods in zebrafish by using the Tol2 transposable element. Methods Cell Biol 77:201–222

Kawakami K, Shima A (1999) Identification of the Tol2 transposase of the medaka fish *Oryzias latipes* that catalyzes excision of a nonautonomous Tol2 element in zebrafish *Danio rerio*. Gene 240:239–244

Kimchi T, Xu J, Dulac C (2007) A functional circuit underlying male sexual behaviour in the female mouse brain. Nature 448:1009–1014

Knaut H, Werz C, Geisler R, Nuslein-Volhard C (2003) A zebrafish homologue of the chemokine receptor Cxcr4 is a germ-cell guidance receptor. Nature 421:279–282

Knaut H, Blader P, Strahle U, Schier AP (2005) Assembly of trigeminal sensory ganglia by chomokine signaling. Neuron 47:653–666

Korsching SI (2002) Olfactory maps and odor images. Curr Opin Neurobiol 12:387–392

Kwan KM, Fujimoto E, Grabher C, Mangum BD, Hardy ME, Campbell DS, Parant JM, Yost HJ, Kanki JP, Chen CB (2007) Construction kit for Tol2 transposon transgenesis constructs. Dev Dyn 236:3088–3099

Kyle AL, Sorensen PW, Stacey NE, Dulka JG (1987) Medial olfactory tract pathways controlling sexual reflexes and behavior in teleosts. Ann N Y Acad Sci 519:97–107

Laberge F, Hara TJ (2001) Neurobiology of fish olfaction: a review. Brain Res Rev 36:41–59

Laberge F, Hara TJ (2003) Non-oscillatory discharges of an F-prostaglandin responsive neuron population in the olfactory bulb-telencephalin transition area in lake whitefish. Neuroscience 116:1089–1095

Larsson MC, Domingos AI, Jones WD, Chiappe E, Amrein H, Vosshall LB (2004) *Or83b* encodes a broadly expressed odorant receptor essential for *Drosophila* olfaction. Neuron 43:703–714

Lewcock JW, Reed RR (2004) A feedback mechanism regulates monoallelic odorant receptor expression. Proc Natl Acad Sci USA 101:1069–1074

Leypold BG, Yu CR, Leinders-Zufall T, Kim MM, Zufall F, Axel R (2002) Altered sexual and social behaviors in trp2 mutant mice. Proc Natl Acad Sci USA 99:6376–6381

Li J, Mack JA, Souren M, Yaksi E, Higashijima S, Mione M, Fetcho JR, Friedrich RW (2005a) Early development of functional spatial maps in the zebrafish olfactory bulb. J Neurosci 25:5784–5795

Li Q, Shirabe K, Thisse C, Thisse B, Okamoto H, Masai I, Kuwada JY (2005b) Chemokine signaling guides axons within the retina in zebrafish. J Neurosci 25:1711–1717

Li W, Scott AP, Siefkes MJ, Yan H, Liu Q, Yun SS, Gage DA (2002) Bile acid secreted by male sea lamprey that acts as a sex pheromone. Science 296:138–141

Liberles SD, Buch LB (2006) A second class of chemosensory receptors in the olfactory epithelium. Nature 442:645–650

Lipschitz DL, Michel WC (2002) Amino acid odorants stimulate microvillar sensory neurons. Chem Senses 27:277–286

Luu P, Acher F, Bertrand HO, Fan J, Ngai J (2004) Molecular determinants of ligand selectivity in a vertebrate odorant receptor. J Neurosci 24:10128–10137

Malnic B, Hirono J, Sato T, Buck LB (1999) Combinatorial receptor codes for odors. Cell 96:713–723

Martini S, Silvotti L, Shirazi A, Ryba NJP, Tirindelli R (2001) Co-expression of putative pheromone receptors in the sensory neurons of the vomeronasal organ. J Neurosci 21:843–848

Meijerink J, Carlsson MA, Hansson BS (2003) Spatial representation of odorant structure in the moth antennal lobe: a study of structure–response relationships at low doses. J Comp Neurol 467:11–21

Meister M, Bonhoeffer T (2001) Tuning and topography in an odor map on the rat olfactory bulb. J Neurosci 21:1361–1360

Michel WC (1999) Cyclic nucleotide-gated channel activation is not required for activity-dependent labeling of zebrafish olfactory receptor neurons by amino acids. Biol Signals Recept 8:338–347

Michel WC, Derbidge DS (1997) Evidence of distinct amino acid and bile salt receptors in the olfactory system of the zebrafish, *Danio rerio*. Brain Res 764:179–187

Michel WC, Sanderson MJ, Olson JK, Lipschitz DL (2003) Evidence of a novel transduction pathway mediating detection of polyamines by the zebrafish olfactory system. J Exp Biiol 206:1697–1706

Miyasaka N, Sato Y, Yeo SY, Hutson LD, Chien CB, Okamoto H, Yoshihara Y (2005) Robo2 is required for establishment of a precise glomerular map in the zebrafish olfactory system. Development 132:1283–1293

Miyasaka N, Knaut H, Yoshihara Y (2007) Cxcl12/Cxcr4 chemokine signaling is required for placode assembly and sensory axon pathfinding in the zebrafish olfactory system. Development 134:2459–2468

Mombaerts P (2004) Genes and ligands for odorant vomeronasal and taste receptors. Nat Rev Neurosci 5:263–278

Mombaerts P, Wang F, Dulac C, Chao SK, Nemes A, Mendelsohn M, Edmondson J, Axel R (1996) Visualizing an olfactory sensory map. Cell 87:675–686

Monti Graziadei GA, Stanley RS, Graziadei PPC (1980) The olfactory marker protein in the olfactory system of the mouse during development. Neuroscience 5:1239–1252

Mori K, Mataga N, Imamura K (1992) Differential specificities of single mitral cells in rabbit olfactory bulb for a homologous series of fatty acid odor molecules. J Neurophysiol 67:786–789

Mori K, Nagao H, Yoshihara Y (1999) The olfactory bulb: coding and processing of odor molecule information. Science 286:711–715

Mori K, Takahashi YK, Igarashi KM, Yamaguchi M (2006) Maps of odorant molecular features in the mammalian olfactory bulb. Physiol Rev 86:409–433

Morita Y, Finger TE (1998) Differential projections of ciliated and microvillous olfactory receptor cells in the catfish, *Ictalurus punctatus*. J Comp Neurol 398:539–550

Nasevisius A, Ekker SC (2000) Effective targeted gene 'knockdown' in zebrafish. Nat Genet 26:216–220

Neuhaus EV, Gisselmann G, Zhang W, Dooley R, Stortkuhl K, Hatt H (2005) Odorant receptor heterodimerization in the olfactory system of *Drosophila melanogaster*. Nat Neurosci 8:15–17

Neuhauss SCF (2003) Behavioral genetic approaches to visual system development and function in zebrafish. J Neurobiol 54:148–160

Ngai J, Alioto TS (2007) Genomics of odor receptors in zebrafish. In: Firestein S, Beauchamp GK (eds) The senses: a comprehensive reference, vol. 4. Olfaction and taste. Academic, Oxford, pp 553–560

Nguyen-Ba-Charvet KT, Chedotal A (2002) Role of slit proteins in the vertebrate brain. J Physiol (Paris) 96:91–98

Nikonov AA, Caprio J (2001) Electrophysiological evidence for a chemotopy of biological relevant odors in the olfactory bulb of the channel catfish. J Neurophysiol 86:1869–1876

Nikonov AA, Caprio J (2004) Odorant specificity of single olfactory bulb neurons to amino acids in the channel catfish. J Neurophysiol 92:123–134

Nikonov AA, Caprio J (2007) Highly specific olfactory receptor neurons for types of amino acids in the channel catfish. J Neurophysiol 98:1909–1918

Nikonov AA, Finger TE, Caprio J (2005) Beyond the olfactory bulb: an odoropic map in the forebrain. Proc Natl Acad Sci USA 102:18688–18693

Nishizumi H, Kumasaka K, Inoue N, Nakashima A, Sakano H (2007) Deletion of the core-H region in mice abolishes the expression of three proximal odorant receptor genes in cis. Proc Natl Acad Sci USA 104:20067–20072

Pfister P, Rodriguez I (2005) Olfactory expression of a single and highly variable V1r pheromone receptor-like gene in fish species. Proc Natl Acad Sci USA 102:5489–5494

Ressler KJ, Sullivan SL, Buck LB (1994) Information coding in the olfactory system: evidence for a stereotyped and highly organized epitope map in the olfactory bulb. Cell 79:1245–1255

Rolen SH, Caprio J (2007) Processing of bile salt odor information by single olfactory bulb neurons in the channel catfish. J Neurophysiol 97:4058–4068

Rolen SH, Sorensen PW, Mattson D, Caprio J (2003) Polyamines as olfactory stimuli in the goldfish *Carassius auratus*. J Exp Biol 206:1683–1696

Rossi D, Zlotnik A (2000) The biology of chemokines and their receptors. Annu Rev Immunol 19:23–45

Rubin BD, Katz LC (1999) Optical imaging of odorant representations in the mammalian olfactory bulb. Neuron 23:499–511

Saraiva LR, Korsching SI (2007) A novel olfactory receptor gene family in teleost fish. Genome Res 17:1448–1457

Sato T, Hamaoka T, Aizawa H, Hosoya T, Okamoto H (2007a) Genetic single-cell mosaic analysis implicates ephrinB2 reverse signaling in projections from the posterior tectum to the hindbrain in zebrafish. J Neurosci 27:5271–5279

Sato Y, Miyasaka N, Yoshihara Y (2005) Mutually exclusive glomerular innervation by two distinct types of olfactory sensory neurons revealed in transgenic zebrafish. J Neurosci 25:4889–4897

Sato Y, Miyasaka N, Yoshihara Y (2007b) Hierarchical regulation of odorant receptor gene choice and subsequent axonal projection of olfactory sensory neurons in zebrafish. J Neurosci 27:1606–1615

Serizawa S, Miyamichi K, Nakatani H, Suzuki M, Saito M, Yoshihara Y, Sakano H (2003) Negative feedback regulation ensures the one receptor-one olfactory neuron rule in mouse. Science 302:2088–2094

Shykind BM, Rohani SC, O'Donnell S, Nemes A, Mendelsohn M, Sun Y, Axel R, Barnea G (2004) Gene switching and the stability of odorant receptor gene choice. Cell 117:801–815

Sood R, English MA, Jones M, Mullikin J, Wang DM, Anderson M, Wu D, Chandrasekharappa SC, Yu J, Zhang J, Paul Liu P (2006) Methods for reverse genetic screening in zebrafish by resequencing and TILLING. Methods 39:220–227

Sorensen PW, Caprio J (1998) Chemoreception. In: Evans DH (ed) The physiology of fishes, 2nd edn. CRC, Boca Raton, pp 375–405

Sorensen PW, Hara TJ, Stacey NE (1991) Sex pheromones selectively stimulate the medial olfactory tracts of male goldfish. Brain Res 558:343–347

Sorensen PW, Fine JM, Dvornikovs V, Jeffrey CS, Shao F, Wang J, Vrieze LA, Anderson KR, Hoye TR (2005) Mixture of new sulfated steroids functions as a migratory pheromone in the sea lamprey. Nat Chem Biol 1:324–328

Speca DJ, Lin DM, Sorensen PW, Isacoff EY, Ngai J, Dittman AH (1999) Functional identification of a goldfish odorant receptor. Neuron 23:497–498

Spehr M, Leinders-Zufall T (2005) One neuron-mulitple receptors: increased complexity in olfactory coding? Sci STKE 285:pe25. doi:10.1126/stke.2852005pe25

Stacey NE, Kyle AL (1983) Effects of olfactory tract lesions on sexual and feeding behavior in the goldfish. Physiol Behav 30:621–628

Steele CW, Owens DW, Scarfe AD (1990) Attraction of zebrafish *Brachydanio rerio* to alanine and its suppression by copper. J Fish Biol 36:341–352

Steele CW, Scarfe AD, Owens DW (1991) Effects of group size on the responsiveness of zebrafish *Brachydanio rerio* to alanine, a chemical attactant. J Fish Biol 38:553–564

Stowers L, Holy TE, Meister M, Dulac C, Koentges G (2002) Loss of sex discrimination and male–male aggression in mice deficient for TRP2. Science 295:1493–1500

Strotmann J, Levai O, Fleischer J, Schwarzenbacher K, Breer H (2004) Olfactory receptor proteins in axonal processes of chemosensory neurons. J Neurosci 24:7754–7761

Takahashi YK, Kurosaki M, Hirono S, Mori K (2004) Topographic representation of odorant molecular features in the rat olfactory bulb. J Neurophysiol 92:2413–2427

Thermes V, Grabher C, Ristratore F, Bourrat F, Choulika A, Wittbrodt J, Joly JS (2002) *I-SceI* meganuclease mediates highly efficient transgenesis in fish. Mech Dev 118:91–98

Tran PB, Miller RJ (2003) Chemokine receptors: signposts to brain development and disease. Nat Rev Neurosci 4:444–455

Troemel ER, Chou JH, Dwyer ND, Colbert HA, Bargmann CI (1995) Divergent seven transmembrane receptors are candidate chemosensory receptors in *C. elegans*. Cell 83:207–218

Uchida N, Takahashi YK, Tanifuji M, Mori K (2000) Odor maps in the mammalian olfactory bulb: domain organization and odorant structural features. Nat Neurosci 3:1035–1043

Valentincic T, Metelko J, Ota D, Pirc V, Blejec A (2000) Olfactory discrimination of amino acids in brown bulhead catfish. Chem Senses 25:21–29

Vassar R, Chao SK, Sitcheran R, Numez JM, Vosshall LB, Axel R (1994) Topographic organization of sensory projections to the olfactory bulb. Cell 79:981–991

Vitebsky A, Reyes R, Sanderson MJ, Michel WC, Whitlock KE (2005) Isolation and characterization of the *laure* olfactory behavioral mutant in the zebrafish, *Danio rerio*. Dev Dyn 234:229–242

Vosshall LB, Stocker RF (2007) Molecular architecture of smell and taste in Drosophila. Annu Rev Neurosci 30:505–533

Vosshall LB, Wong AM, Axel R (2000) An olfactory sensory map in the fly brain. Cell 102:147–159

Wang F, Nemes A, Mendelsohn M, Axel R (1998) Odorant receptors govern the formation of a precise topographic map. Cell 93:47–60

Wang JW, Wong AM, Flores J, Vosshall LB, Axel R (2003) Two-photon calcium imaging reveals an odor-evoked map of activity in the fly brain. Cell 112:271–282

Weltzien FA, Hoglund E, Hamdani EH, Doving KB (2003) Does the lateral bundle of the medial olfactory tract mediate reproductive behavior in male crucian carp? Chem Senses 28:293–300

Westerfield M (1995) The zebrafish book, 3rd edn. University of Oregon Press, Eugene, Oregon

Wienholds E, van Eeden F, Kosters M, Mdde J, Plasterk RHA, Cuppen E (2003) Efficient target-selected mutagenesis in zebrafish. Genome Res 13:2700–2707

Yang Z, Jiang H, Chachainasakul T, Gong S, Yang XW, Heintz N, Lin S (2006) Modified bacterial artificial chromosome for zebrafish transgenesis. Methods 39:183–188

Yu CR, Power J, Barnea G, O'Donnell S, Brown HE, Osborne J, Axel R, Gogos JA (2004) Spontaneous neural activity is required for the establishment and maintenance of the olfactory sensory map. Neuron 42:553–566

Zielinski BS, Hara TJ (2007) Olfaction. In: Hara T, Zielinski B (eds) Fish physiology, vol 25. Sensory systems neuroscience. Academic, San Diego, pp 1–43

Zippel HP, Voigt R, Knaust M, Luan Y (1993) Spontaneous behaviour, training and discrimination training in goldfish using chemosensory stimuli. J Comp Physiol A 172:81–90

Zou Z, Buck LB (2006) Combinatorial effects of odorant mixes in olfactory cortex. Science 311:1477–1481

Insect Olfaction: Receptors, Signal Transduction, and Behavior

K. Sato and K. Touhara

Abstract The insect olfactory system is a suitable model for exploring molecular function of odorant receptors, axonal projection of olfactory receptor neurons onto secondary neurons, and the neural circuit for odor perception. Recent progress in the study of insect olfaction revealed that the heteromeric insect olfactory receptor complex forms a cation nonselective ion channel directly gated by odor or pheromone ligands independent of known G-protein signaling pathways. Despite fundamental differences in transduction machineries between insects and vertebrates, the anatomical and functional features of insect odor-coding strategy are similar and thus justify any consideration of mammalian olfaction in the study of insects. The understanding of the molecular mechanism of insect olfaction will help in the development of insect repellents for controlling insect pest and vector populations for a wide range of pathogens.

1 Introduction

Insects are the most diverse group animals on earth, with approximately five million species described to date (Novotny et al. 2002). Amidst this great diversity are adaptations common to all insects that maximize inclusive fitness in their respective habitats. One such fundamental adaptation is the ability to respond to cues in the environment, in particular the ability to detect external biological compounds via a chemical sensor. The sophisticated olfactory system of insects is able to sense volatile odorants derived from prey, host plants, and conspecific individuals. These compounds are detected by olfactory receptor neurons (ORNs) housed in the antennae, and these ORNs relay information about food sources, oviposition sites, and mates that leads to behavior based on neural responses mediated by the ORNs. The binding

---

K. Touhara (✉)
Department of Integrated Biosciences, The University of Tokyo, 5-1-5 Kashiwanoha, Kashiwa, Chiba 277-8562, Japan
e-mail: touhara@k.u-tokyo.ac.jp

of a ligand to an odorant receptor (OR) is the initial event of olfactory transduction, the process that converts information from chemical signals into electrical events. Recent progress in the molecular analysis of ORs has revealed fundamental differences between insects and mammalians in the structures and functional properties of their ORs. Although there are mechanistic differences, the physiological properties of insect olfactory organs and those of mammals have similarities. Thus, the insect olfactory system is a suitable model for comparing regulation of OR gene expression, axonal projection of ORNs onto secondary neurons and odor perception, and exploring novel molecular mechanisms. It is clear that any improvements in our understanding of insect olfaction will likewise improve our knowledge of how insects behave as agricultural pests and as vectors for a wide range of pathogens.

2 Olfactory Organ and Sensory Neurons

An insect's morphology and size change dramatically during development, a process called metamorphosis. For example, the morphology of the hatching larvae of Diptera and Lepidoptera is generally wormlike, but the larvae transform into sessile pupae before emerging as adult insects. Naturally the foods of adults and larvae differ, both in terms of the dietary source and also the distance the insect must travel to its food. The larvae of Diptera and Lepidoptera hatch and live directly on their food sources, which can be bits of refuse or the leaves of host plants, for example. In contrast, adults must forage for food as well as find suitable sites to lay eggs. Although both larval and adult olfactory organs respond to volatile chemicals from food sources, a subset of ORNs in the olfactory organ of adults is specialized to detect semiochemicals such as pheromones. Thus, the adult olfactory organ of any given insect is anatomically and physiologically much more complex than that of its larvae.

2.1 Adult Olfactory Organ

The diverse morphology of the olfactory organ among insect species provides a basis upon which classification has been based. Basically, the olfactory organ in the adult insect comprises two components on the head: the antenna and the maxillary palp (Fig. 1a). Numerous sensilla cover the surface of the antennae and prevent direct contact of ORNs with the external environment. Each sensillum is filled with a potassium- and protein-rich fluid called sensillum lymph and houses one to four ORN dendrites. The small pits on the cuticle surfaces of sensilla allow contact of the ORN dendrite with volatile odorants that dissolve in the lymph. In Drosophila, the third segment of the antenna and of the maxillary palp possess approximately 1,200 and 120 ORNs, respectively.

The sensilla are divided into three groups on the basis of morphology. In Drosophila, there are approximately 200 basiconic (long, conical, large or small

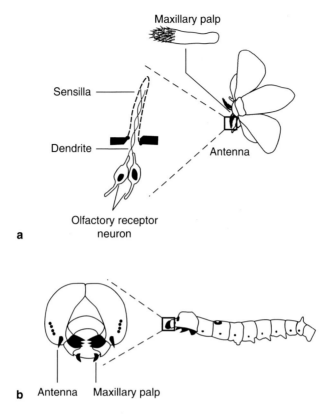

Fig. 1 Olfactory organ of insects. Insects have two types of olfactory organ: antenna and maxillary palp. Antennae and maxillary palps both bear numerous sensilla. An individual sensillum houses between one and four dendrites of olfactory receptor neurons. **a** Morphology of the olfactory organ of the adult silkmoth. **b** Morphology of the olfactory organ of silkmoth larvae

subtypes), 150 trichoid (largest, hairlike), and 60 coeloconic (smallest, tiny pointed pin) sensilla on the surface of antennae. Maxillary palps contain only basiconic sensilla. Individual sensilla on the antennae each house between one and four ORNs, whereas maxillary palps bear 60 basiconic sensilla, each housing two ORNs. All ORNs are typical bipolar neurons that project a single axon that forms a synapse with a projection neuron in a glomerulus in the antennal lobe. Projection neurons are second-order olfactory neurons that transmit the information of odorants to the higher olfactory centers. The anatomy of neural termination of the ORNs is functionally homologous to that seen in vertebrate glomeruli, in which the ORN axon synapses with a mitral cell in a glomerulus of the olfactory bulb. A Drosophila antennal lobe contains only approximately 50 glomeruli, almost 1/20 the number found in rodents. Thus, the relative simplicity of the insect model is well suited for the study of odor-information processing at a second-order-neuron level (Wilson et al. 2004; Olsen and Wilson 2008).

2.2 Larval Olfactory Organ

The morphology of the larval olfactory organ is much simpler than that of the adult. This simplicity is reflective of the larval stage that does not forage beyond its host plant. The Lepidoptera larval olfactory organ is also distributed to its antennae and maxillary palps (Fig. 1b). Any morphological differences between larval and adult maxillary palps are not obvious. Both types of chemosensory organ possess sticklike extensions on the head. As in adults, these larval sensors consist of multiple sensilla, each comprising one to several ORNs. The Drosophila olfactory organ is a dorsal organ that functions in both smell and taste and is composed of a central olfactory "dome" and six peripheral sensilla (Heimbeck et al. 1999; Python and Stocker 2002). The colocalization of both olfactory and gustatory sensory neurons in the same chemosensory organ is one of the larva-specific features of Drosophila. The dome is innervated by seven triplet dendrites originating from 21 ORNs, each expressing a different OR. Similar to adult ORNs, larval ORNs strongly respond to many volatile odorants derived from food sources (Fishilevich et al. 2005; Kreher et al. 2005). The fundamental difference between the adult and larval olfactory organ is a dramatic difference in the number of ORNs: 21 ORNs in the larva vs. approximately 1,200 ORNs in the adult. The greater number of ORNs in adults highlights the importance of detection of diverse, long-range chemical signals in flying adults.

2.3 Odorant Response by Olfactory Neurons

The first step for any olfactory organ is reception of odor ligands by individual ORNs. Extracellular single-unit electrical activity of individual ORNs has been recorded in individual sensilla using fine glass, crystal, or tungsten electrodes inserted into the base. With use of this recording technique, odor coding of basiconic sensilla in antennae and maxillary palps has been examined (de Bruyne et al. 1999, 2001). Interestingly, ORNs in these sensilla exhibit a spontaneous action potential. Stimulation with particular odorants elicited either an increase or a decrease in the firing rates of individual ORNs, but ORNs also exhibit a marked desensitization to longer olfactory stimulation. An electrophysiological survey of particular odorants revealed that ORNs in both antennae and maxillary basiconic sensilla are strongly activated by food-related odors such as ester and acetate (de Bruyne et al. 1999, 2001). A subset of ORNs in the antenna basiconic sensilla also respond to carbon dioxide (de Bruyne et al. 2001). ORNs in coeloconic sensilla respond to water vapor, ammonia, diaminobutane, and wide range of food sources (Yao et al. 2005). However, none of the ORNs in trichoid sensilla respond to food-related odors, but they do respond to hexane extracts of fly bodies (van der Goes van Naters and Carlson 2007). Moreover, some food odors strongly inhibit the electrical activity of ORNs in the trichoid sensilla. This olfactory specificity suggests that trichoid sensilla function as pheromone sensors. Indeed, a male-specific volatile compound,

11-cis-vaccenyl acetate, has been identified as the unisex pheromone detected by specific ORNs in the trichoid sensilla (Ha and Smith 2006).

3 Odorant Receptors

Since the discovery of rodent OR genes (Buck and Axel 1991), large OR gene families have been identified in both vertebrates and invertebrates. In vertebrates, it is thought that olfactory receptors comprise at least four types of G-protein-coupled receptors (GPCRs): ORs, type I vomeronasal receptors (Dulac and Axel 1995), type II vomeronasal receptors (Matsunami and Buck 1997), and the trace amine-associated receptor family (Liberles and Buck 2006). The total number of functional ORs varies widely across species; fishes have about 100 ORs (Alioto and Ngai 2005), whereas mice have about 900 and humans have about 350 (Niimura and Nei 2005). The primitive lamprey, Lamptera, also expresses seven-transmembrane receptors in the olfactory epithelium that share a low sequence identity with other vertebrate ORs (Freitag et al. 1999). However, the OR gene family has not been found in Ciona, a common ancestor of chordates and vertebrates (Dehal et al. 2002; Satoh 2005).

3.1 OR Genes

To date, insect OR gene families have been identified in the fruit fly Drosophila melanogaster (Clyne et al. 1999; Gao and Chess 1999; Vosshall et al. 1999; Robertson et al. 2003), the malaria vector mosquito Anopheles gambiae (Hill et al. 2002), the yellow fever vector mosquito Aedes aegypti (Bohbot et al. 2007), the honeybee Apis mellifera (Robertson and Wanner 2006), the silkmoth Bombyx mori (Sakurai et al. 2004; Nakagawa et al. 2005; Wanner et al. 2007), and the flour beetle Tribolium castaneum (Engsontia et al. 2008). Because these OR genes exhibit no known homology to vertebrate ORs, the first identification of insect OR genes came much later than studies of nematode ORs and nearly a decade later than Buck and Axel's report on rodents. Although bioinformatic analysis predicted that these insect ORs would have seven-transmembrane domains, the ORs exhibit a novel membrane topology in which the N-terminus is intracellular and the C-terminus is extracellular (Benton et al. 2006; Lundin et al. 2007). Furthermore, insect ORs lack sequence homology to GPCRs in vertebrates (Wistrand et al. 2006). Therefore, it remains questionable whether insect ORs belong to the GPCRs (see later).

The number of ORs greatly varies among insect species. So far, 62, 79, 131, 157, 48, and 265 ORs have been identified in Drosophila, Anopheles, Aedes, Apis, Bombyx, and Tribolium, respectively. Phylogenetic analysis has revealed several species-specific OR clusters (Robertson and Wanner 2006; Bohbot et al. 2007; Engsontia et al. 2008) and has shown that ORs are highly divergent (approximately 20%

amino acid homology), indicating that the insect OR genes are quite ancient. No information is yet available regarding the origin of the insect OR gene family, since only a few insect genome projects have been established. Although its arthropod ancestors are still unknown (Akam 2000) and members of Insecta occupy a single branch within the arthropods (Dunn et al. 2008), it is likely that insect OR genes developed during the evolution of Arthropoda (Fig. 2). It will be important to elucidate the molecular evolution of the insect OR genes to understand how the invertebrate chemosensory system developed over time.

3.2 OR Expression Patterns

Consistent with the drastic change in food preference after metamorphosis, the expression pattern of ORs in the adult insect olfactory organ differs from that in the larval olfactory organ and also shows sexual dimorphism. In Drosophila, the adult expresses 40 ORs, 30 of which are expressed in antennae. Larvae express 25 ORs (Fishilevich et al. 2005; Kreher et al. 2005), 14 of which are larva-specific. As described below, a subset of adult-specific ORs responds to food odors, whereas ten of the 30 adult-specific ORs are specifically expressed in trichoid sensilla and are strongly inhibited or only weakly activated by food-related odors, consistent with their purported role in pheromone rather than food detection. ORs expressed specifically at each developmental stage are likely to be essential for survival during the stage.

In vertebrates, each ORN expresses a single OR gene from a large gene family in a mutually exclusive manner (one neuron–one receptor rule; Mombaerts 2004). In contrast, individual insect ORNs express one, two, or three different ligand-binding ORs along with an Or83b family coreceptor (see Sect. 3.3) in a single ORN

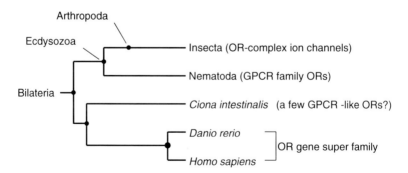

Fig. 2 A phylogenetic tree of the animals, in which the evolution of odorant receptor (OR) genes were explored. In nematodes, Ciona and vertebrates, OR genes belong to the family of seven-transmembrane G-coupled-protein receptors (GPCR). An ancestor of insects possessing a unique OR gene family appeared after branching from the arthropod ancestor. Thus, the origin of insect OR genes is likely to have arisen independently of the GPCR family

in fruit flies (Vosshall et al. 2000). Most antenna ORNs express one type of OR plus Or83b, although eight of the 40 adult ORs have been shown to be coexpressed, thus resulting in up to three ORs plus Or83b being expressed in a single neuron. In one such case, Or65a, Or65b, and Or65c were found to be colocalized in a single ORN in the trichoid sensilla. It has been shown that Or33c and Or85e are coexpressed in the maxillary palp (Couto et al. 2005; Fishilevich et al. 2005). Sometimes an OR is coexpressed with a gustatory receptor: Or10a is coexpressed with the putative gustatory receptor Gr10a in the basiconic sensilla. Although a given combination of coexpressed, canonical OR genes is always the same in an ORN, there are notable exceptions to the one neuron–one receptor rule, and the molecular receptive range of an ORN will depend on the combination of ORs that are coexpressed. The control region for expression of some Drosophila OR genes resides within approximately 0.5 kb upstream/downstream of the OR open reading frame and consists of multiple regulatory elements (Ray et al. 2007). The regulatory factors that determine which OR genes in a particular ORN will be expressed are likely to control the organ-specific expression (either in antenna or in maxillary pads) as well, and the coexpression of specific combinations of ORs in a single ORN appears to be under positive/negative regulation, meaning that some genes are actively expressed while others are actively suppressed.

3.3 Or83b Family

One fundamental difference in how vertebrate and insect ORs function is the presence of the insect-specific Or83b receptor family. Unlike most other insect ORs that share low homology, the Or83b family is one insect-specific receptor family whose molecular structure and function are conserved across diverse insect species (Hill et al. 2002; Nakagawa et al. 2005; Robertson and Wanner 2006; Bohbot et al. 2007; Engsontia et al. 2008). Or83b is coexpressed with conventional ORs in all ORNs (Vosshall et al. 1999; Larsson et al. 2004; Benton et al. 2006). It is thought that Or83b does not function directly in odor recognition. In vivo, Or83b is located on the surface of the dendrite cell membrane and forms a heteromeric complex with conventional ORs (Neuhaus et al. 2005; Benton et al. 2006). Or83b is necessary for electrical activity of ORNs and ciliary targeting of the other "ligand-binding" ORs to the membrane (Larsson et al. 2004; Benton et al. 2006). Coexpression of Or83b is also essential for response of ligand-binding ORs to odorant in heterologous expression systems (Nakagawa et al. 2005; Wanner et al. 2007; Sato et al. 2008; Wicher et al. 2008). In Drosophila, some ORNs in basiconic and coeloconic sensilla do not express Or83b. Instead, these ORNs respond to carbon dioxide and express a pair of gustatory receptors, Gr21a and Gr63a (Jones et al. 2007; Kwon et al. 2007). Coexpression of a pair of gustatory receptors is necessary and sufficient for olfactory carbon dioxide chemosensation.

3.4 OR Function

The molecular receptive range of insect ORs has been well characterized by in vivo and in vitro experiments. The first functional identification of ligands for a candidate Drosophila OR was obtained by expression of the Or43a gene in Xenopus oocytes (Wetzel et al. 2001). Complementary RNA of Or43a was injected into oocytes together with the $G\alpha_{15}$ subunit, which coupled with various GPCRs and thus activated the calcium pathway, leading to opening of endogenous calcium-activated chloride channels. Using the two-electrode voltage-clamp technique, Or43a-expressing oocytes exhibited electrical responses to cyclohexanol, cyclohexanone, benzaldehyde, and benzyl alcohol at nanomolar concentrations. This ligand repertoire was consistent with the result obtained from recording of Or43a-expressing ORNs in antennae in vivo (Störtkuhl and Kettler 2001).

A similar approach was applied in the study of the silkmoth bombykol pheromone receptor (Sakurai et al. 2004; Nakagawa et al. 2005), a receptor for honeybee queen substance p-oxo-2-decenoic acid (Wanner et al. 2007), and a ligand repertoire for Anopheles ORs (Lu et al. 2007). In these cases, however, coexpression of $G\alpha_{15}$ was not necessary, but expression of one member of the Or83b family was absolutely necessary. The molecular mechanisms for signal transduction are discussed in Sect. 4.

Much of the progress in in vivo pairing of a Drosophila OR with its ligand has been made using the Δ halo mutant fly lacking Or22a/b that is normally expressed in ab3A neurons (Dobritsa et al. 2003). With use of the Gal4/UAS expression system, when the Or22a/b gene was expressed ectopically in ab3A neurons under the control of the Or22a promoter Gal4 in the Δ halo mutant, the odorant response of ab3A neurons was rescued. The ectopic expression of Or47a normally expressed by the ab5B neurons into "empty" ab3A neurons resulted in a change in the olfactory responsiveness of ab3A to that similar to ab5B. Thus, ectopic expression of individual OR genes in ab3A neurons of the Δ halo mutant has been utilized as an in vivo expression system for the large-scale screening of molecular receptive range of insect ORs. With use of this approach, nearly all Drosophila ORs expressed in antennae have been characterized (Hallem et al. 2004b; Hallem and Carlson 2006). The odor responsiveness of an ectopically expressed OR was also compared with that of wild-type neurons from which the receptors were derived. With use of the Δ halo mutant system, certain aromatic and aliphatic food odors were identified as the ligands for larva-specific ORs (Kreher et al. 2005). The system is also useful for analyzing ORs of other insects, such as disease vector mosquitoes (Hallem et al. 2004a; Jones et al. 2007; Kwon et al. 2007). These results provide direct functional evidence that an OR gene encodes a bona fide odorant-binding receptor and that functional or ectopic expression of OR genes turns out to be a powerful tool for identifying their ligands.

These studies revealed both similarities and fundamental differences in the odor-coding mechanisms of insect and mammalian ORs. The striking parallel with mammalian ORs is that an individual insect OR can recognize multiple odors and that one odor can activate multiple ORs, although some of the ORs responded specifically to only one of the compounds tested. The key difference in the physiological function of insect and mammalian ORs is that insect ORs possess spontaneous electrical activity that is either activated or inhibited by particular odorants (Hallem et al. 2004b; Hallem and Carlson 2006). Inhibition may narrow an otherwise wide spectrum response to a single odorant in multiple ORs by suppressing the activity of certain ORs. The molecular mechanism and its contribution to odor coding of this inhibitory response are discussed later.

4 Receptor Transduction Mechanism

During the 1980s and 1990s, physiological aspects of vertebrate olfactory transduction were characterized by electrophysiological recordings and biochemical approaches using intact ORNs isolated from main olfactory epithelia. In vertebrates, it is widely accepted that odor transduction in ORNs is mediated by a G-protein ($G\alpha_{olf}$)-mediated second messenger pathway. Odorant binding to G-protein-coupled ORs (Buck and Axel 1991) is the initial event of olfactory transduction, and this in turn results in the activation of type III adenylate cyclase to produce the second messenger cyclic AMP (Lowe et al. 1989). The resulting activation of cyclic-nucleotide-gated (CNG) channels in the ciliary membrane of ORNs subsequently generates a depolarizing receptor potential (Nakamura and Gold 1987). Genetic disruption of $G\alpha_{olf}$, adenylyl cyclase, or the CNG channel subunit causes anosmia detectable as both ORN electrical activity and odor-learning behavior, confirming that these components are involved in peripheral vertebrate olfactory transduction (Brunet et al. 1996; Belluscio et al. 1998; Wong et al. 2000).

4.1 G-Protein Cascade?

In contrast with vertebrates, insect olfactory transduction has been a longstanding mystery in the field of sensory physiology. Insect ORNs are bipolar neurons from which dendrites extend into cuticular structures. Studies of extracellular recordings in antennae and genetic approaches using a G-protein-signaling fly mutant have provided fragmented knowledge of the intracellular signaling mechanism involved in insect olfaction. It has been demonstrated that insect ORs are capable of activating $G\alpha_{15}$ and $G\alpha_q$ (Wetzel et al. 2001; Sakurai et al. 2004). The maxillary palp of norpA phospholipase C mutant flies turned out to be olfactory-defective (Riesgo-Escovar et al. 1995). RNA interference knockdown of $G\alpha_q$ in antenna ORNs results in odor-specific defects in olfactory behavior (Kalidas and Smith 2002). These studies support the hypothesis that binding of odorants and insect ORs activates the $G\alpha_q$ pathway. In contrast, the possibility of involvement of a cyclic-nucleotide pathway has also been suggested (Gomez-Diaz et al. 2004). Lepidoptera antennae produce cyclic GMP (cGMP) upon stimulation by pheromones (Ziegelberger et al. 1990);

however, it is unclear whether cGMP is produced within the ORNs. It has been shown that insect ORNs express a subset of molecules involved in canonical G-protein signaling such as that via the $G\alpha_o$ subunit (Rützler et al. 2006) and a CNG channel (Baumann et al. 1994). The coupling of insect ORs with G-proteins and downstream effector enzymes has not yet been fully confirmed.

4.2 Ion Channel Hypothesis

Although bioinformatics suggest that insect OR gene families belong to the seven-transmembrane receptor family, they lack homology to GPCRs in vertebrates (Wistrand et al. 2006) and possess a distinct topology (Benton et al. 2006; Lundin et al. 2007). Therefore, it seems unlikely that insect ORs in the cell membrane of an ORN dendrite transduce chemical signals into an electrical potential via G-proteins. Recently, an intriguing possibility that insect ORs are themselves heteromeric ligand-gated ion channels was tested (Sato et al. 2008). A combination of canonical OR genes and Or83b families cloned from Drosophila, Anopheles, and Bombyx was transfected into mammalian cell lines, and the resulting odorant responsiveness was characterized by various intracellular recording techniques such as patch-clamp and calcium imaging. The cells expressing insect ORs showed both electrical and calcium responses to a cognate ligand with sensitivity similar to that observed in in vivo antenna recordings. The analysis of waveforms revealed shorter response latency than that of vertebrate ORNs' olfactory response via G-proteins. The sensitivity to ruthenium red, a known calcium and transient receptor potential channel blocker, depends on the combination of an OR and a member of the Or83b family. Swapping Or83b with an ortholog in silkmoth resulted in changes in potassium permeability. These results suggest that a combination of ORs regulates ion permeability independently of G-protein pathways.

More direct evidence that insect ORs are odor-gated ion channels was obtained from an outside-out single-channel recording of cell membrane excised from insect OR-expressing HEK293T cells (Sato et al. 2008). The cell membrane expressing the Anopheles 2-methylphenol receptor complex, GPROR2 + GPROR7 (Hallem et al. 2004a), clearly showed a cluster of single channels opening upon stimulation with a cognate ligand, but not with nonagonist eugenol (Fig. 3). The estimated channel conductance at a holding potential of –60 mV was almost the same as that recorded in an independent experiment using a Xenopus oocyte recording system (Sato et al. 2008). Interestingly, both HEK293T and oocyte cell membranes expressing insect ORs showed spontaneous channel opening. The spontaneous activity of ORs accounts for the spontaneous firing of ORNs that exhibit bipolar electrical activity (Hallem et al. 2004b; Hallem and Carlson 2006) and become electrically negative upon deletion of Or83b in vivo (Larsson et al. 2004).

Consistent with the findings of Sato et al. (2008), Wicher et al. (2008) also reported ion-channel-like properties of insect ORs; however, they also reported $G\alpha_s$-dependent cyclic-nucleotide activation of Or83b itself. Sato et al. (2008) also reported the cGMP sensitivity of silkmoth pheromone receptor BmOR-1 although

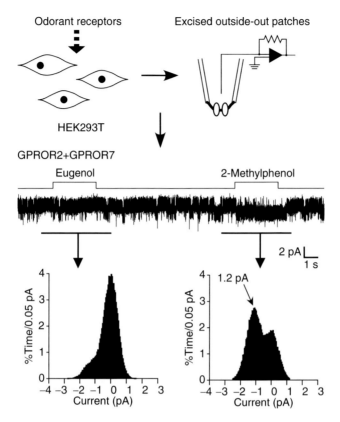

Fig. 3 Single-channel conductance of the complex of Anopheles 2-methylphenol receptor GPROR2 with the Or83b receptor GPROR7 in vitro. The pair of receptor genes was transfected into an HEK293T cell line using a pME18S expression vector together with green fluorescent protein or monomeric red fluorescent protein. Outside-out patch membranes were excised from cells expressing green fluorescent protein or monomeric red fluorescent protein. The ligands were applied focally to the excised patch membrane using a pressure ejection system. Although nonagonist eugenol had no effect on spontaneous channel opening, its cognate ligand 2-methylphenol evoked a cluster of channel openings. The single-channel conductance of 1.2 pS at –60 mV was obtained from the current-amplitude distribution

they failed to demonstrate the elevation of cyclic nucleotides in BmOR-1-expressing cells. Ziegelberger et al. (1990) reported that bombykol stimulation resulted in increases in cGMP in antennae of male silkmoths and that cyclic AMP levels were unchanged upon bombykol stimulation; however, the electrical activity of pheromone-sensing ORNs was independent of the level of cGMP. Thus, the cyclic-nucleotide-mediated pathway is not likely to be involved in the primary olfactory transduction process that generates the receptor potential, although some insect OR complexes do have cyclic-nucleotide sensitivity.

A new model has been proposed for olfactory signal transduction by a heteromeric insect OR complex (Sato et al. 2008; Fig. 4). In this model, a conventional insect OR and Or83b family coreceptor form a cation nonselective ion channel

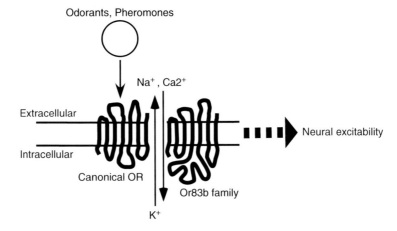

Fig. 4 The mechanism of insect olfactory transduction. Odorants and pheromones regulate the channel opening probability of the insect OR–Or83b receptor complex. An influx of cations depolarizes the olfactory receptor neuron (ORN), resulting in an increase in the firing rate. Depression of channel activity suppresses neural electrical activity. See Sato et al. (2008) for more details

directly gated by odor or pheromone ligands. The G-protein pathway is negligible in producing the current elicited by odor activation of such insect OR heteromultimers. How odorants inhibit the OR complex ion channel to suppress electrical neural activity remains to be investigated.

5 Sending Peripheral Odor Information to the Brain

The receptor potential produced by the binding of a ligand to ORs is transduced into action potentials, which are propagated to the central nervous system through the axon. The axons of individual ORNs terminate in an olfactory glomerulus. In rodents, each glomerulus in the olfactory bulb receives the synaptic input from ORNs expressing the same OR (Ressler et al. 1994; Vassar et al. 1994; Mombaerts et al. 1996; Fig. 5). Similarly, axons of Drosophila ORNs that express the same combination of ORs converge upon a single glomerulus (Couto et al. 2005; Fishilevich et al. 2005). Thus, single glomeruli receive input from the neurons that possess the same molecular receptive range, which is a common rule for odor recognition at second-order olfactory neuron levels in both vertebrates and insects.

The axon wiring in the insect antennal lobe is anatomically and physiologically similar to that in the mammalian olfactory bulb. Insect ORNs target and synapse with two neurons: projection neurons and GABA-mediated local interneurons. Projection neurons are second-order olfactory interneurons that transmit the information of odorants into the mushroom body calyx and lateral horn, which is the center for olfactory-related memory formation (Heisenberg

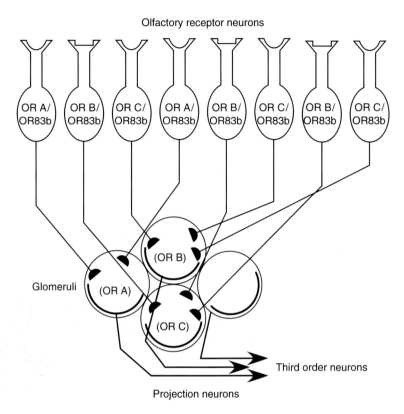

Fig. 5 Synaptic convergence onto a single glomerulus of ORN axons expressing the same ORs. Insect ORNs generally express two ORs on the surface of dendrites, a single OR plus a receptor of the Or83b family. The axons of ORNs expressing the same receptor combination synapse with projection neurons in unique glomeruli in the antennal lobe

2003). Local interneurons make the lateral interglomerular excitatory and inhibitory connections (Wilson et al. 2004; Olsen and Wilson 2008). An alternative model of odor coding at second-order neurons proposes that synaptic integration between local interneurons and the ORNs expressing the same ORs sharpens the tuning of projection neurons to particular odorants (Olsen and Wilson 2008). Although the origin of insect OR genes is likely to be evolutionarily distinct from that of mammalian ORs, the anatomical and functional features of insect glomeruli and the odor-coding strategy are similar and thus justify any consideration of mammalian olfaction in the study of insect.

6 Biologically Relevant ORs and Behavior

Since the discovery of the first pheromone in silkmoth, Bombyx mori (Butenandt et al. 1959), knowledge of the structures of chemical signals that correspond to insect-specific reactions has accumulated. So far, chemical information regarding more than 3,900 insect pheromones and semiochemicals in more than 7,000 species has been made available in The Pherobase (http://www.pherobase.net/). The compounds are catalogued into two groups: pheromones interacting within species, and allelochemicals mediating interspecific communication. Pheromones are further categorized on the basis of biological function, such as sex, aggregation, and trail and alarm pheromones. The best-studied insect pheromone is bombykol, which is released from sexually mature female silkmoths and activates male reproductive behavior. The presence of this pheromone was demonstrated by Fabre (1886) and finally identified by Butenandt almost a half century later. Butenandt et al. (1959) documented that a few molecules of bombykol molecules are sufficient to activate male-specific behavior.

The antennae of male silkmoths possess bombykol-specific pheromone receptor neurons expressing the receptor for bombykol, BmOR-1. BmOR-1 shows sexual dimorphism in its expression pattern (Sakurai et al. 2004; Nakagawa et al. 2005). Bombykol released from the female activates male sexual behavior with high specificity and sensitivity via the male-specific BmOR-1. The axons of BmOR-1-expressing neurons extend to the macroglomerular complex in males. Further searching in the silkmoth genome database for additional male-specific ORs revealed BmOR-3, a receptor for bombykal that acts as an inhibitory pheromone for wing vibration behavior (Nakagwa 2005). Female silkmoth antennae also show biased expression of BmOR-19 and BmOR-30, whose ligands are unknown (Wanner et al. 2007). These aspects of OR expression and function suggest the fundamental role of ORs in recognizing biologically relevant odorants and pheromones from the diverse chemical environment.

The molecular mechanism and neural circuit for the behavioral response in fruit flies to the male-specific volatile compound 11-cis-vaccenyl acetate have been well studied (Xu et al. 2005; Ha and Smith 2006; Kurtovic et al. 2007; Benton et al. 2007; Datta et al. 2008). 11-cis-Vaccenyl acetate is released from the mature male body and acts as an aggregation pheromone in both males and females (Bartelt et al. 1985). This pheromone detected by Or67d-expressing ORNs appears to regulate reproductive behavior in both sexes. At the periphery, activation of Or67d-expressing ORNs by 11-cis-vaccenyl acetate requires the expression of the family of sensory neuron membrane proteins (Xu et al. 2005; Benton et al. 2007). In males, activation of Or67d by 11-cis-vaccenyl acetate inhibits male–male courtship behavior, whereas female Or67d promotes mating behavior (Kurtovic et al. 2007). Or67d-expressing ORNs activate a sexually dimorphic third-order olfactory neuron, resulting in different behaviors in males and females (Kurtovic et al. 2007; Datta et al. 2008).

Carbon dioxide released as a result of animal respiration is the most widely utilized chemical signal for host- and prey-seeking behavior in insects, especially in blood-feeding insects. CO_2 also acts as a repellent in fruit flies to avoid stressful environments (Suh et al. 2004). As a repellent, CO_2 is detected by antennal ORNs coexpressing both Gr21a and Gr63a (Jones et al. 2007; Kwon et al. 2007). The axons of the CO_2-sensing neurons converge upon the V glomerulus in the antennal lobe (Jones et al. 2007). Mosquitoes also coexpress the homologs GPRGR22 and GPRGR24 in ORNs of the maxillary palp connecting to medial glomeruli (Anton et al. 2003). These data suggest that the localization of CO_2-sensing ORNs in either antennae or maxillary palps determines the neural circuit for species-specific behavioral response to CO_2. Although Drosophila-specific microRNAs downregulate the expression of CO_2 receptors in maxillary palps (Cayirlioglu et al. 2008), the neural circuit and any additional molecular mechanism regulating species-specific behavior is unknown. At least four functionally distinct chemosensors, including Or35a, are present in coeloconic sensilla, and detect water vapor, ammonia, diaminobutane, and a wide range of food sources (Yao et al. 2005).

7 OR-Based Design of Insect Regulators

Every year, more than one million people die from deadly tropical diseases transmitted by blood-feeding insects. Insects also are the most serious agricultural pests. Thus, insect control is a worldwide health, agricultural, and therefore economic problem, and the use of odorants that target ORs could help control insects. The common prevention strategies for mosquitoes and flies are application of repellents and mass traps that emit attractants. N,N-Diethyl-3-methylbenzamide (DEET) is a highly effective mosquito repellent and is the active ingredient in the most widely used insect repellents applied to the skin (Brown and Hebert 1997). DEET works by selectively blocking the odor-evoked activation of insect OR–Or83b complexes, resulting in a reduced perception of the smell of attractants (Ditzen et al. 2008). Other natural organic compounds such as eucalyptus and citronella that interact with ORs have traditionally been used as repellents. Methyl eugenol released from the flowers of papaya and mango is a highly potent male fly attractant (Metcalf et al. 1975) and also acts as a pheromonal precursor (Nishida et al. 1988). Since the first application of methyl eugenol in combination with insecticide in Hawaii in the 1960s, box traps and methyl eugenol have been used worldwide and have helped to eradicate oriental fruit flies. The molecular mechanism for the sex-specific effect of methyl eugenol has not yet been fully elucidated.

8 Conclusion

Recent progress in the study of the molecular and neural basis of insect olfaction has revealed both similarities and fundamental differences between vertebrates and invertebrates. Although insect ORs were identified as seven-transmembrane receptors, they are distinct from vertebrate ORs in that they are part of heteromeric multimers that include one member of the Or83b receptor family and act as odor-gated cation-nonselective channels. This molecular feature may represent the largest single family of ion-channel-like proteins in any organism. Despite fundamental functional differences between the ORs of insects and vertebrates, they share similar neural circuitry for odor coding at the secondary olfactory neuron level. The tight connection of interneurons and olfactory-driven behaviors in insects allows us to understand how stereotypic behavior is induced by stimulation of the olfactory organ with a single compound. Olfaction-based repellents should be useful for controlling insect pest populations. Further analysis of insect chemosensory systems will be important for the advancement of both sensory biology and health science.

References

Akam M (2000) Arthropods: developmental diversity within a (super) phylum. Proc Natl Acad Sci USA 97:4438–4441

Alioto TS, Ngai J (2005) The odorant receptor repertoire of teleost fish. BMC Genomics 6:173

Anton S et al (2003) Central projections of olfactory receptor neurons from single antennal and palpal sensilla in mosquitoes. Arthropod Struct Dev 32:319–327

Bartelt RJ et al (1985) cis-Vaccenyl acetate as an aggregation pheromone in Drosophila melanogaster. J Chem Ecol 11:1747–1756

Baumann A et al (1994) Primary structure and functional expression of a Drosophila cyclic nucleotide-gated channel present in eyes and antennae. EMBO J 13:5040–5050

Belluscio L et al (1998) Mice deficient in G(olf) are anosmic. Neuron 20:69–81

Benton R et al (2006) Atypical membrane topology and heteromeric function of Drosophila odorant receptors in vivo. PLoS Biol 4:e20

Benton R et al (2007) An essential role for a CD36-related receptor in pheromone detection in Drosophila. Nature 450:289–293

Bohbot J et al (2007) Molecular characterization of the Aedes aegypti odorant receptor gene family. Insect Mol Biol 16:525–537

Brown M, Hebert AA (1997) Insect repellents: an overview. J Am Acad Dermatol 36:243–249

Burnet LJ et al (1996) General anosmia caused by a targeted disruption of the mouse olfactory cyclic nucleotide-gated cation channel. Neuron 17:681–693

Buck LB, Axel R (1991) A novel multigene family may encode odorant receptors: a molecular basis for odor recognition. Cell 65:175–187

Butenandt A et al (1959) N-Acetyl tyramine, its isolation from Bombyx cocoons and its chemical and biological properties. Arch Biochem Biophys 83:76–83

Cayirlioglu P et al (2008) Hybrid neurons in a microRNA mutant are putative evolutionary intermediates in insect CO_2 sensory system. Science 319:1256–1260

Clyne PJ et al (1999) A novel family of divergent seven-transmembrane proteins: candidate odorant receptors in Drosophila. Neuron 22:327–338

Couto A et al (2005) Molecular, anatomical, and functional organization of the Drosophila olfactory system. Curr Biol 15:1535–1547

Datta SR et al (2008) The Drosophila pheromone cVA activates a sexually dimorphic neural circuit. Nature 45:473–477

de Bruyne M et al (1999) Odor coding in a model olfactory organ: the Drosophila maxillary palp. J Neurosci 19:4520–4532

de Bruyne M et al (2001) Odor coding in the Drosophila antenna. Neuron 30:537–552

Dehal P et al (2002) The draft genome of Ciona intestinalis: insights into chordate and vertebrate origins. Science 298:2157–2167

Ditzen M et al (2008) Insect odorant receptors are molecular targets of the insect repellent DEET. Science 319:1838–1842

Dobritsa A et al (2003) Integrating the molecular and cellular basis of odor coding in the Drosophila antenna. Neuron 37:827–841

Dulac C, Axel R (1995) A novel family of genes encoding putative pheromone receptors in mammals. Cell 83:195–206

Dunn CW et al (2008) Broad phylogenomic sampling improves resolution of the animal tree of life. Nature 452:745–749

Engsontia P et al (2008) The red flour beetle's large nose: An expanded odorant receptor gene family in Tribolium castaneum. Insect Biochem Mol Biol 38:387–397

Fabre JH (1886) Souvenirs entomologiques (Troisiémè serie). Librairie Ch. Delagrave, Paris

Fishilevich E et al (2005) Chemotaxis behavior mediated by single larval olfactory neurons in Drosophila. Curr Biol 15:2086–2096

Freitag J et al (1999) On the origin of the olfactory receptor family: receptor genes of the jawless fish (Lampetra fluviatilis).Gene 226:165–174

Gao Q, Chess A (1999) Identification of candidate Drosophila olfactory receptors from genomic DNA sequence. Genomics 60:31–39

Gomez-Diaz G et al (2004) cAMP transduction cascade mediates olfactory reception in Drosophila melanogaster. Behav Genet 34:395–406

Ha TS, Smith DP (2006) A pheromone receptor mediates 11-cis-vaccenyl acetate-induced responses in Drosophila. J Neurosci 26:8727–8733

Hallem EA et al (2004a) Olfaction: mosquito receptor for human-sweat odorant. Nature 427:212–213

Hallem EA et al (2004b) The molecular basis of odor coding in the Drosophila antenna. Cell 117:965–979

Hallem EA, Carlson JR (2006) Coding of odors by a receptor repertoire. Cell 125:143–160

Heimbeck G et al (1999) Smell and taste perception in Drosophila melanogaster larva: toxin expression studies in chemosensory neurons. J Neurosci 19:6599–6609

Heisenberg M (2003) Mushroom body memoir: from maps to models. Nat Rev Neurosci 4:266–275

Hill CA et al (2002) G-protein-coupled receptors in Anopheles gambiae. *Science* 298:176–178

Jones WD et al (2007) Two chemosensory receptors together mediate carbon dioxide detection in Drosophila. Nature 445:86–90

Kalidas S, Smith DP (2002) Novel genomic cDNA hybrids produce effective RNA interference in adult Drosophila. Neuron 33:177–184

Kreher SA et al (2005) The molecular basis of odor coding in the Drosophila larva. Neuron 46:445–456

Kurtovic A et al (2007) A single class of olfactory neurons mediates behavioural responses to a Drosophila sex pheromone. Nature 446:542–546

Kwon JY et al (2007) The molecular basis of CO2 reception in Drosophila. Proc Natl Acad Sci USA 104:3574–3578

Larsson MC et al (2004) Or83b encodes a broadly expressed odorant receptor essential for Drosophila olfaction. Neuron 43:703–714

Liberles SD, Buck LB (2006) A second class of chemosensory receptors in the olfactory epithelium. Nature 442:645–650

Lowe G et al (1989) Adenylate cyclase mediates olfactory transduction for a wide variety of odorants. Proc Natl Acad Sci USA 86:5641–5645

Lu T et al (2007) Odor coding in the maxillary palp of the malaria vector mosquito Anopheles gambiae. Curr Biol 17:1533–1544

Lundin C et al (2007) Membrane topology of the Drosophila OR83b odorant receptor. FEBS Lett 581:5601–5604

Matsunami H, Buck LB (1997) A multigene family encoding a diverse array of putative pheromone receptors in mammals. Cell 90:775–784

Metcalf RL et al (1975) Attraction of the oriental fruit fly, Dacus dorsalis, to methyl eugenol and related olfactory stimulants. Proc Natl Acad Sci USA 72:2501–2505

Mombaerts P et al (2004) Genes and ligands for odorant, vomeronasal and taste receptors. Nat Rev Neurosci 5:263–278

Mombaerts P et al (1996) Visualizing an olfactory sensory map. Cell 87:675–678

Nakagawa T et al (2005) Insect sex-pheromone signals mediated by specific combinations of olfactory receptors. Science 307:1638–1642

Nakamura T, Gold GH (1987). A cyclic nucleotide gated conductance in olfactory receptor neurons. Nature 325:442–444

Neuhaus EM et al (2005) Odorant receptor heterodimerization in the olfactory system of Drosophila melanogaster. Nat Neurosci 8:15–17

Niimura Y, Nei M (2005) Comparative evolutionary analysis of olfactory receptor gene clusters between humans and mice. Gene 346:13–21

Nishida R et al (1988) Accumulation of phenypropanoids in the rectal glands of males of the oriental fruit fly, Dacus dorsalis. Experientia 44:534–536

Novotny V et al (2002) Low host specificity of herbivorous insects in a tropical forest. Nature 416:841–844

Olsen SR, Wilson RI (2008) Lateral presynaptic inhibition mediates gain control in an olfactory circuit. Nature 452:956–960

Python F, Stocker RF (2002) Adult-like complexity of the larval antennal lobe of D. melanogaster despite markedly low numbers of odorant receptor neurons. J Comp Neurol 445:374–387

Ray A et al (2007) Mechanisms of odor receptor gene choice in Drosophila. Neuron 53:353–369

Ressler KJ et al (1994) Information coding in the olfactory system: evidence for a stereotyped and highly organized epitope map in the olfactory bulb. Cell 79:1245–1278

Riesgo-Escovar J et al (1995) Requirement for a phospholipase C in odor response: overlap between olfaction and vision in Drosophila. Proc Natl Acad Sci USA 92:2864–2868

Robertson HM et al (2003) Molecular evolution of the insect chemoreceptor gene superfamily in Drosophila melanogaster. Proc Natl Acad Sci USA 100:14537–14542

Robertson HM, Wanner KW (2006) The chemoreceptor superfamily in the honeybee, Apis mellifera: expansion of the odorant, but not gustatory, receptor family. Genome Res 16:1395–1403

Rützler M et al (2006) Gα encoding gene family of the malaria vector mosquito Anopheles gambiae: Expression analysis and immunolocalization of a Gαq and a Gαo in female antennae. J Comp Neurol 499:533–545

Sakurai T et al (2004) Identification and functional characterization of a sex pheromone receptor in the silkmoth Bombyx mori. Proc Natl Acad Sci USA 101:16653–16658

Sato K et al (2008) Insect olfactory receptors are heteromeric ligand-gated ion channels. Nature 452:1002–1006

Satoh G (2005) Characterization of novel GPCR gene coding locus in amphioxus genome: gene structure, expression, and phylogenetic analysis with implications for its involvement in chemoreception. Genesis 41:47–57

Störtkuhl KF, Kettler R (2001) Functional analysis of an olfactory receptor in Drosophila melanogaster. Proc Natl Acad Sci USA 98:9381–9385

Suh GS et al (2004) A single population of olfactory sensory neurons mediates an innate avoidance behaviour in Drosophila. Nature 431:854–859

van der Goes van Naters W, Carlson JR (2007) Receptors and neurons for fly odors in Drosophila. Curr Biol 17:606–612

Vassar R et al (1994) Topographic organization of sensory projections to the olfactory bulb. Cell 79:981–978

Vosshall LB et al (1999) A spatial map of olfactory receptor expression in the Drosophila antenna. Cell 96:725–736

Vosshall LB et al (2000) An olfactory sensory map in the fly brain. Cell 102:147–159

Wanner KW et al (2007) A honeybee odorant receptor for the queen substance 9-oxo-2-decenoic acid. Proc Natl Acad Sci USA 104:14383–14388

Wetzel CH et al (2001) Functional expression and characterization of a Drosophila odorant receptor in a heterologous cell system. Proc Natl Acad Sci USA 98:9377–9380

Wicher D et al (2008) Drosophila odorant receptors are both ligand-gated and cyclic-nucleotide-activated cation channels. Nature 452:1007–1011

Wilson RI et al (2004) Transformation of olfactory representations in the Drosophila antennal lobe. Science 303:366–370

Wistrand M et al (2006) A general model of G protein-coupled receptor sequences and its application to detect remote homologs. Protein Sci 15:509–521

Wong ST et al (2000) Disruption of the type III adenylyl cyclase gene leads to peripheral and behavioral anosmia in transgenic mice. Neuron 27:487–497

Xu P et al (2005) Drosophila OBP LUSH is required for activity of pheromone-sensitive neurons. Neuron 45:193–200

Yao CA et al (2005) Chemosensory coding by neurons in the coeloconic sensilla of the Drosophila antenna. J Neurosci 25:8359–8378

Ziegelberger et al (1990) Cyclic GMP levels and guanylate cyclase activity in pheromone-sensitive antenna of the silkmoths Antheraea polyphemus and Bombyx mori. J Neurosci 10:1217–1225

Smelling, Tasting, Learning: *Drosophila* as a Study Case

B. Gerber, R.F. Stocker, T. Tanimura, and A.S. Thum

Abstract Understanding brain function is to account for how the sensory system is integrated with the organism's needs to organize behaviour. We review what is known about these processes with regard to chemosensation and chemosensory learning in *Drosophila*. We stress that taste and olfaction are organized rather differently. Given that, e.g., sugars are nutrients and should be eaten (irrespective of the *kind* of sugar) and that toxic substances should be avoided (regardless of the *kind* of death they eventually cause), tastants are classified into relatively few behavioural matters of concern. In contrast, what needs to be done in response to odours is less evolutionarily determined. Thus, discrimination ability is warranted between different kinds of olfactory input, as any difference between odours may potentially be or become important. Therefore, the olfactory system has a higher dimensionality than gustation, and allows for more sensory–motor flexibility to attach acquired behavioural 'meaning' to odours. We argue that, by and large, larval and adult *Drosophila* are similar in these kinds of architecture, and that additionally there are a number of similarities to vertebrates, in particular regarding the cellular architecture of the olfactory pathway, the functional slant of the taste and smell systems towards classification versus discrimination, respectively, and the higher plasticity of the olfactory sensory–motor system. From our point of view, the greatest gap in understanding smell and taste systems to date is not on the sensory side, where indeed impressive advances have been achieved; also, a satisfying account of associative odour-taste memory trace formation seems within reach. Rather, we lack an understanding as to how sensory and motor formats of processing are centrally integrated, and how adaptive motor patterns actually are selected. Such an understanding, we believe, will allow the analysis to be extended to the motivating factors of behaviour, eventually leading to a comprehensive account of those systems which make *Drosophila* do what *Drosophila's* got to do.

B. Gerber (✉)
Universität Würzburg, Biozentrum, Am Hubland, 97074 Würzburg, Germany
e-mail: bertram.gerber@biozentrum.uni-wuerzburg.de

1 Introduction

There are more things in the world than there are possible behaviours. Thus, in order to fulfil the needs of life, the things in the outside world need to be 'funnelled' into far fewer behavioural matters of concern. Integrating the sensory system with the biological needs to come up with appropriate behaviour is what brains have evolved for. It is this triad of things, needs and actions that neurobiology needs to understand.

Notably, it cannot be known in advance which sensory–motor match would be the most fitting one; thus, both during evolution and during learning, possible matches need to be tried out, by taking chances, and the ones with the relatively best fit are stabilized. As a study case, we focus on the functional architecture of the fruit fly chemosensory–motor system to see with which kind of circuitry these problems have evolutionarily been solved regarding smell and taste. We then move on to chemosensory associative learning, to see which degrees of freedom remain for the individual to seize upon the opportunities, and cope with the perils, of life.

Taste is more closely entangled with immediate behaviour control than olfaction. That is, the behavioural 'meaning' of tastants is evolutionarily obvious, in that, for example, energy-rich foods should be eaten and toxic substances should be avoided. Accordingly, tastants seem to be classified into relatively few behavioural matters of concern (edible/sweet, non-edible/bitter, to mention two of them), leaving largely superfluous discrimination between, e.g., different kinds of sweetness. In contrast, it seems much less obvious how to behave towards a given odour. This not only requires flexibility in the sensory–motor 'switchboard', but also requires the ability to discriminate between as many different odours as possible. This has two corollaries, namely that the olfactory system has a higher dimensionality on the sensory side, and that it possesses a dedicated subsystem which allows acquired behavioural 'meaning' to be attached to them. Owing to its cellular simplicity and genetic accessibility, the fruit fly *Drosophila* is a suitable study case to understand how these processes come about.

2 Smelling

Olfactory systems help to track down matters of concern, such as food sources, shelters, oviposition sites or social interaction partners. How does this work in a fly? (For classical accounts see Rodrigues and Siddiqi 1978 and Rodrigues 1980.) Are the mechanisms similar to those in mice or in humans? Indeed, there are surprising parallels between these phylogenetically distant kinds of animal (Ache and Young 2005; Hildebrand and Shepherd 1997; Strausfeld and Hildebrand 1999). These similarities do not necessarily postulate a common origin of olfactory systems, however; rather, to the extent that these systems are not of common origin, similarities and discrepancies between them point to common versus specific functional demands of olfactory systems in different animals.

Common to both phyla is that odorants need to travel through an extracellular matrix ('lymph' in insects) to the olfactory receptor neurons (ORNs). Their dendritic membranes carry olfactory receptor proteins (ORs), which determine the spectrum of odours that can activate the cell. Similar is also that all and only those cells that express the same OR converge in one spherical 'glomerulus' structure in the primary olfactory centre (called 'olfactory bulb' in vertebrates and 'antennal lobe' in insects) (Fig. 1). Lateral connections between the glomeruli contribute to the establishment of specific patterns of activated glomeruli for each particular odour. For further processing, output neurons typically sample one glomerulus each and establish divergent, combinatorial connections to higher-order brain centres. Such architecture seems suitable to achieve both a good signal-to-noise ratio (convergence) and high discriminability (combinatorial divergence). Higher centres thus appear to increase the distinctiveness of 'odour images', and in addition act as a switchboard to refer different odours to distinct behavioural programmes. In other words, they act as a 'watershed' along the sensory–motor pathway, transforming olfactory information ('Which odour?') into motor commands ('What should be done?'). However, this reformatting and in particular the premotor processes themselves are poorly understood. What seems plausible is that the pathways underlying innate behaviour are simpler, more direct and certainly more stereotypic than those which mediate learning-related changes.

2.1 Olfactory Organs of Adult Drosophila

While mammalian ORNs are densely clustered in an epithelium deep inside the nose, in insects one to four ORNs are housed in hairlike structures on the body surface, called 'sensilla' (Fig. 2). The dendrites of the ORNs, expressing the ORs, extend into the lymph of the sensillum shaft. At least for certain pheromones, ORN activation requires the presence of an odorant-binding protein in the lymph (Ha and Smith 2006). The stereotyped assembly of ORNs in sensilla has proven useful to record from identified neurons and to define the range of odours to which they respond.

Olfactory sensilla of *Drosophila* are located at two sites, the third antennal segment and the maxillary palp (Fig. 1). The palp carries approximately 60 morphologically uniform basiconic sensilla, each housing two ORNs. These ORNs fall into six different functional classes with respect to their odour spectra (de Bruyne et al. 1999) reflecting different combinations of expressed ORs (Couto et al. 2005; Goldman et al. 2005). The third antennal segment is covered by three major morphological types of sensilla – basiconic, trichoid and coeloconic – each comprising several subtypes (Shanbhag et al. 1999) (Fig. 2). Every subtype is found in a specific spatial arrangement on the antenna. Trichoid and basiconic sensilla are sexually dimorphic in number, with 30% more trichoids and 20% fewer basiconics in males than in females (Stocker 1994). Antennal basiconic sensilla house two or four neurons, trichoid sensilla house one, two or three neurons and coeloconic sensilla house two or three neurons (Shanbhag et al. 1999). For the majority of antennal ORNs from all of these sensilla types, ligand ranges have been determined

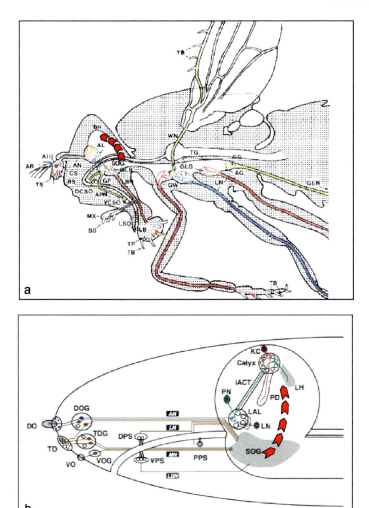

Fig. 1 Overview of the *Drosophila* chemosensory system. **a** Overview of the adult chemosensory pathways. Olfactory pathways project to the brain proper, whereas gustatory afferents are collected in various regions of the suboesophageal and thoracic ganglion. The chevrons indicate the proposed pathway to short-circuit a taste-driven reward signal carried by octopaminergic neurons from the suboesophageal ganglion towards the brain. **b** Overview of the larval chemosensory pathways. As in adults, olfactory pathways project into the brain proper, whereas gustatory afferents are collected in various regions of the suboesophageal ganglion. The chevrons have the same meaning as in a. **c** Scanning electron microscopy (SEM) overview of the larval head. One can discern the dome-shaped dorsal organ, and the wart-like terminal organ. The cirri surround the mouth opening (triangle) and, in the third row of cirri, cover the tiny ventral organ. **d** SEM overview of the adult head and appendages in labellum-opened state. Medial from the large complex eyes, one can discern the arista and the third antennal segment, as well as the maxillary palps and the labellum. **e** Comparison of the approximate number of, from left to right, olfactory receptor neurons, antennal lobe glomeruli, projection neurons, calycal glomeruli in the mushroom bodies and mushroom body Kenyon cells. Note that the local interneurons in the antennal lobe, which shape olfactory activity, are present in both larva and adult, but are omitted in this figure. AIII third antennal

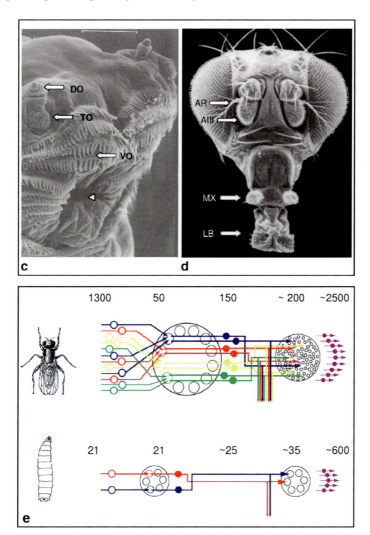

Fig. 1 (continued) segment, AR arista, AL antennal lobe, *AN* antennal nerve, APN accessory pharyngeal nerve, BR brain, BS basiconic sensilla, CS coeloconic sensilla, DO dorsal organ, DOG dorsal organ ganglion, DPS dorsal pharyngeal sense organ, DCSO dorsal cibarial sense organ, GEN genitalia, GG gustatory centre of genitalia, GLB gustatory centre of the labellum, GLG gustatory centres of the leg, GP gustatory centre of the pharynx, GW gustatory centre of the wing, iACT inner antennocerebral tract, KC Kenyon cells, LAL larval antennal lobe, LB labellum (labial palps), *LBN* labial nerve, LH lateral horn, LN local interneurons, *LN* labral nerve, LSO labral sense organ, *MN* maxillary nerve, MX maxillary palp, PD pedunculus, PN projection neuron, PPS posterior pharyngeal sense organ, SOG suboesophageal ganglion, TO terminal organ, TOG terminal organ ganglion, TB taste bristle, TG thoracic ganglion, TP taste peg, TS trichoid sensilla, VCSO ventral cibarial sense organ, VO ventral organ, VOG ventral organ ganglion, VPS ventral pharyngeal sense organ, WN wing nerve. (**a** From Stocker 1994, copyright Springer. **b** From Stocker 2006, copyright Landes Bioscience. **c** Copyright K. Neuser, Universität Würzburg. **d** From "The Interactive Fly", http://www.sdbonline.org/fly/aimain/1aahome.htm", copyright F.R. Turner, Indiana University. **e** From Ramaekers et al. 2005, copyright Elsevier)

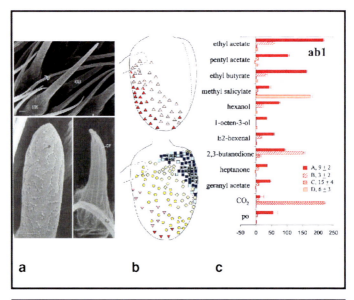

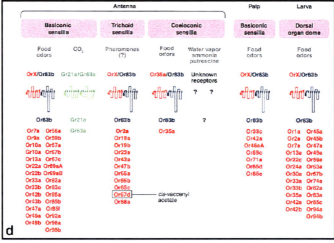

Fig. 2 Major features of the *Drosophila* olfactory system. **a** Scanning electron micrographs of trichoid, basiconic and coeloconic antennal sensilla. **b** Distribution of trichoid (upper panel) and basiconic (lower panel) sensilla on the anterior surface of the third antennal segment. The different symbols refer to morphological subtypes of these sensilla. The arista (stippled) is located on the lateral side of the antenna. **c** Response profiles of the four olfactory receptor neurons (ORNs) comprised within the basiconic sensillum type ab1 to a set of 11 volatile compounds and the solvent (paraffin oil, po). The data present the increase of spikes/s relative to the spontaneous firing frequency. **d** Inventory of *Drosophila* olfactory receptor proteins (ORs) expressed in the different olfactory organs of the adult and the larva, subdivided by sensillum type and possible activating odours. OR83b is an obligate coreceptor for all ORNs except the CO_2-sensitive neurons expressing the gustatory receptor genes *Gr21a* and *Gr63a*. OR67d is strongly activated by the aggregation pheromone 11-*cis*-vaccenyl acetate. **e** Distribution of 31 classes of ORNs expressing

Smelling, Tasting, Learning: *Drosophila* as a Study Case

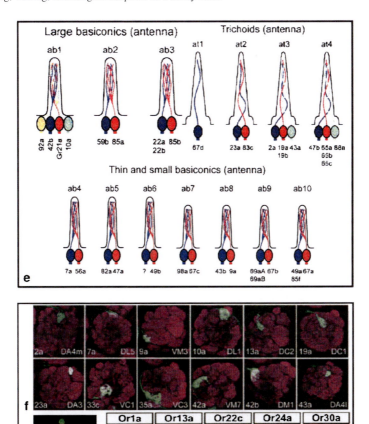

Fig. 2 (continued) specific ORs (and *Gr21a*) in 14 subtypes of antennal basiconic and trichoid sensilla. **f** Terminals of selected *Or*-green fluorescent protein (GFP) lines (green) in specific glomeruli of the adult antennal lobe, which is counterstained with the neuropile marker nc82 (magenta). The OR expressed in each type of ORN is indicated at the bottom left of each panel, the glomerular terminology is indicated on the right. Lateral is to the right. **g** Dorsal organ (DO) and terminal organ (TO) of a third instar larva labelled by the 4551-Gal4 driver line (green). Neuronal nuclei are tagged by α-Elav staining (red). **h** Terminals of selected *Or*-Gal4 or *Or*-GFP lines (green) in specific glomeruli of the larval antennal lobe, counterstained with the neuropile marker nc82 (magenta). Lateral is to the left. (**a, b** From Shanbhag et al. (1999), copyright Elsevier. **c** From de Bruyne et al. (2001), copyright Elsevier. **d** From Vosshall and Stocker (2007), copyright Annual Reviews. **e f** From Couto et al. (2005), copyright Elsevier. **g** From Grillenzoni et al. (2007), copyright Springer. **h** From Fishilevich et al. (2005), copyright Elsevier)

(de Bruyne et al. 2001; Goldman et al. 2005; Hallem and Carlson 2006; van der Goes van Naters and Carlson 2007). While basiconic sensilla are activated by food odours (de Bruyne et al. 1999, 2001), trichoid sensilla respond mainly to fly odours (van der Goes van Naters and Carlson 2007). Coeloconic sensilla comprise ORNs, as well as neurons that respond to humidity changes (Yao et al. 2005). About 50 additional, ill-characterized sensilla are found in the sacculus, a pit on the posterior side of the antenna.

The afferent fibres of the 1,100–1,250 sensory neurons from the third antennal segment (Stocker 2001) each project into single glomeruli of the antennal lobe (Stocker et al. 1983) (Figs. 1, 2). Most of the projections are bilateral, extending into corresponding glomeruli in both ipsilateral and contralateral lobes. However, about 200 fibres that terminate in five specific glomeruli (see later) remain strictly ipsilateral (Stocker et al. 1983). The antennal lobe also is the target of the 120 maxillary ORNs; their projections are bilateral throughout.

2.2 Larval Olfactory Organs

Adults and larvae of insects that undergo full metamorphosis display very different ways of life. Adult flies roam over considerable distance to locate nutrients or mates, while larvae, which live on the food, stay within rather limited territories. Does this entail corresponding differences in complexity of the olfactory circuitry? (For a comprehensive summary of the earlier literature, also on non-*Drosophila* species, see Cobb 1999.) Indeed, both olfactory organs and the central olfactory pathway are much simpler in larvae than in adults, at least in terms of cell number. The tiny larval antenna, the 'dorsal organ', and specifically its prominent 'dome' sensillum, is innervated by only 21 ORNs (Heimbeck et al. 1999; Kreher et al. 2005) (Figs. 1, 2). Larvae in which these neurons were selectively silenced by transgenic toxin expression did not respond behaviourally to odours anymore, suggesting that these neurons are the exclusive larval ORNs (Fishilevich et al. 2005; Larsson et al. 2004). They are arranged in seven triplets, corresponding to a developmental fusion of seven three-neuron-type sensilla (Grillenzoni et al. 2007). Six other sensilla that surround the dome are thought to be gustatory, rendering the dorsal organ a mixed organ for smell and taste, a situation which is not paralleled in adults.

Both the dramatic reduction of ORNs from almost 1,300 in the adult to a mere 21 in the larva and the mixed modality of the dorsal organ suggest that long-range chemosensory signals and the distinction between olfactory and gustatory cues may be less important for a substrate feeder than for a flying insect. As another discrepancy from the adult, all larval ORN projections remain exclusively ipsilateral in the brain. Interestingly, although larvae with a single functional ORN still are attracted by odorants, the accuracy of navigation is enhanced when the larva can use two identical ORNs, one on each side and both expressing the same OR (see later) (Louis et al. 2007); whether a similar improvement would be seen if two different kinds of ORN are functional on the same body side, however, remains to be tested. Common to both adult and larval stages is that the central targets for smell and taste are well separated: Olfactory afferents project into a glomerulus-type antennal lobe (Python and Stocker 2002; Stocker et al. 1983), whereas taste information bypasses the brain proper and rather is sent to various target regions in the suboesophageal ganglion (Colomb et al. 2007; see later) (Fig. 1).

2.3 Odorant Receptors

Homology-based screens for fly genes resembling vertebrate *Or* genes had failed for many years. *Drosophila* ORs were ultimately detected by searching for a family of seven-transmembrane-domain proteins that are selectively expressed in ORNs (Clyne et al. 1999; Gao and Chess 1999; Vosshall et al. 1999). The *Drosophila Or* gene family thus identified comprises 60 genes which encode 62 ORs (Robertson et al. 2003) (Fig. 2). Although fly ORs are characterized by seven-transmembrane domains like their mammalian counterparts (Buck and Axel 1991), the two families are not homologous. Indeed, the membrane topology of fly ORs appears to be inverted relative to that of other ORs (Benton et al. 2006), a feature whose functional implications will have to be elucidated (Sato et al. 2008; Wicher et al. 2008).

Fly ORs fall into two distinct classes: conventional, ligand-specifying ORs and the atypical OR83b, which is expressed in the large majority of adult ORNs and all 21 larval ORNs (Benton et al. 2006; Larsson et al. 2004; Vosshall et al. 1999) (Fig. 2). OR83b is an obligatory coreceptor that associates with the conventional ORs; the OR/OR83b complex is then targeted to the dendrite (Benton et al. 2006). By contrast, the conventional, ligand-specifying ORs are expressed each in a specific subpopulation of ORNs in the antenna or palp and/or the larval dorsal organ (Clyne et al. 1999; Couto et al. 2005; Fishilevich et al. 2005; Gao and Chess 1999; Goldman et al. 2005; Kreher et al. 2005, 2008; Vosshall et al. 1999). As shown by RNA in situ hybridization, 48 of these ORs are detected in adults and 25 in the larvae (Fig. 2). Twelve ORs are expressed in both larva and adult, while the rest are specific for their stage (Fishilevich et al. 2005). Why a given ORs is expressed at a given stage or in a particular olfactory organ or sensillum type is not understood. In general, each ORN expresses only one ligand-specifying OR, but there are at least seven documented cases of OR coexpression (Couto et al. 2005; Fishilevich and Vosshall 2005; Goldman et al. 2005) (Fig. 2).

Imaging of ORNs expressing a given OR or recording from 'empty' adult ORNs in which single ORs were misexpressed allows the range of ligands that can act via each OR to be identified and ORs to be assigned to a specific class of ORN and sensillum type (Dobritsa et al. 2003; Goldman et al. 2005; Hallem and Carlson 2006; Hallem et al. 2004; Kreher et al. 2005, 2008; Pelz et al. 2006) (Fig. 2). Thus, ORs expressed in antennal and palp basiconic ORNs tend to be strongly activated by general food odours (Goldman et al. 2005; Hallem and Carlson, 2006) (Fig. 2). Larval ORs are tuned either to aromatic or aliphatic food components (Kreher et al.

2005, 2008). Distinct from these receptors, four ORs expressed in two subtypes of trichoid sensilla respond to pheromonal components, such as the aggregation pheromone 11-*cis*-vaccenyl acetate (Ha and Smith 2006; Kurtovic et al. 2007; van der Goes van Naters and Carlson 2007). Recently, a CD36-related receptor has been identified as a putative cofactor of ORs for pheromone detection (Benton et al. 2007). ORNs in coeloconic sensilla are mainly tuned to amines, ammonia and putrescine (Yao et al. 2005), but the identity of most of their ORs remains to be discovered. A very distinct receptor arrangement is found in a subset of antennal basiconic ORNs that are specialized for the detection of CO_2. Notably, detection of this particular substance requires neither OR83b nor any of the conventional ORs, but the coexpression of the gustatory receptor genes *Gr21a* and *Gr63a* (Benton et al. 2006; Jones et al. 2007; Kwon et al. 2007; Suh et al. 2004) (Fig. 2).

2.4 Target Glomeruli of Odorant Receptors

The groundbreaking discovery in mice that ORNs expressing the same OR converge upon discrete glomeruli in the olfactory bulb (Mombaerts et al. 1996; Ressler et al. 1994; Vassar et al. 1994) prompted researchers to ask whether the fly uses the same logic of connectivity. Given that the adult olfactory system does its job with about 50 ORs and about 50 antennal lobe glomeruli (Couto et al. 2005; Laissue et al. 1999), *Drosophila* is a particularly suitable model for studying the principles of ORN wiring at the cellular level. Indeed, fly ORNs expressing a given OR were shown to target one glomerulus or exceptionally two glomeruli (Gao et al. 2000; Vosshall et al. 2000), which allowed the establishment of an almost complete OR-to-glomerulus map (Couto et al. 2005; Fishilevich and Vosshall 2005) (Fig. 2). This map comprises 46 different ORs; it assigns glomerular identity to every antennal basiconic and trichoid ORN, every palp ORN and provides indirect evidence on eight glomeruli that are targeted by ORNs from coeloconic sensilla. Six particular glomeruli deserve special attention. Three lateral, large glomeruli may be implicated in mating behaviour, because they are innervated by neurons that express *fruitless*, a gene which is involved in shaping the circuitry of male courtship (Manoli et al. 2005; Stockinger et al. 2005). Two of these glomeruli are good candidates for processing pheromonal cues, as they are targets of trichoid sensilla (see earlier) and are larger in males than in females (Kondoh et al. 2003; Stockinger et al. 2005). Furthermore, the most ventral glomerulus in the antennal lobe comprises the terminals of the CO_2-sensitive ORNs which coexpress *Gr21a* and *Gr63a* (see earlier). Interestingly, prolonged exposure to CO_2 induces a reversible volume increase in this glomerulus (Sachse et al. 2007). Finally, as shown by previous studies, two other glomeruli are the targets of six putative thermosensory or hygrosensory neurons in the featherlike antennal 'arista' (Foelix et al. 1989; Lienhard and Stocker 1987). In the present map, the identity of the innervating sensory neurons remains unknown for two glomeruli only. Interestingly, one of the target glomeruli of coeloconic ORNs (Vosshall and Stocker 2007), as well as the target glomerulus of the CO_2-sensitive neurons and the two aristal target glomeruli receive innervation exclusively from the ipsilateral antenna; almost all other glomeruli are bilaterally innervated (Stocker 1994).

A number of conclusions can be drawn from this sensory map (Couto et al. 2005; Fishilevich and Vosshall 2005; Hallem and Carlson 2006; Stocker 1994):

1. The afferents from the antenna and the palp segregate into different glomeruli, suggesting an ability to distinguish between the two types of signals.
2. The majority of glomeruli receive bilateral inputs; however, a group of five ventral glomeruli are exclusive targets of the ipsilateral antenna.
3. Most of the glomeruli appear to be responsive to a variety of odorants, i.e. those recognized by their proper OR. These types of glomeruli are very likely involved in the processing of food odours.
4. Other glomeruli may accomplish more specialized functions, as suggested by the putative pheromone glomeruli, the CO_2 glomerulus and the two aristal glomeruli.
5. Target glomeruli of basiconic, trichoid and coeloconic sensilla tend to cluster in different areas of the antennal lobe.
6. At least seven glomeruli are targeted by two types of OR, owing to coexpression in the corresponding ORNs. One particular OR is coexpressed with either of two different ORs in two types or ORNs; accordingly it has two target glomeruli.
7. A possible chemotopic arrangement of glomeruli, i.e. a clustering of glomeruli that are activated by similar odours, remains controversial (Couto et al. 2005; Fishilevich and Vosshall 2005; Hallem and Carlson 2006).

The functional significance of many of these observations will have to be shown.

2.5 Central Olfactory Pathway in Adult Flies

The odour information that each glomerulus receives from its corresponding ORNs is significantly processed in the antennal lobe, regarding both quantitative and qualitative parameters, such as detection threshold and odour discrimination, respectively.

The two major target neurons of the ORNs are local interneurons, which interconnect many or even all glomeruli, and projection neurons, which mostly link single glomeruli with higher olfactory centres, the mushroom bodies and the lateral horn (Stocker 1994) (Figs. 1, 4). Many of the local interneurons are GABAergic (Wilson and Laurent 2005). They receive excitatory input from ORNs and – via recurrent synapses – from projection neurons and establish inhibitory synapses with both ORNs and projection neurons. A possible role of this inhibitory network may be to synchronize projection neuron activity, within a given glomerulus and/or between projection neurons innervating different glomeruli (Ng et al. 2002). Recently, a second class of cholinergic, excitatory local interneurons was identified (Olsen et al. 2007; Shang et al. 2007). These neurons likely provide the substrate for another long-known property of projection neurons, their significantly broadened odour

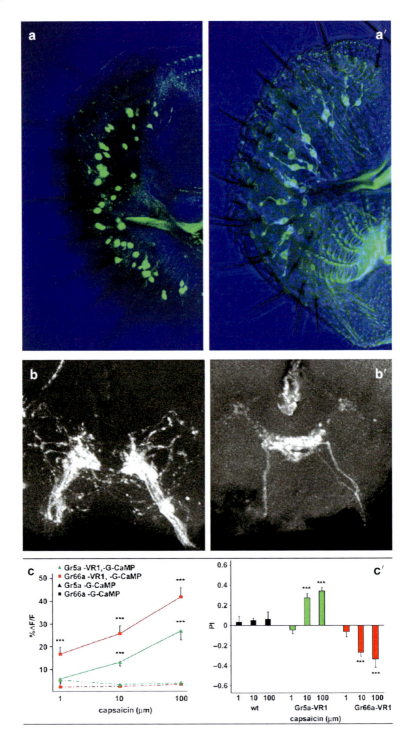

tuning compared with ORNs (Ng et al. 2002; Wilson et al. 2004). Although projection neurons get their major input from ORNs that project to their "own" glomerulus (Root et al. 2007), excitatory local interneurons allow projection neurons to respond to signals from ORNs that target neighbouring glomeruli.

The 'odour image' (Laurent 1996) represented by patterned temporal and combinatorial activity of an estimated 150 projection neurons (Stocker et al. 1997) is then transferred onto third-order neurons in the mushroom bodies and the lateral horn (Figs. 1, 4). These two higher centres are thought to control distinct olfactory functions. The mushroom bodies represent key regions for olfactory learning (see later), whereas the lateral horn appears to be involved in innate odour recognition (de Belle and Heisenberg 1994; Heimbeck et al. 2001; Jefferis et al. 2007; Tanaka et al. 2004). Uniglomerular projection neurons establish terminals in both of these centres. Their output synapses onto the about 2,500 intrinsic mushroom body neurons (the Kenyon cells) are located in the so-called calyx (Crittenden et al. 1998; Ito et al. 1997; Lee et al. 1999; Strausfeld et al. 2003; Yasuyama et al. 2002) (Fig. 1). Projection neurons deriving from specific glomeruli were shown to establish synaptic boutons preferentially in moderately specific, relatively broad zones of the calyx (Jefferis et al. 2007; Lin et al. 2007; Tanaka et al. 2004). Consistent with this observation, odour stimulation evokes spatially distinct, stereotyped activity in the calyx (Fiala et al. 2002; Wang et al. 2004a). Calycal zones were reported to correspond to the clonally and developmentally segregated dendritic arborizations of five Kenyon cell subtypes (Lin et al. 2007). Comparing the projection neuron-to-Kenyon cell map with electrophysiological data from ORNs (Hallem and Carlson 2006; Wilson et al. 2004) reportedly suggests that Kenyon cell responses in the different zones may be correlated with chemical classes of odour (Lin et al. 2007) (see, however, Murthy et al. 2008).

A prominent feature of calycal connectivity is that projection neurons synapse onto multiple Kenyon cells, and that Kenyon cells receive input from multiple projection neurons, generating an intricate local divergence–convergence network. Accordingly, Kenyon cells may act as coincidence detectors, which integrate the

◄──

Fig. 3 Two receptor genes, *Gr5a* and *Gr66a*, are expressed in different subsets of gustatory receptor neurons of the labellar chemosensilla; their axons project to separate regions in the suboesophageal ganglion; their activation induces attractive or aversive behaviour, respectively. (**a**) *Gr5a* expression at the labellum as approximated by GFP expression from the promoter-Gal4 line *Gr5a*-Gal4. GFP-expressing neurons are observed in all taste sensilla. While s-type sensilla have only one GFP-positive neuron, half of the l-type sensilla have more than one GFP-positive neuron. (**a'**) GFP expression from the *Gr66a*-Gal4 strain. One GFP-positive neuron is observed per s- and l-type sensillum. (**b**) Projection patterns of *Gr5a*-Gal4-positive and of *Gr66a*-Gal4-positive neurons (**b'**) in the suboesophageal ganglion. (**c**) When a capsaicin receptor (VR1) is transgenically expressed in either the *Gr5a*-Gal4 or the *Gr66a*-Gal4 pattern, application of capsaicin can drive the respective neurons; without the transgene, no such activation is found. (**c'**) If animals expressing VR1 in the *Gr5a*-Gal4 expression pattern are presented with capsaicin, flies prefer capsaicin, whereas if VR1 is expressed in the *Gr66a*-Gal4 expression pattern, flies avoid capsaicin; control genotypes are behaviourally indifferent towards capsaicin. (a, a', b, b' copyright T. Inoshita and T. Tanimura, Kyushu University. c, c' from Marella et al. 2006, copyright Elsevier)

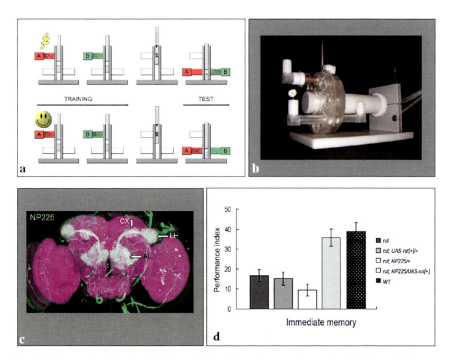

Fig. 4 Olfactory learning in adult *Drosophila*. **a** Learning experiments in adult *Drosophila* use a T-maze. Either electric shock as punishment (upper part) or sugar as reward (lower part) can be used as a reinforcer. In either version, about 100 flies are put in a training tube (red) where one odour (A) is paired with reinforcement. Then, a second odour (B) is applied alone (green) (a second group of flies is trained reciprocally, i.e. odour A is presented without reinforcement and odour B is presented with reinforcement). Finally, the flies are forced into an elevator and moved to a choice point where both odours are presented. Counting the number of flies on either side and comparing the distribution between reciprocally trained groups allows the learning index to be calculated. **b** The 'revolver' device for measuring learning with high throughput; the apparatus is partially disassembled to highlight the training tube (top) and the two testing tubes (bottom). **c** Expression pattern of the Gal4 line NP225 visualized by driving UAS-mCD8::GFP (green); anti-Synapsin staining as a neuropile marker (magenta). About 75 projection neurons innervating 35 glomeruli in the antennal lobe (AL) and projecting to both the mushroom body calyx (CX) and the lateral horn (LH) are labelled. **d** rut^{2080} mutants are impaired in appetitive olfactory learning. Both control genotypes (both in the mutant genetic background: rut^{2080}; UAS-rut and rut^{2080}; NP225 Gal4) also show the memory impairment. If in the mutant genetic background the *rutabaga* cDNA is expressed in the NP225-Gal4 pattern, learning scores are restored to wild-type (WT) level. (**a**, **c** Copyright A.S. Thum, University of Fribourg. **b** Copyright A. Yarali, Universität Würzburg. **d** Modified from Thum et al. 2007, copyright Society for Neuroscience)

odour information carried by parallel channels of projection neurons (Heisenberg 2003; Perez-Orive et al. 2002; Wang et al. 2004a). The activity patterns across the Kenyon cells are then read out by relatively few classes of mushroom body output neurons (Ito et al. 1998; Tanaka et al. 2008), which ultimately activate specific motor channels. Yet, exactly how this odour-to-behaviour switchboard at the mushroom body output is organized remains a mystery.

Regarding the connectivity in the lateral horn, the most striking feature observed is a segregation of terminals between putative pheromone-representing projection neurons – which get their inputs from the two putative pheromone glomeruli (see earlier) – and 'normal' projection neurons (Jefferis et al. 2007; Schlief and Wilson 2007). Interestingly, the candidate pheromone region in the lateral horn receives both excitatory and inhibitory signals from the two glomeruli, the former via cholinergic projection neurons and the latter via GABA-positive projection neurons (Jefferis et al. 2007). Balanced excitation and inhibition of these pathways may allow lateral horn neurons to mediate behavioural alternatives, depending on the attractive or repulsive nature of the pheromone. Not unexpectedly, sexually dimorphic connectivity has been observed in the pheromone region (Datta et al. 2008). Normal projection neurons, which are mostly activated by food odours, establish stereotypic, largely overlapping patterns of terminals (Jefferis et al. 2007; Marin et al. 2002; Wong et al. 2002). Thus, information about food appears to become intensely integrated across antennal input channels within the mushroom body and the lateral horn, whereas pheromones may be signalled via discrete channels all the way from the sensory periphery to the lateral horn. This may correspond to the evolutionarily fixed and discrete behavioural 'meaning' of different pheromones, as contrasted with a requirement for an integrated sensory–motor switchboard for general odours.

A number of putative third-order neurons have been identified in the lateral horn (Jefferis et al. 2007; Tanaka et al. 2004); however, their roles in mediating odour-driven behaviours are hard to predict, because their dendritic fields overlap with many classes of projection neurons and because they target different brain areas. Establishing a complete neuronal circuit diagram of such behavioural programmes will require the identification of the as yet unknown downstream premotor and motor neurons as well as of biologically meaningful behavioural 'modules'.

2.6 Larval Olfactory Pathway

The larval central olfactory pathway largely shares the layout and the types of neurons of its adult counterpart, but is much simpler in terms of cell numbers (Python and Stocker 2002). Similar to the situation in adults, the 21 larval ORNs target single glomeruli in the antennal lobe (Figs. 1, 2). However, larval glomeruli do not represent sites of ORN convergence. Rather, every ORN (each expressing its proper OR) has its own glomerulus among a total of 21 glomeruli (Fishilevich et al. 2005; Kreher et al. 2005; Ramaekers et al. 2005) (Figs. 1, 2). Recently, Kreher et al. (2008) reported that the similarity in ORN activation pattern allows a partial prediction of behavioural odour similarity, based on masking experiments. In any event, as in the adult, local interneurons establish horizontal connections between glomeruli (Ramaekers et al. 2005) and most of the larval projection neurons are of the uniglomerular type (Marin et al. 2005; Ramaekers et al. 2005). Each glomerulus appears to be innervated only by one or a few projection neurons (Ramaekers et al. 2005), suggesting that their total number may not be much higher than the number of glomeruli.

Similar to the situation in the adult, projection neurons target both the mushroom bodies and the lateral horn (Fig. 1). The lateral horn circuitry has not been addressed so far. Studying the output connectivity of larval projection neurons in the mushroom body was simplified by the fact that the larval calyx comprises about 30–40 relatively large, identifiable structures, called 'calyx glomeruli' (Marin et al. 2005; Masuda-Nakagawa et al. 2005; Ramaekers et al. 2005). Projection neurons choose mostly single, exceptionally two calyx glomeruli as targets (Marin et al. 2005; Ramaekers et al. 2005). Each of them is innervated by only one or a few projection neurons. Many of these neurons were shown to each stereotypically connect a specific antennal lobe glomerulus with a specific calyx glomerulus (Ramaekers et al. 2005).

While some of the larval Kenyon cells innervate a single calyx glomerulus (Ramaekers et al. 2005), many establish dendritic arbours in usually six, apparently randomly selected glomeruli (Masuda-Nakagawa et al. 2005; Murthy et al. 2008). Thus, as in adults, projection neurons diverge onto multiple Kenyon cells, and most Kenyon cells receive input from multiple projection neurons, again providing a local divergence–convergence connectivity (Masuda-Nakagawa et al. 2005; Murthy et al. 2008). Finally, it should be noted that the two types of Kenyon cells, i.e. the ones receiving input in one or multiple calyx glomeruli, may allow different modes of signal transfer, acting either in elementary odour coding or as coincidence detectors for interpreting combined activity (Heisenberg 2003; Perez-Orive et al. 2002).

In conclusion, the general organization of the central olfactory pathway in the larva is similar to that in the adult (Fig. 1) and still shares the essential layout of the vertebrate olfactory system. Yet, the larval circuit displays a number of specific properties. Firstly, every larval ORN and probably most of the larval projection neurons are unique (Ramaekers et al. 2005). Any loss of these cells should theoretically affect olfactory function more severely than in the adult system. However, silencing of single or even multiple ORNs had surprisingly little effect on larval odour-driven behaviour, implying that the ligand receptive ranges of the different ORs must be largely overlapping (Fishilevich et al. 2005). Secondly, the presence of only 21 antennal lobe glomeruli suggests that the number of primary olfactory dimensions is reduced in the larva compared with adult flies comprising about 50 glomeruli. Thirdly, given that the numbers of ORNs, antennal lobe glomeruli, projection neurons and calyx glomeruli are almost the same, the larval olfactory pathway lacks convergent and divergent connectivity up to the mushroom bodies (Ramaekers et al. 2005) (Fig. 1). This contrasts with the adult olfactory circuit, in which 1,300 ORNs converge onto 50 glomeruli, which diverge again to an estimated 150 projection neurons, each of which innervates a broad zone of the calyx. The lack of cellular redundancy, the reduced number of primary olfactory dimensions and the lack of convergent connectivity in the antennal lobe are likely to reduce the signal-to-noise ratio. Hence, larvae can be expected to be relatively poorer in odour discrimination than adult flies. Yet, for the simple discrimination tasks of a substrate feeder this may not be a too serious drawback. Given its mere 21 primary olfactory dimensions, the larva is an ideally suited comprehensive model for analyzing the translation of olfactory input into behavioral output (Kreher et al. 2008; Hoare et al. 2008).

3 Tasting

After having been tracked down using visual and/or olfactory cues, contact chemosensation serves to immediately handle things in physical contact with the animal. In insects it contributes to diverse behaviour functions: selection of oviposition sites in butterflies (Feeny et al. 1983); kin and/or nestmate recognition to support nepotism in ants (Ozaki et al. 2005); and the pursuit of courtship (Ferveur 2005; Lacaille et al. 2007). Most obviously, however, contact chemosensation organizes eating and drinking behaviour (the 'taste' system). Here, we chose not to mention much of the fascinating biology of contact chemosensation in insects and restrict ourselves largely to how taste function is organized in *Drosophila*.

To start with the most striking difference to vertebrates, gustatory receptor neurons in *Drosophila*, as in insects in general, are primary sensory neurons, in contrast to the situation in vertebrates where taste cells originate from the epidermis and only are innervated by neurons. Also, the gene family coding for sugar- and bitter-sensitive gustatory receptor proteins is not apparently homologous to the functionally corresponding gene family in vertebrates (Clyne et al. 2000; Robertson et al. 2003; Scott et al. 2001); this, as in the case for olfaction, argues for quite some degree of evolutionary divergence in the chemosensory systems of insects versus vertebrates. Still, if even under such conditions functional similarities are found, these may be particularly good hints towards common functional constraints on taste processing.

Indeed, there are a number of similarities. The taste system of *Drosophila* seems to categorize sensory inputs into relatively few modalities, including sweet, salt and bitter. Regarding sweet, this is achieved by coexpression of gustatory receptor proteins with distinct sugar-ligand profiles within the same gustatory sensory neuron; a similar architecture likely applies to bitter as well. Sour taste may be detected by a depression of the sugar response. In addition, water-sensitive neurons are included in many taste organs, which in vertebrates is not typically the case. Still, the relatively few dimensions of taste, in particular the lower dimensionality of taste compared with olfaction, and the logic of coexpression of multiple gustatory receptor genes in a given sensory neuron for either sweet or bitter taste seem to generally conform with the situation in vertebrates.

The taste sensilla of *Drosophila* are cuticular, hairlike structures with a single pore at the tip into which two or four gustatory receptor neurons send their dendrites (Falk et al. 1976; Ishimoto and Tanimura 2004). Taste sensilla typically include, in addition to three non-neuronal cells with homeostatic function, also one mechanosensory neuron, serving to integrate the 'what' with the 'where'. Thus, taste is most closely entangled with touch, also in development (Awasaki and Kimura 1997). Such organization is similarly found in mammals, as taste neurons are grouped into taste 'buds', as taste and touch sensory neurons are intermingled on the tongue, and as gustatory and somatosensory cortex are closely entangled functionally (Kaas 2005).

Taste sensilla can be found both on multiple external sense organs, used to probe the environment before ingestion, and at internal sense organs, used to monitor the quality of already ingested food (Fig. 1); such an architecture is also seen in mammals. The projections from both kinds of organ typically bypass the brain proper, and rather send their axons to the suboesophageal ganglion; here, in concert with centrifugal interneurons, ingestion behaviour is thought to be organized. This triad of contact-chemosensory input, ingestion-related motor output and central motivating factors comprises the 'taste system' of *Drosophila*. How does this system work?

3.1 Adult

Adult flies taste with their 'feet' (i.e. tarsi), with their 'tongue' (i.e. labellum), with taste neurons along their pharynx, and additionally have contact chemosensory neurons at their wing margin (Stocker 1994; Singh 1998) (Fig. 1). The external taste organs of the adult comprise hair-shaped sensilla and conically shaped pegs.

After initial contact of a tastant via the tarsi, flies initiate extension of the proboscis (such proboscis extension can also happen by direct stimulation of labellar sensilla). This brings the tastant into contact with labellar sensilla and makes the flies open the labellar lobes. This exposes a set of taste pegs buried in the ridges of the opened labellum to the tastant (Fig. 1), to finally trigger ingestion. The quality of ingested food can then be monitored with three different taste organs along the pharynx (i.e. labral, ventral cibarial and dorsal cibarial sense organ).

Inputs from labellum, pharynx and the tarsi are collected in the suboesophageal ganglion, where taste information is integrated with centrifugal interneurons to organize ingestion behaviour.

3.1.1 Taste Neurons at Tarsi

On the fore-, mid-, and hindlegs, contact chemosensilla are located on the tarsal segments (Fig. 1). On the forelegs, the numbers of such electrophysiologically confirmed sensilla differ between sexes (females have 18 sensilla, males have 28) (Meunier et al. 2000; Meunier et al. 2003), likely related to males using their forelegs to touch the female abdomen just before copulation. On the basis of their electrophysiological profiles, the tarsal sensilla are classified into A- B- and C-type sensilla (see later).

Different from the situation on the labellum (see later), not all tarsal sensilla contain a sugar-sensitive sensory neuron (Meunier et al. 2000). Also, the response spectra of the tarsal sensory neurons differ from those of labellar sensilla. That is, labellar contact chemosensory neurons are classically called S, W, L1 or L2 neurons (Ishimoto and Tanimura. 2004), on the basis of their electrophysiological response spectra. The L neurons are activated by salt, in the case of L1 with a low threshold (typically no electrophysiological responses are seen for concentrations below 10 mM; Fujishiro et al. 1984) and in the case of L2 neurons with a high threshold; L2 neurons can in some cases also be activated by bitter compounds.

W cells are activated by pure water, a response which typically can be inhibited by high osmolarity, i.e. higher than a few hundred millimolar sugars and salts. S cells are activated by sugars and can be inhibited by bitter substances. Are these kinds of cells also found on the tarsi?

For the tarsal A-type sensilla (Meunier et al. 2000), spikes from S, W and L1 cells can be discerned, and the water response can inhibited by high osmolarity. In B-type sensilla, responses to sugars are observed only phasically, during 100–200 ms after stimulation, while otherwise such responses are more sustained. Also, in the B-type sensilla, the W cell is not inhibited by high osomolarity. In C-type sensilla, only a W cell has so far been identified electrophysiologically, which however is not inhibited by high osmolarity. As the neurons in a number of tarsal sensilla do not respond to any compound examined so far, they cannot be classified yet as A-, B- or C-type. Furthermore, there also are bitter-sensitive neurons on the tarsi, as the proboscis extension reflex to sugars can be suppressed by bitter compounds applied to the tarsi. Indeed, electrophysiological studies revealed that in specific tarsal sensilla there are L2 neurons which do respond to bitter compounds (Meunier et al. 2003). Interestingly, the initiation of spikes in these bitter-sensitive neurons has a delay of up to 200 ms; this delay is shortened as the concentration of the bitter substance increases. Strikingly, the same long delay is seen for the inhibitory effect of bitter substances on the S and W cells, even in sensilla that do not contain an L2 neuron. Similarly long delays likely are typical for electrophysiological bitter responses in labellar sensilla as well. One explanation for these long latencies may be that bitter compounds need to diffuse *into* the receptor cells to activate receptor sites; given that many bitter substances are hydrophobic, such a process may take some time. To summarize, tarsal chemosensilla are special in quite some respects.

3.1.2 Taste Neurons at Labellar Taste Pegs

The labellum of *Drosophila* is decorated on its bottom surface with numerous ridges called pseudotrachea (Shanbhag et al. 2001). Along these pseudotrachea, multiple rows of a total of about 30–40 taste pegs are located (Fig. 1). When flies are sucking liquid food, the fluid passes this pseudotracheal ridge system, which eventually merges to the actual mouth opening. Thus, the taste pegs are guideposts for ingestion, triggering pumping behaviour. Taste pegs are distinct from regular taste sensilla in that each taste peg is innervated by only one gustatory sensory neuron together with one mechanosensory neuron. The number of taste pegs is variable among individuals and differs between sexes. Interestingly, in *poxn* mutants (CG8246, *poxn*$^{70-23}$; Awasaki and Kimura 1997) only the neurons in the external taste sensilla, but not in the internal taste sensilla and labellar pegs, are transformed into mechanosensory neurons (Awasaki and Kimura 1997). In such mutants, sugar still can enhance and bitter substances can still reduce food uptake. This suggests that chemoreception via internal taste organs and/or taste pegs provides the necessary information for these kinds of behaviour effect. No electrophysiological studies have been performed on the respective taste sensory neurons, as it is difficult to fixate the labellum such that the taste pegs in the pseudotracheal ridges, not to mention the internal sense organs, are accessible (but see Dethier and Hanson 1964 for such recordings in the blowfly).

However, with transgenic techniques hints towards the function of the taste pegs could be obtained. With use of the enhancer trap line E409, which supports transgene expression in the taste pegs (and in the mushroom bodies), a novel functional class of gustatory sensory neurons, distinct from *Gr5a*-Gal4- and *Gr66a*-Gal4-positive neurons (see later) was identified (Fischler et al. 2007). With use of Ca^{2+} imaging at the target region of these neurons in the suboesophageal ganglion, it was found that they can be activated by carbonated water. Behaviourally, stimulation with carbonated water elicits feeding. Silencing the E409-positive neurons abolishes these behavioural responses, whereas driving these cells (by means of ectopically expressing a capsaicin receptor and then stimulating with capsaicin) triggers proboscis extension (note, however, that airborne CO_2 is a repellent for *Drosophila*; Faucher et al. 2006; Suh et al. 2004).

3.1.3 Taste Neurons at Labellar Sensilla

On the labellum, there are 31 contact chemosensilla, each containing two to four gustatory and one mechanosensory neuron. They are classified into three types (Shanbhag et al. 2001; Hiroi et al. 2002). The s-type sensilla are short, house four gustatory sensory neurons and are located near the opening of the labellum. The i-type sensilla are intermediate in size and contain two gustatory sensory neurons; they are located mostly on the anterior and posterior part of the labellum. The l-type sensilla are long, possess four gustatory sensory neurons and are located such that they can contact the substrate even when the labellum is closed, suggesting a role in initiating labellar opening. Each individual labellar sensillum can be identified across subjects by its specific location. The axons from all labellar gustatory sensory neurons project towards the suboesophageal ganglion.

Drosophila possesses nine l-type sensilla. They house the W, S, L1 and L2 neurons, classified on the basis of their electrophysiological characteristics (Fujishiro et al. 1984). Activation of W, S and L1 cells can trigger ingestion, while activity in L2 neurons inhibits it. Bitter substances, which – just as high NaCl concentrations – can activate L2 neurons in i- and s-type sensilla (see later), do not do so in L2 neurons of l-type sensilla; whether and which non-NaCl compounds might stimulate these cells remains to be investigated. In all l-type sensilla, sugar responses can be inhibited by bitter compounds.

Neither of the two gustatory sensory neurons in the nine to ten i-type sensilla is water-sensitive (Hiroi et al. 2004). One type of neuron responds to sugar as well as to NaCl with low threshold. Given that low NaCl concentrations are behaviourally attractive, these cells would seem to indiscriminatively report 'edible'. The other neuron type responds to NaCl with high threshold, as well as to bitter compounds. As both kinds of substance are potentially toxic, these cells seem to indiscriminatively report 'non-edible'. Interestingly, in the so far examined i-type sensilla on the proboscis this type of neuron also responsd to the pheromone (Z)-7-tricosene at subnanomolar concentration (Lacaille et al. 2007). This compound from the male cuticle, just as the bitter substances which activate these cells, inhibits male–male courtship. Furthermore, if one leg is stimulated with sugar, both kinds of compound can inhibit proboscis extension when applied to the other leg. Finally, adaptation to (Z)-7-tricosene reduces subsequent electrophysiological responses also to bitter substances, collectively suggesting that both kinds of stimulus may use the same input channel.

Finally, regarding the 12–13 s-type sensilla, electrophysiological recordings are scarce. Hiroi et al. (2002) reported responses to sucrose and other sugars from a few sensilla, but obviously the range of activating compounds for the four gustatory sensory neurons in these sensilla still needs to be examined in more detail.

3.1.4 Taste Neurons at Pharyngeal Taste Organs

There are five internal sense organs along the pharynx: the labral sense organ, the ventral and dorsal cibarial sense organs, the ventral sense organ and one dorsal row of 'fishtrap' bristles (Fig. 1). Whereas most of the neurons in these organs may be gustatory, the monoinnervated fishtrap bristles and many of the neurons of the labral sense organ appear to be mechanosensory (Nayak and Singh 1983). Interestingly, most of these sensory neurons have persisted from the larval period (Gendre et al. 2004), suggesting some persistence of taste function between larva and adult. In any event, for all pharyngeal sensilla, the compounds to excite these neurons and the behaviours relying on their input are unknown.

3.1.5 Receptor Genes: Sweet and Bitter

At present, the functional architecture of gustatory receptor gene expression, in particular for sweet and bitter, is being unravelled; however, the functional configuration of these receptors as monomer, dimers or oligomers is not clear, and neither are their downstream intracellular signalling cascades, the transmitter used by their host sensory neurons and the precise connectivity of these cells to second-order interneurons.

Recently, an about 60-member family of putative gustatory receptor genes was found (the *Gr* family; Clyne et al. 2000). Promoter-Gal4 strains are widely used to approximate their expression patterns and ligand profiles. We largely restrict our discussion to the three best understood *Gr* genes, namely *Gr5a*, *Gr64a-f* and *Gr66a* (Fig. 3). This ignores the richness of *Gr* processing, in particular with regard to the emerging understanding of pheromone function including the role of ligand-binding proteins in this respect (Shanbhag et al. 2001; Park et al. 2006; Matsuo et al. 2007).

Gr5a codes for the trehalose receptor identified earlier on the basis of classical genetics and electrophysiology (Tanimura et al. 1982, 1988; Dahanukar et al. 2001; Ueno et al. 2001). As seen in Fig. 3, it is expressed in all S cells of all three sensillar types on the labellum (Wang et al. 2004b). Central projections of the *Gr5a*-Gal4-

positive neurons target the suboesophageal ganglion in a relatively lateral and anterior region (Wang et al. 2004b). In addition, a subset of contact chemosensilla on the legs is included in the *Gr5a*-Gal4 expression pattern, which send projections to their cognate thoracic ganglion (Wang et al. 2004b). Notably, ectopic expression in cultured cells combined with Ca^{2+} imaging reveals that this protein specifically binds to trehalose at micromolar ranges (Chyb et al. 2003). In vivo, *Gr5a*-Gal4-positive neurons are activated by all sugars tested (arabinose, fructose, galactose, glucose, maltose, sucrose, trehalose) and reportedly also by NaCl with low (10 mM) threshold, but not by bitter compounds (caffeine, denantonium) (Marella et al. 2006). Strikingly, in null mutants for *Gr5a*, phenotypes are more specific: electrophysiological responses are abolished for only four out of 14 sugars tested (trehalose, methyl-α-glucoside, glucose, melezitose) (Dahanukar et al. 2007); this defect can be restored by expression of the Gr5a protein from a *Gr5a*-Gal4 driver. Preliminary data (Slone et al. 2007; Jiao et al. 2007) may suggest that *Gr5a* and members of the *Gr64b-f* gene group produce dimers for trehalose detection; indeed, Dahanukar et al. (2007) propose concordant expression of *Gr64f* and *Gr5a*.

As the Gr5a protein is dispensable for the electrophysiological responses to many sugars, but as *Gr5a*-Gal4-positive neurons have a broader activation profile than the requirement of the Gr5a protein suggests, one wonders which other members of the *Gr* gene family, expressed in the same set of neurons, might be responsible for this discrepancy. Deleting the *Gr64a* gene abolishes (maltotriose, stachyose, raffinose, leucrose, fructose) or partially reduces (sucrose, maltose, turanose, maltitol, palatinose) the *Gr5a*-independent effects (Dahanukar et al. 2007). Rescue expression of *Gr64a* driven by *Gr5a*-Gal4 notably restores these deficits. All electrophysiological responses to all the sugars tested were fully abolished in *Gr5-Gr64a* double mutants. Most importantly, behavioural analyses using the proboscis extension response conform with the complementary requirement of *Gr5a* and *Gr64a* for detecting different kinds of sugars; if both genes are deleted, proboscis extension responses to all sugars tested (note that this analysis did not include trehalose), but not towards very low concentration NaCl (5 mM), are fully abolished (Dahanukar et al. 2007). Consistently, if the *Gr5a*-Gal4-positive neurons are disabled by transgenic toxin expression (Wang et al. 2004b), proboscis extension to all sugars tested (trehalose, low [5 mM] concentration sucrose, glucose) as well as to very low NaCl concentrations is abolished, but the suppression of high [100 mM] concentration sucrose responses by added high-concentration NaCl or bitter compounds remains intact; this is consistent with *Gr5a* and *Gr64a* being expressed in the same set of cells.

The complementary involvement of *Gr5a* and *Gr64a* for detecting different kinds of sugar, together with their proposed concordant expression (Dahanukar et al. 2007), suggests they act as independent sensors within the same cell. Such an architecture is a good example for the functional logic of the taste system: it is as if differential behaviour to both classes of sugar were deliberately precluded.

In contrast, *Gr66a*-Gal4-positive neurons seem to be devoted to processing 'bad' (Moon et al. 2006). The *Gr66a*-Gal4 pattern covers one neuron each in the i- and s-type labellar sensilla (Fig. 3), and several sensilla on the legs; in all cases, these cells are non-overlapping with *Gr5a*-Gal4 (Wang et al. 2004b; Dahanukar et al. 2007). Different from *Gr5a*-Gal4, *Gr66a*-Gal4 also shows expression in the pharyngeal sense organs (Wang et al. 2004b). The central projections of *Gr66a*-Gal4-positive versus *Gr5a*-Gal4-positive neurons also are non-overlapping, in that *Gr66a*-Gal4-positive neurons from the labellum project to more posterior and medial portions of the suboesophageal ganglion. Finally, projections from the legs reportedly target the suboesophageal ganglion in the case of *Gr66a*-Gal4, but the thoracic ganglia in the case of *Gr5a*-Gal4 (Wang et al. 2004b). Disabling *Gr66a*-Gal4-positive neurons leaves proboscis extension to all sugars tested (trehalose, sucrose, glucose) as well as to very low (5 mM) NaCl intact; in contrast, these flies cannot suppress proboscis extension to sucrose when bitter compounds (berberine, caffeine, denantonium, quinine) are added to the sucrose solution, whereas such suppression by high concentrations (100–1,000 mM) of NaCl remains intact (Wang et al. 2004b). As shown by in vivo imaging, *Gr66a*-Gal4-positive neurons are activated by these and other bitter compounds (aristolochic acid, azadirachtin, limonin, lobeline, papaverine, quassin), but not by any of the sugars tested (Marella et al. 2006). Furthermore, flies transgenically expressing a capsaicin receptor in the *Gr66a*-Gal4 pattern show avoidance of capsaicin, a substance to which normal flies reportedly are indifferent; in turn, such capsaicin expression in *Gr5a*-Gal4 neurons induces attraction (Marella et al. 2006).

Interestingly, *Gr66a*-Gal4-positive neurons can also be activated by NaCl, with high threshold (above 10 mM) (Marella et al. 2006). This suggests additional expression of a high-threshold NaCl sensor in these cells. As high concentrations of NaCl still are behaviourally active even when these cells are disabled (Wang et al. 2004b), one such salt sensor may well be expressed outside the *Gr66a*-Gal4 pattern. Furthermore, it seems as if genetically defined subsets of *Gr66a*-Gal4-positive neurons were all activated by the same kinds of bitter ligands (Marella et al. 2006), lending at present no support for a functional heterogeneity within these neurons.

3.1.6 Sensor Genes: NaCl

Processing of low and high NaCl concentrations is distinct: (1) low salt concentrations are attractive, but high salt concentrations suppress proboscis extension and lead to avoidance; (2) L1 cells respond best to low concentrations, whereas L2 neurons respond to high concentrations only; (3) *Gr66a*-Gal4 cells are activated only by high concentration. Thus, there must be two kinds of NaCl sensor, one with low threshold, expressed in L1 neurons, and one with high threshold, expressed in L2/*Gr66a*-Gal4 neurons plus possibly in some additional as yet uncharacterized non-*Gr66a*/non-L2 cells. Both processes likely involve discrete molecular sensor mechanisms, as one member of the *pickpocket* (*ppk*) gene family (*ppk11*) is necessary for the behavioural responses to low salt, but is dispensable for the aversive responses to high salt; high NaCl responses may be mediated by another *ppk* gene, *ppk19* (Liu et al. 2003a). The *ppk* gene family is homologous to the vertebrate epithelial Na^+-channel/degenerin gene family (ENaC), different members of which supposedly act as sensors for salt in vertebrates (Lindemann 2001).

3.1.7 Interplay: Combinatorial Coding of Taste?

Strikingly, Marella et al. (2006) report that neurons covered by *Gr5a*-Gal4 are activated not only by sugars, but also by both low (10 mM) and high (1 M) salt concentrations; correspondingly, Wang et al. (2004b) report that disabling *Gr5a*-Gal4-positive neurons abolishes behavioural responses not only to sugars but also to very low (5 mM) concentrations of salt. As it is possible that at least in some sensilla *Gr5a*-Gal4 labels more than one neuron (Inoshita and Tanimura, unpublished results), these data speak to the set of *Gr5a*-Gal4-positive neurons as a whole; that is, it remains unclear whether indeed one and the same cell can be activated by sugars and low and high salt concentrations. Actually, electrophysiological sensilla recordings do not support this notion. Still, if this were so (as is the case in i-type sensilla which house neurons activated by both high salt and bitter, and neurons which are activated by sugars and low salt; Hiroi et al. 2004), a discrimination between these three kinds of tastant would need to rely on combinatorial coding downstream of the gustatory sensory neurons. Obviously, looking at the connectivity towards and the physiological function of gustatory interneurons now is highly warranted. Such studies are still in their infancy, and the few ones available (e.g. Bader et al. 2007; Hammer 1993; Melcher and Pankratz 2005) have so far not addressed the issue of gustatory coding, but rather have focused on the 'valuation' of tastants.

Still, in a completely different sense, combinatorial activity patterns obviously are used by flies, e.g. when combining chemosensory information from various legs to locate a food source, or when monitoring the stage of ingestion by combining taste information from the sense organs located at the various stages of ingestion.

3.1.8 Watery

At present, nothing is known about the molecular mechanism of water sensation. However, the NP1017-Gal4 strain covers sensory neurons likely responsible for watery taste (Inoshita and Tanimura 2006). This strain marks one sensory neuron per taste sensillum in s- and l-type sensilla on the labellum; it expresses Gal4 in taste pegs of the labellar pseudotrachea, as well as in taste sensilla on the tarsi, and in contact chemosensory neurons on the wing margin. If these neurons are disabled, proboscis extension towards water stimulation is severely reduced, but responses to glucose as well as the suppression of proboscis extension by adding high-concentration NaCl to the sugar stimulus remain intact. Ablating these cells abolishes the electrophysiological responses of labellar l-type sensilla to water, but leaves sugar and salt responses in these sensilla intact. The projections of NP1017-Gal4-positive neurons from the labellum target the suboesophageal ganglion; however, the projections from the labellar sensilla and the pegs of the labellar pseudotracheae have distinct target sites in the central versus the lateral anterior region. The projection from labellar NP1017-Gal4-positive neurons overlaps with that from *Gr5a*-Gal4-positive cells, indicating that inputs from water- and sugar-sensing cells may to some extent be funnelled into a common pathway.

3.2 Larva

Larval behaviour towards tastants is very similar to what is observed in adults. Larvae show preference for sugars, avoidance of various bitter substances and dose-dependent responses to salt: at low concentrations, larvae are attracted and at high concentrations they are repelled by NaCl, the concentration of draw being 0.2 M (Miyakawa 1982; Liu et al. 2003a; Niewalda et al. 2008). Interestingly, Miyakawa (1982) reported preference for low-concentration (0.01 M) NaCl remaining intact even in situations where glucose is presented at saturated concentration, suggesting at least some functional independence between glucose and low-salt processing. In contrast, low concentrations of sucrose or fructose reportedly could not be detected by the larvae in the presence of high-concentration glucose (compare Dahanukar et al. 2008).

The chemosensory equipment of the larval head comprises three external sense organs – dorsal, terminal and ventral organs–and three pharyngeal organs (Gendre et al. 2004; Python and Stocker 2002; Singh and Singh 1984) (Fig. 1). Each organ includes several multineuronal sensilla. The dorsal organ is composed of the olfactory dome sensillum (see earlier) and six smaller sensilla. Five of them and most of the terminal, ventral and pharyngeal sensilla are characterized by a distal pore, suggesting gustatory function (for *Musca*, see Chu and Axtell 1971; Chu-Wang and Axtell 1972). However, thermosensory (Liu et al. 2003b), hygrosensory or mechanosensory neurons may also be present. The estimated 90 per body side taste neurons of the larva (Colomb et al. 2007) outnumber the 21 ORNs (see earlier), consistent with an expected predominant short-range chemical orientation and proverbial (Carle 1969) feeding obsession of the larva. By contrast, in the flying adults, about 1,300 ORNs (Stocker 2001) outnumber approximately 600 taste neurons (Stocker 1994).

The patterns of expression of *Gr* genes in larval sensilla–studied using promoter-Gal4 strains–are only partially described (Colomb et al. 2007; Fishilevich et al. 2005; Scott et al. 2001). Notably, none of the *Gr5a*-Gal4 strains available show any expression in the larva (Colomb et al. 2007). *Gr2a*, *Gr21a*, *Gr22e*, *Gr28be*, *Gr32a* and *Gr66a*, known to be expressed in the adult, are also expressed in the larva, that is in the terminal organ. *Gr2a*-Gal4 labels in addition two neurons in the dorsal organ. In adults, *Gr22e*, *Gr28be*, *Gr32a* and *Gr66a* are suspected to encode bitter receptors, as they are coexpressed in many taste neurons (Thorne et al. 2004; Wang et al. 2004b); however, in the larva, no coexpression is observed for *Gr32a* and *Gr66a* (Colomb et al. 2007; Scott et al. 2001). Interestingly, *Gr21a*, which mediates CO_2 responses in adults (see earlier), is expressed in neurons of the terminal organ that are necessary for the behavioural response to CO_2 (Faucher et al. 2006). Remarkably also, several *Or*-Gal4 lines (*Or30a*, *Or42a*, *Or49a*, *Or63a*) label neurons in both dorsal and terminal organs (Fishilevich et al. 2005; Kreher et al. 2005; Scott et al. 2001). However, whether the Gal4 expression patterns in the terminal organ faithfully reflect gene expression has to be verified.

Regarding salt processing, one member of the *ppk* gene family (*ppk11*) is exclusively expressed in three pairs of neurons of the terminal organ and is necessary for the appetitive behavioural responses to low salt, but is dispensable for the aversive responses to high salt (Liu et al. 2003a). An involvement of *ppk* genes for aversive responses to high salt in the larva, as suggested by Liu et al. (2003a) regarding *ppk19*, must remain tentative, however, as Colomb et al. (2007) did not find *ppk19* expression in the larva.

Similar to the situation in adult flies, taste information is sent to multiple areas in the suboesophageal ganglion (Fig. 1). Four major target subregions have been identified via single-cell labelling in various Gal4 driver lines (Colomb et al. 2007; Scott et al. 2001). They seem to be correlated primarily with the nerve through which the afferents travel and less with the *Gr* gene expressed. Consequently, as in the adult, neurons in different sense organs but expressing the same gene, for example *Gr2a*, may have different central targets (Colomb et al. 2007; Scott et al. 2001). Gustatory afferents from external sense organs, such as those from the terminal organ labelled by *Gr66a*-Gal4 (Scott et al. 2001), generally establish ipsilateral projections, in contrast to the bilateral *Gr66a* projections in the adult. Afferents involved in attractive responses (Heimbeck et al. 1999) were suggested to project to a region slightly different from the four subregions mentioned (Colomb et al. 2007). Moreover, the neuron from the terminal organ expressing *Gr21a* (see earlier; Faucher et al. 2006) appears to have its own, specific suboesophageal target region (Colomb et al. 2007). Interestingly, the neurons of the terminal organ expressing *Or30a*-Gal4, *Or42a*-Gal4 and *Or49a*-Gal4 project into the suboesophageal ganglion rather than the antennal lobe (Fishilevich et al. 2005; Kreher et al. 2005).

Little information is available about potential target neurons of larval taste afferents. Intriguing candidates are a set of 20 neurons in the suboesophageal ganglion that express the *hugin* gene (Bader et al. 2007; Melcher and Pankratz 2005). They establish dendritic arborizations that partially overlap with the terminals of taste receptor neurons (Bader et al. 2007; Colomb et al. 2007) and send processes to the protocerebrum, the ventral nerve cord, the ring gland and the pharyngeal apparatus. In adults, blocking synaptic output from *hugin* neurons increases feeding. Hence, these interneurons may integrate taste processing, the endocrine system, higher-order brain centres and motor output in order to modify feeding. Also, octopaminergic interneurons receiving their input in the soboesopaheal ganglion are suspects for receiving appetitive gustatory input and then distributing an internal reward signal to the brain (see the discussion later).

4 Associating Smell and Taste

Why learn? Well, it does not hurt, and may even help. In other words, associative plasticity is a basic feature of nervous systems: Activity-dependent, associative mechanisms are engaged in developmental processes and thus are at disposal for behaviour control. Further, being able to use past experience to predict the future is an obvious advantage, for example when it comes to predicting food.

We discuss associative, Pavlovian learning between odours and food reward in *Drosophila*. We argue that flies (just like insects in general) posses a discrete side branch in their olfactory pathway to accommodate experience-dependent changes in olfactory behaviour. This side branch diverts from the antennal lobes and forms a loop via the so-called mushroom bodies towards premotor centres. These centres thus receive both direct and indirect olfactory input (from antennal lobes and mushroom bodies, respectively) to organize behaviour. We review what is known about this system in larval and adult *Drosophila*.

4.1 Adult *Drosophila*

In their seminal 1974 study, Quinn et al. (1974) showed that adult *Drosophila* can be differentially conditioned to odours. The initial version of the experiment was later (Tully and Quinn 1985) simplified into a purely Pavlovian conditioning paradigm: flies receive electric shock in the presence of one odour, and subsequently are exposed to another odour without shock (to average-out non-associative as well as odour-specific effects, the chemical identity of the odours is reversed in a reciprocally trained set of flies). In a final choice test, flies avoid the previously punished over the previously non-punished odour (Fig. 4). This paradigm is used for integrative analyses of learning and memory (reviewed in Davis 2005; Gerber et al. 2004; Heisenberg 2003; Heisenberg and Gerber in press; Keene and Waddell 2007; Margulies et al. 2005; McGuire et al. 2005; Zars 2000). In a pioneering study (Dudai et al. 1976), 'learning mutants' were found with this kind of assay, including mutants in the *dunce* (CG 32498, dnc^1) and *rutabaga* (CG 9533, rut^1) genes, marking the discovery of the role of the cyclic AMP/protein kinase A (cAMP/PKA) cascade for associative learning, which was later confirmed in vertebrates as well. Subsequently, a plethora of further mutants were characterized as impaired in this kind of task (reviewed in Davis 2005; Keene and Waddell 2007; McGuire et al. 2005), again providing educated guesses for research in vertebrates.

Importantly for the current purpose, Tempel and co-workers (1983) showed that an appetitive version of the paradigm is possible as well, using sugar as a reward (Fig. 4); however, this appetitive version of olfactory learning had received considerably less attention until recently (Keene et al. 2006; Kim et al. 2007; Krashes et al. 2007; Schwaerzel et al. 2003; Schwaerzel et al. 2007; Thum et al. 2007), rendering the focus of the current review timely. In the following we want to ask how appetitive learning works in adult flies, comparing the underlying mechanisms with the ones known for aversive learning.

4.1.1 Bridging the Gap

As detailed already, the olfactory system conveys odour information initially to the antennal lobe and then further via the projection neurons to the lateral horn and the mushroom bodies. From both these centres, premotor commands are thought to originate. The gustatory system, in contrast, carries sugar information to the suboesophageal ganglion, from where premotor commands likely can be triggered

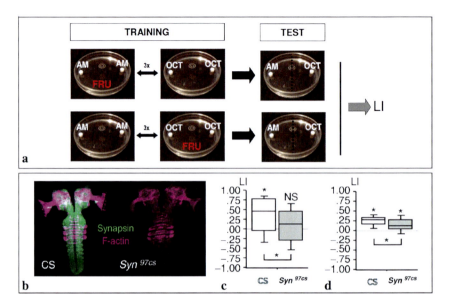

Fig. 5 Olfactory learning in larval *Drosophila*. **a** Learning experiments in larval *Drosophila* use agarose-filled Petri dishes. The agarose can be torn to reward (by adding fructose) or punishment (by adding high-concentration sodium chloride or quinine). Odours are supplied by evaporation from small, perforated Teflon containers. A two-group, reciprocal training design is used (Scherer et al. 2003). In one of the groups, isoamylacetate (AM) is presented with sugar reward (+) and alternately 1-octanol (OCT) is presented either without any reinforcer (as in this figure, AM+/OCT) or with high-concentration salt or quinine as punishment. The other group receives reciprocal training (AM/OCT+). Subsequently, animals are tested for their choice between AM versus OCT. Relatively higher preferences for AM after AM+/OCT training than after AM/OCT+ training reflect associative learning and can be expressed as a learning index. **b** Antibody staining reveals localization of synapsin throughout the neuropile regions of the brain in the wild-type CS strain (green), and total absence of synapsin in the deletion mutant syn^{97CS}; both strains were outcrossed for 13 generations to effectively adjust genetic background. The frontal two brain hemispheres are to the top, the caudal ventral nerve cord is to the bottom. In magenta, F-actin is labelled by phalloidin to orient within the preparations; towards the top one can discern the F-actin-rich fibre bundles of the developing adult eyes. **c, d** In two independent experiments, syn^{97CS} show a reduction of appetitive learning scores by approximately 50%. In **c**, larvae are tested individually, whereas in **d** larvae are tested in cohorts of 30; clearly, scores are not higher when testing cohortwise (arguing against a stamped effect), but scatter is much reduced. Control experiments testing for sensory or motor defects have revealed no difference between syn^{97CS} and the wild-type CS strain (see the text and Michels et al. 2005). Box plots represent the median as the middle line and 25, 75 as well as 10, 90% quantiles as box boundaries and whiskers, respectively. (**a, b** Copyright B. Michels, Universität Würzburg. **c, d** From Michels et al. 2005, copyright Cold Spring Harbor Laboratory Press)

directly. Thus, gustatory information seems to bypass the brain proper and to not converge directly with the olfactory pathway. How, then, can an association of smell and taste ever come about? Does the insect brain contain neurons to short-circuit smell and taste?

Indeed, Hammer (1993) in the honeybee identified the so-called VUM_{mx1} neuron; the cell body of this unpaired neuron is positioned in the maxillary segment of the suboesophageal ganglion, ventrally near the midline. It likely receives input in the suboesophageal ganglion and provides output to the antennal lobe, the mushroom body calyx and the lateral horn. Sucrose application to the antennae and proboscis, which elicits the proboscis-extension feeding reflex, drives this neuron; however, driving this neuron does not elicit the feeding reflex. Strikingly, if an odour is presented together with an activation of VUM_{mx1}, bees learn appetitively about that odour (Hammer 1993). Thus, the VUM_{mx1} neuron is not sufficient to substitute for sugar, as its activation does not trigger the feeding reflex; rather it is specifically sufficient to mediate the rewarding function of sugar, i.e. its role as something 'good'.

In *Drosophila*, both in the adult and in the larva, such a VUM_{mx1} neuron is present as well (H. Tanimoto, Universität Würzburg, and A. Thum, Université Fribourg, respectively; personal communications). As in the honeybee, it is located medioventrally at the midline and innervates the suboesophageal ganglion, the antennal lobe, the mushroom body calyx and the lateral horn. It is part of a cluster of ventral unpaired median neurons, which also in the fly likely are octopaminergic (Sinakevitch and Strausfeld 2006). Adult flies lacking octopamine (owing to a lack of the synthesizing enzyme tyramine β-hydroxylase, CG 1543, in the $T\beta H^{M18}$ mutant) are impaired in odour–sugar learning, but not in odour–shock learning (Schwaerzel et al. 2003). In the larva, driving octopaminergic/tyraminergic neurons as covered by the TDC-Gal4 line (Cole et al. 2005) can reportedly substitute for the sugar reward in olfactory learning (Schroll et al. 2006); whether the VUM_{mx1} neuron as an individual neuron can also in *Drosophila* mediate this appetitively reinforcing function is as yet unknown. If this were so, the target areas of the VUM_{mx1} neuron, namely the antennal lobe, the mushroom bodies and the lateral horn, would be prime suspects for housing memory traces for odour–sugar learning in *Drosophila*.

4.1.2 Interplay: Localizing Memory Traces?

Is it possible to localize memory? No. This is because having a memory is a psychological property of a person, or an animal, as a whole. In other words, as brains do not have memories, one cannot localize them in the brain. However, maybe one can localize those neuronal changes necessary and sufficient for a particular change in behaviour? Thus, the trick is to not try to localize a psychological process ('memory'), but the substrate of its behaviour corollary (the 'memory trace').

In adult *Drosophila*, the mushroom bodies arguably are the site of the short-term odour–shock associative memory trace (reviewed in Gerber et al. 2004; Heisenberg 2003; Heisenberg and Gerber in press). The working model is that whenever the activation of a Kenyon cell, as part of the pattern of Kenyon cells activated by a given odour, coincides with a shock-triggered, likely dopaminergic, reinforcement signal impinging onto the Kenyon cells, future output from this Kenyon cell (and from its concomitantly activated fellow Kenyon cells) onto mushroom body output neurons is modulated. This modulated output then is thought to mediate future conditioned avoidance in response to the odour. Which data are the bases for this working model? (for a detailed discussion of two recent functional imaging studies by Yu et al. 2004, 2005, see Heisenberg and Gerber in press):

1. Proteins required for synaptic plasticity, such as the type I adenylate cyclase coded for by the *rutabaga* gene (CG 9533, *rut*), are preferentially expressed in the mushroom bodies (Crittenden et al. 1998). This cyclase is required for cAMP production in neurons and can be activated by both G-proteins and the Ca^{2+}/calmodulin signalling cascade (Han et al. 1992; Levin et al. 1992); importantly, in vitro studies suggest that only a simultaneous activation by both these mechanisms leads to overadditive cAMP production (Abrams et al. 1998; Dudai et al. 1988). Given that Kenyon cell activation by odours leads to Ca^{2+} influx (Wang et al. 2004a), and that shock application likely leads to the activation of G-protein-coupled dopamine receptors (Han et al. 1996; Kim et al. 2003; Schwaerzel et al. 2003; Riemensperger et al. 2005; Schroll et al. 2006; reviewed in Blenau and Baumann 2001), this cyclase could act to molecularly detect the coincidence of odour and shock to then trigger the cAMP/PKA cascade. Consistent with such a role of the cyclase, mutations in the *rutabaga* gene (*rut^1*, *rut^{2080}*, *rut^{2769}*) entail learning defects in all associative learning tasks reported to date (Duerr and Quinn 1982; Liu et al. 2006; Perisse et al. 2007; Tempel et al. 1983; Wustmann et al. 1996). Notably, restoring the cyclase in the mushroom bodies restores odour–shock learning (Mao et al. 2004; McGuire et al. 2003; Zars et al. 2000), but does not restore learning in other tasks such as visual pattern learning (Liu et al. 2006). Pattern learning, however, can be rescued by restoring the cyclase in the central complex (Liu et al. 2006). Central complex expression, in turn, does not appear to rescue odour–shock learning (Zars et al. 2000). Also, cyclase expression in the projection neurons does not rescue odour–shock learning (Thum et al. 2007). Importantly, the cyclase seems to act acutely during the learning process, as acute expression is sufficient to rescue learning, arguing against a purely developmental role of the cyclase for establishing a properly functioning mushroom body (McGuire et al. 2003).
2. Connolly et al. (1996) transgenically expressed a mutant $G_{\alpha s}$ protein (CG 2835, using the $G_{\alpha s}^*$ mutant) in the mushroom bodies which constitutively activates the cyclase, hence presumably rendering any modulation of cyclase activity impaired. This leads to an abolishment of memory scores after odour–shock learning. Whether a knockdown of the *rut*-cyclase by means of RNA interference would lead to a similar abolishment of short-term odour–shock memory is unknown.
3. Three groups independently found that output from chemical synapses of the mushroom body is required at test, but is dispensable during training (Dubnau et al. 2001; McGuire et al. 2001; Schwaerzel et al. 2002). If output from the projection neurons is blocked during training, however, flies cannot establish an odour–shock memory trace (Schwaerzel, 2003).
4. Dopaminergic neurons innervating the mushroom bodies are activated by shock (Riemensperger et al. 2005), and blocking synaptic output from dopaminergic neurons as part of the *TH*-Gal4 pattern prevents acquisition but not retrieval of odour–shock memory (Friggi-Grelin et al. 2003; Schwaerzel et al. 2003).

Thus, synaptic plasticity in the mushroom bodies is sufficient (*rut*-rescue) and necessary ($G_{\alpha s}^*$) to establish a short-term memory trace during odour–shock training. Furthermore, olfactory information needs to enter the mushroom bodies during training but does not have to leave them; during test, in turn, both input to and output from the mushroom bodies is required to support normal memory scores. Reinforcement signalling through dopaminergic neurons, on the other hand, seems to be required only during training, but not at test. With due caveats in mind (Gerber et al. 2004; Heisenberg 2003; Heisenberg and Gerber in press), it therefore seems a reasonable working hypothesis that the short-term memory trace for odour–shock learning is located in the mushroom bodies. Does this also apply for odour–sugar learning?

4.1.3 Odour–Sugar Learning

A first hint towards a role of the mushroom bodies for appetitive learning was provided by the *mushroom body deranged* mutant (*mbd*; no CG number can as yet be assigned to the affected gene; Heisenberg et al. 1985): olfactory learning is abolished in adult *mbd* mutants, regardless of whether shock or sugar reinforcement is used. Recently, analyses of memory trace localization for odour–sugar learning are being pursued more systematically (Schwaerzel et al. 2003; Thum et al. 2007). Transgenic expression of the *rutabaga* adenylate cyclase in the mushroom bodies is sufficient to rescue the sugar-learning defect of the *rutabaga* mutant (rut^{2080}; Schwaerzel et al. 2003); this is also the case for an acute expression in the mushroom body (Thum et al. 2007). Furthermore, if output from the mushroom body is possible during training but blocked at test, flies show no appetitive olfactory memory score (Schwaerzel et al. 2003). If, however, mushroom body output is blocked during training, but is possible at test, flies show normal learning scores (Schwaerzel et al. 2003). Thus, all odour–sugar memory trace(s) must be located upstream of mushroom body output and require processing through the mushroom body for retrieval; one such memory trace, as addressed by *rut* function, is located within the mushroom bodies themselves. This situation matches the findings for odour–shock learning.

However, research on honeybees suggests that the antennal lobes may house a memory trace for odour–sugar associations as well (Erber et al. 1980; Faber et al. 1999; Farooqui et al. 2003; Hammer and Menzel 1998; but see Peele et al. 2006). This prompted the question of whether acute expression of the *rut*-cyclase in the projection neurons of *rut* mutants (rut^{2080}) would restore learning performance; this is indeed the case (Thum et al. 2007) (Fig. 4)! Does this prove that there are two fully redundant memory traces in projection neurons and mushroom bodies? Such a conclusion would require showing that expression of the constitutively active $G_{\alpha s}$* protein, or knockdown of the cyclase, in either the projection neurons or the mushroom bodies does not affect reward learning. Also, a block of input towards the mushroom bodies during training, e.g. by a temperature-sensitive, dominant-negative acetylcholine receptor, should leave appetitive memory scores unaffected but should abolish aversive learning. Finally, testing whether the projection-neuron versus mushroom-body memory traces are different in terms of their specificity and/or their role during the various phases of memory (for the situation in honeybees, see Hammer and Menzel 1998) now is warranted.

4.1.4 Longer-Term Appetitive Memory

For sugar-reward learning, relatively little is known about the organization of longer-term memory (for reviews concerning longer-term memory of aversive memories, see McGuire et al. 2005; Keene and Waddell 2007; but see the recent account by Krashes and Waddell 2008). Mutants in the *amnesiac* gene (CG 11937, *amn¹* and *amn^X*) have no substantial defect in short-term memory after appetitive learning; a memory impairment, however, becomes particularly prominent for longer (more than 60 min) retention intervals (Keene et al. 2006; Waddell et al. 2000). This defect can be rescued by transgenic expression of the *amnesiac*-encoded protein in the dorsal paired median (DPM) neurons (Tamura et al. 2003; Waddell et al. 2000). *Drosophila* posses one such neuron per hemisphere, large neurons innervating the mushroom body lobes in a mesh-like way (Waddell et al. 2000). They receive input from, and provide output to, the mushroom bodies. If output from these neurons is blocked, appetitive short-term memory is unaffected, regardless of when the block is induced; this is consistent with the lack of phenotype of *amn* mutants (*amn¹*; *amn^X8*) in this regard. Strikingly, however, if output from the DPM neurons is transiently blocked during the break between training and a longer-term (3-h) retention test, memory scores are reduced; this is true both for sugar learning and for shock learning (Keene et al. 2006). Also, for retention 3 h after training, output from the DPM neurons at the moment of test is dispensable for both appetitive and aversive memory retrieval (Keene et al. 2006). Thus, it seems that off-line processing along a mushroom body–DPM neuron–mushroom body loop is required to support 3-h memory. If this were so, then certainly one would predict that blocking output from the mushroom bodies towards the DPM neurons during the retention period should affect 3-h-memory – as is indeed the case for both appetitive and aversive longer-term memory (for a more detailed account, see Keene and Waddell, 2007). Thus, the situation regarding the role of *amn* in aversive learning is strikingly similar to what has been mentioned here for appetitive learning (Keene et al. 2004; Keene et al. 2006; Waddell et al. 2000).

In summary, appetitive and aversive olfactory learning are similar in terms of the involvement of the PKA/cAMP cascade as addressed by adenylate cyclase function; also, for both kinds of paradigm, a short-term memory trace is localized in the mushroom bodies. Finally, the similar role of the mushroom body–DPM neuron–mushroom body loop for stabilizing both appetitive and aversive longer-term memories is particularly noteworthy. However, as the mushroom body houses both an appetitive and the aversive memory trace for short-term retention, how can similar molecular mechanisms (the PKA/cAMP cascade) in the same set of neurons establish different memories (appetitive versus aversive)? Interestingly, Schwaerzel et al. 2007 showed that a specific pool of PKA-RII (CG 15862, defining the A kinase-anchoring protein anchored pool of PKA) is required for aversive learning, but is dispensable for appetitive learning. This suggests that different memory traces within the same cell may use the same molecular pathway, but at different subcellular compartments.

In any event, also at other levels there are major dissociations between appetitive and aversive olfactory learning. Before discussing these dissociations, however, one should mention that for appetitive learning flies have to be starved before the experiment, whereas such starvation is neither necessary nor is usually performed for aversive learning; thus, discrepancies between appetitive and aversive learning may in part result from differences in motivational state. With this caveat in mind, the major discrepancies between both kinds of learning appear to be

- Appetitive but not aversive training establishes an additional short-term memory trace in the projection neurons.
- Regarding the short-term memory domain, either dopamine or octopamine is necessary and sufficient for reinforcement in aversive and appetitive learning, respectively [Schroll et al. 2007; Schwaerzel et al. 2003; but see Kim et al. 2007 regarding a common role of the *D1-like dopamine receptor* gene (CG 9652 coding for *dumb1* and *dumb2*)].
- Memory scores for odour–shock learning decay characteristically faster than for odour–sugar learning (Tempel et al. 1983) and
- initial retention (less than 5 min) of aversive but not of appetitive memory reportedly is impaired in the *dunce* mutant (CG 32498, using the dnc^1 allel) (Tempel et al. 1983).

4.2 Larval Drosophila

The taste and smell systems of larvae are much reduced in terms of cell number as compared with adults, but by and large follow the same functional architecture (Fig. 1). But do larvae also show the same potency for learning as adults do, and if so, are the same molecular processes and the same sets of cells involved? As will be argued below, the shared characters to us appear to outweigh the discrepancies.

Learning experiments are performed with larvae crawling on an agarose surface in standard Petri dishes. In the pioneering studies of Aceves-Pina and Quinn (1979), Tully et al. (1994) and Heisenberg et al. (1985), electric shock was used as an aversive reinforcer (but see Forbes 1993 for a failure to replicate these results). Here, we review the more recent literature using gustatory reinforcement. That is, the agarose can be torn to reward (by adding sugar) or punishment (by adding high-concentration salt or bitter) for association with odours (isoamylacetate, AM; or 1-octanol, OCT) evaporating from custom-made odorant containers (Fig. 5). As in adult flies, a two-group, reciprocal experimental design is used (Scherer et al. 2003); experimental designs which do not use reciprocal training (Honjo and Furukubo-Tokunaga 2005) confound associative and non-associative effects and therefore are not reviewed here (for a detailed discussion, see Gerber and Stocker 2007).

4.2.1 Appetitive Learning

Employing a reward-only paradigm (AM+/OCT and AM/OCT+), Neuser et al. (2005) showed that learning success increases with the number of learning trials and reaches an asymptote after three trials each with the rewarded and the unrewarded odour. Learning scores increase with reward strength, 2 M fructose supporting asymptotic levels of learning (see also Schipanski et al. 2008). The ensuing memory is stable for at least 30 min; after 90 min, there is no measurable learning effect left. In addition, there does not appear to be any effect of larval gender or age, although regarding age one may notice a trend for best scores at an intermediate larval age (5 days after egg laying, as compared with 4 and 6 days).

In the tradition of the olfactory learning experiments in adult flies, larval learning experiments had initially been performed in darkness (i.e. using red light, which does not allow for vision in *Drosophila*), but as Yarali et al. (2007) reported, these experiments work just as well under normal illumination conditions; even changes of visual context between training and test leave olfactory memory scores unaffected. Finally, an en mass version of the assay is possible, such that animals are trained and tested in groups of 30; under such conditions, learning scores are not higher than in the individual-animal version (arguing against a stampede effect), but the scatter of the data is reduced (Neuser et al. 2005).

What is known about the genetic and cellular bases of appetitive learning in the larva? Four recent studies offer the first hints, but before going into detail, we would like to briefly discuss the kinds of behavioural control procedures for 'learning mutants' (for a more detailed discussion, see Gerber and Stocker 2007). That is, one may wonder whether the mutant is able to taste, to smell, and whether it can crawl fast enough during the test to move among the sources of the different odours in the allotted time (typically 3 min). For odour–sugar learning, one therefore often compares naïve animals from the different genotypes in terms of (1) their preference between the fructose reward and plain agarose, (2) their preference between an AM-scented and an unscented control side as well as (3) their preference between an OCT and a control side. However, a learning defect, logically, can only be detected after training, i.e. after animals had undergone extensive handling, exposure to reinforcers and exposure to odours. Thus, one may in addition want to test for those olfactory and motor abilities that the mutants need at the very moment of test. In other words, can the larvae still respond to odour after 'sham training' that involves the same handling and general procedure as for training, but (1) omits the reinforcer, exposing the larvae to only the odours, and (2) omits the odours, exposing the larvae to only the reinforcer? Finally, in some cases, the mutation in question may entail a developmental delay; to the extent to which developmental stage matters for learning, one may therefore want to allow the mutant more time for development so that it can mature to the same stage as the wild type.

Michels et al. (2005) investigated the role of the *synapsin* gene (CG 3985, *syn*), the single fly homolog of the vertebrate *synapsin* genes. Synapsin is a brain-wide-expressed, evolutionarily conserved presynaptic phosphoprotein (Godenschwege et al. 2004; Hilfiker et al. 1999; Klagges et al. 1996). It is associated with the cytoskeleton and the cytoplasmic side of synaptic vesicles and regulates the balance between the readily releasable versus the reserve pool of vesicles in a phosphorylation-dependent way, thus contributing to the regulation of synaptic output. Mutations in the human *synapsin 1* gene can cause seizures and, in a subset of patients, learning defects (Garcia et al. 2004; see also Gitler et al. 2004 for similar phenotypes in mice), and psychotic symptoms (Chen et al. 2004). In *Drosophila*, the *syn*^{97CS} deletion mutant (Godenschwege et al. 2004) lacks the synapsin protein and shows a reduction in learning ability by about 50% as compared with an effectively isogenized (13 generations) wild-type control strain (Fig. 5) (Michels et al. 2005). This phenotype is not due to an impairment of those sensory and motor skills required in the learning paradigm, as both naïve responses towards the odours and to the reward as well as odour responses after 'sham training' are indistinguishable between *syn*^{97CS} and wild-type control (Michels et al. 2005).

A second paper focused on the *foraging* gene (CG 10033, *for*) (Kaun et al. 2007). This gene shows a polymorphism which seems to underlie a behavioural polymorphism: while in the absence of food locomotion is the same for both kinds of genotype, in the presence of food larvae carrying the so-called sitter (*for-s*) allele forage largely within their food patch, whereas larvae with the rover (*for-r*) allele move between patches. Arguably, either of these strategies may be beneficial, depending on the spatial and temporal distribution of food sources. The *for* gene codes for a protein kinase G, and sitters and rovers indeed differ in protein kinase G activity (low for sitter and high for rover; Osborne et al. 1997). Kaun et al. (2007) reported that larvae carrying either allele do not differ in visual learning (Gerber et al. 2004), but do differ in olfactory learning, such that rover larvae show higher initial, but lower later retention. It thus seems as if rovers learn and forget faster than sitters; naïve responsiveness to odours as well as to the reward, however, do not differ between them. Notably, the reduced initial learning scores in sitters can be increased to rover levels by boosting expression of the protein kinase G in the mushroom bodies (driver strains 201Y-Gal4, H24-Gal4, and c739-Gal4); whether this would also lead to rover-like small learning scores at later retention intervals is not known.

Thirdly, two papers focused on the function of the *neurexin* gene. In vertebrates (Dean and Dresbach 2006), neurexins are found to be presynaptic transmembrane proteins. Together with their postsynaptic binding partners of the neurolignin protein family, they act to induce and maintain synaptic contacts, and to organize the molecular machinery at active zone and postsynaptic density, respectively. Zeng et al. (2007) and Li et al. (2007) now report that in *Drosophila* there is but one homolog to the vertebrate *neurexin* genes (CG 7050, *dnrx*), which is expressed throughout the neuropile regions of the larval and adult brains. Specifically, some overlap of immunostaining for the neurexin protein with the active zone protein bruchpilot (CG 34146, *brp*; Kittel et al. 2006; Wagh et al. 2006) but not with a marker of the postsynaptic density (*Drosophila* p21-activated kinase, encoded by CG 10295, *dpak*; Sone et al. 2000) may suggest a presynaptic localization at the neuromuscular junction (Li et al. 2007). Regarding learning, Zeng et al. (2007) reported that while learning was intact in their genetic control strain (*white*^{1118}, which also had been used for five generations of outcrossing with the mutant), a lack of the neurexin protein in the deletion mutant *nrx-1*$^{\Delta 83}$ entails a complete abolishment of learning. This phenotype is not due to an impairment of those sensory and motor skills required in the learning paradigm, as responses to the odours and the reward in naïve animals, as well as odour responses after 'sham training', are normal. Furthermore, the learning defect can at least partially be rescued by spatially extended transgenic expression of neurexin (*elav*-Gal4 driver strain).

Fourthly, Knight et al. (2007) investigated the effect of deleting the *presenilin* gene (CG 18803, *psn*; Boulianne et al. 1997) on both visual and olfactory learning, using the individual-animal version of the assay and employing differential conditioning with appetitive–aversive reinforcement. In the deletion mutant psn^{W6}, the presenilin protein is absent, and in a study nicely controlled for developmental delay, the authors found that both olfactory and visual learning are completely abolished. The $white^{1118}$ strain, which serves as genetic control (and which had been used for five generations of outcrossing), performs fine in both tasks. Responsiveness of naïve animals towards the reinforcers and towards the odours is the same between both genotypes; in the case of olfactory learning, performance can at least partially be restored by using spatially extensive transgenic expression of presenilin (*elav*-Gal4 driver strain). These findings may contribute to an understanding of Alzheimer disease, provided the relation between presenilin function and familial Alzheimer disease is resolved (see the discussion in De Strooper 2007; Wolfe 2007); such analyses may profit from the fact that there is but one *presenilin* gene in flies, and that the amyloid β peptide, to the best of current knowledge, is absent in the fly. Notably, Knight et al. (2007) discussed that, different from the situation in vertebrates, the site of action for presenilin in the fly may be presynaptic. Indeed, it seems noteworthy that three of these four reviewed studies suggest presynaptic mechanisms of plasticity (synapsin, neurexin, presenilin). This is in accordance with a working model proposed for aversive learning in the adult by Heisenberg (2003) which suggests a presynaptic modification of the mushroom body output synapses to underlie short-term associative changes in olfactory behaviour. In the larva, the critical experiments to identify the cells and subcellular site(s) of action for all these three genes remain to be done.

4.2.2 Aversive Learning

Initially, Hendel et al. (2005) suggested that memory was exclusively due to reward, because reward-only but not punishment-only training – using either high-concentration salt or quinine – yields learning effects. Moreover, learning indices after reward-only training are as high as after reward–punishment training. However, larvae do show aversion to high-concentration salt and quinine, and both suppress feeding. Thus, high-concentration salt and quinine, although aversive, did not seem to have any effect as reinforcers. Educatively, this turned out to be wrong:

Both salt and quinine actually are effective as reinforcers, but the respective memories are not automatically expressed in behaviour (Gerber and Hendel 2006). That is, behaviours are expressed if their outcomes offer a benefit (Dickinson 2001; Elsner and Hommel 2001; Hoffmann 2003). Consider that after training with sugar, the test offers the larvae a choice with one odour suggesting 'over there you will find sugar' and the other suggesting 'over there you will not find sugar'. In the absence of sugar, larvae should thus search for the predicted reward. If sugar already is present, however, such a search does not offer any improvement and would not seem warranted. In contrast, after aversive training, one odour may suggest 'over there you will suffer from quinine' whereas the alternative suggests 'over there you will not suffer from quinine'. In the presence of quinine, therefore, the no-quinine-associated odour can give direction to the escape from the aversive reinforcer, while if quinine actually is absent, such a flight response is not warranted to begin with. And this is indeed what is found (Gerber and Hendel 2006). In other words, the behavioural expression of memory is not an automated, but is a regulated process. First, irrespective of the test situation, the odour activates its memory trace. In a second, previously unrecognized evaluative step a comparison is made between the value of this memory trace and the value of the test situation. Only if the value of the memory trace is higher than that of the test situation, tracking down the odour can be expected to improve the situation. It is this expectation of outcome, rather than the activated memory trace per se, which drives conditioned behaviour.

4.2.3 'Remote Control' of Reinforcement

As discussed earlier, olfactory projections target the brain before projecting to motor centres, whereas taste information remains suboesophageal and bypasses the central brain, being transmitted more directly to motor systems. Given that there does not seem to be any convergence between olfactory and taste processing, one may ask how odours can be associated with gustatory reinforcement.

As in adult flies, the solution likely is provided by aminergic interneurons which receive input in the suboesophageal ganglion and provide output to the brain, establishing a short circuit between olfactory and gustatory processing. Specifically, the function of dopaminergic and octopamineric/tyraminergic neurons in the *Drosophila* larva has lately been addressed by remote-controlling neurons (Schroll et al. 2006). With the Gal4/UAS system, the blue-light-gated ion channel channelrhodopsin-2 is expressed in octopaminergic/tyraminergic cells (as covered by *TDC*-Gal4). Owing to the transparency of the larval cuticle, these cells can then non-invasively be driven by switching on the blue light. If light stimulation is paired with one odour, and another odour is presented in darkness, the larvae will subsequently prefer the former, 'virtually' rewarded odour. Thus, light-induced activation of octopaminergic/tyraminergic neurons is sufficient to substitute for appetitive reinforcement (but see Schipanski, 2007 for a failure to replicate these results). In turn, associatively driving dopaminergic neurons (as covered by *TH*-Gal4) reportedly induces aversive learning. The necessity of these neurons for appetitive and aversive learning, respectively, is at present unclear.

5 Outlook

It seems that chemosensation and chemosensory learning in *Drosophila* are beginning to be understood fairly well, in particular in the genuinely sensory aspects, and in terms of odour-taste memory trace formation. The remaining terra incognita, we believe, is how sensory and motor processing formats are integrated, and how adaptive motor patterns are being selected. Only with such an understanding will it be possible to search for the motivating factors of behaviour, the systems which make a *Drosophila* do what *Drosophila's* got to do.

Acknowledgements We express cordial thanks to the members and colleagues at our research institutions, and to the students in our groups, for the critique and discussions that shaped this review. Our research programmes are supported by the Volkswagen Foundation, the German-Israeli Foundation and the Deutsche Forschungsgemeinschaft (Heisenberg Fellowship, SFBs 554 and TR 58, GK 1156 to B.G.), by the Swiss National Funds (grants nos. 31-63447.00 and 3100A0-105517 to R.F.S. and A.S.T.) and by grants from the Ministry of Education, Culture, Sports, Science and Technology of Japan (to T.T.). E. Balamurugan from the Springer production team deserves our gratitude for extended patience.

References

Abrams TW, Yovell Y, Onyike CU, Cohen JE, Jarrard, HE (1998) Analysis of sequence-dependent interactions between transient calcium and transmitter stimuli in activating adenylyl cyclase in *Aplysia*: possible contribution to CS-US sequence requirement during conditioning. Learn Mem 4:496–509

Aceves-Pina EO, Quinn WG (1979) Learning in normal and mutant *Drosophila* larvae. Science 206:93–96

Ache BW, Young JM (2005) Olfaction: diverse species, conserved principles. Neuron 48:417–430

Awasaki T, Kimura K (1997) Pox-neuro is required for development of chemosensory bristles in *Drosophila*. J Neurobiol 32:707–721

Bader R, Colomb J, Pankratz B, Schröck A, Stocker RF, Pankratz MJ (2007) Genetic dissection of neural circuit anatomy underlying feeding behavior in *Drosophila*: distinct classes of hugin expressing neurons. J Comp Neurol 502:848–856

Benton R, Sachse S, Michnick SW, Vosshall LB (2006) Atypical membrane topology and heteromeric function of *Drosophila* odorant receptors in vivo. PLoS Biol 4:e20.

Benton R, Vannice KS, Vosshall LB (2007) An essential role for a CD36-related receptor in pheromone detection in *Drosophila*. Nature 450:289–293

Blenau W, Baumann A (2001) Molecular and pharmacological properties of insect biogenic amine receptors: lessons from *Drosophila melanogaster* and *Apis mellifera*. Arch Insect Biochem Physiol 48:13–38

Boulianne GL, Livne-Bar I, Humphreys JM, Liang Y, Lin C, Rogaev E, St George-Hyslop, P (1997) Cloning and characterization of the *Drosophila* presenilin homologue. Neuroreport 8:1025–1029

Buck L, Axel R (1991) A novel multigene family may encode odorant receptors: a molecular basis for odor recognition. Cell 65:175–187

Carle E (1969) The very hungry caterpillar. Penguin, New York

Chen Q, He G, Qin W, Chen QY, Zhao XZ, Duan SW, Liu XM, Feng GY, Xu YF, St Clair D, Li M, Wang JH, Xing YL, Shi JG, He L (2004) Family-based association study of synapsin II and schizophrenia. Am J Hum Genet 75:873–877

Chu IW, Axtell RC (1971) Fine structure of the dorsal organ of the house fly larva, *Musca domestica* L. Z Zellforsch Mikrosk Anat 117:17–34

Chu-Wang IW, Axtell RC (1972) Fine structure of the terminal organ of the house fly larva, *Musca domestica* L. Z Zellforsch Mikrosk Anat 127:287–305

Chyb S, Dahanukar A, Wickens A, Carlson, JR (2003) *Drosophila* Gr5a encodes a taste receptor tuned to trehalose. Proc Natl Acad Sci USA 100:14526–14530

Clyne PJ, Warr CG, Carlson JR (2000) Candidate taste receptors in *Drosophila*. Science 287:1830–1834

Clyne PJ, Warr CG, Freeman MR, Lessing D, Kim J, Carlson JR (1999) A novel family of divergent seven-transmembrane proteins: candidate odorant receptors in *Drosophila*. Neuron 22:327–338

Cobb M (1999) What and how do maggots smell? Biol Rev 74:425–459

Cole SH, Carney GE, McClung CA, Willard SS, Taylor BJ, Hirsh J (2005) Two functional but noncomplementing *Drosophila* tyrosine decarboxylase genes: distinct roles for neural tyramine and octopamine in female fertility. J Biol Chem 280:14948–14955

Colomb J, Grillenzoni N, Ramaekers A, Stocker RF (2007) Architecture of the primary taste center of *Drosophila melanogaster* larvae. J Comp Neurol 502:834–847

Connolly JB, Roberts IJ, Armstrong JD, Kaiser K, Forte M, Tully T, O'Kane CJ (1996) Associative learning disrupted by impaired Gs signaling in *Drosophila* mushroom bodies. Science 274:2104–2107

Couto A, Alenius M, Dickson BJ (2005) Molecular, anatomical, and functional organization of the *Drosophila* olfactory system. Curr Biol 15:1535–1547

Crittenden JR, Skoulakis EM, Han KA, Kalderon D, Davis RL (1998) Tripartite mushroom body architecture revealed by antigenic markers. Learn Mem 5:38–51

Dahanukar A, Foster K, van der Goes van Naters WM, Carlson JR (2001) A Gr receptor is required for response to the sugar trehalose in taste neurons of *Drosophila*. Nat Neurosci 4 :1182–1186

Dahanukar A, Lei Y-T, Kwon JY, Carlson JR (2007) Two Gr genes underlie sugar reception in *Drosophila*. Neuron 56:503–516

Datta, S.R., Vasconcelos, M.L., Ruta, V., Luo, S., Wong, A., Demir, E., Flores, J., Balonze, K., Dickson, B.J., and Axel, R. (2008). The *Drosophila* pheromone cVA activates a sexually dimorphic neural circuit. Nature 452, 473–477

Davis RL (2005) Olfactory memory formation in *Drosophila*: from molecular to systems neuroscience. Annu Rev Neurosci 28:275–302

Dean C, Dresbach T (2006) Neuroligins and neurexins: linking cell adhesion, synapse formation and cognitive function. Trends Neurosci 29:21–29

De Belle JS, Heisenberg M (1994) Associative odor learning in *Drosophila* abolished by chemical ablation of mushroom bodies. Science 263:692–695

De Bruyne M, Clyne PJ, Carlson JR (1999) Odor coding in a model olfactory organ: the *Drosophila* maxillary palp. J Neurosci 19:4520–4532

De Bruyne M, Foster K, Carlson JR (2001) Odor coding in the *Drosophila* antenna. Neuron 30:537–552

De Strooper B (2007) Loss-of-function presenilin mutations in Alzheimer disease. Talking point on the role of presenilin mutations in Alzheimer disease. EMBO Rep 8:141–146

Dethier VG, Hanson FE (1964) Taste papillae of the blowfly. J Cell Comp Physiol 65:93–100

Dickinson A (2001) Causal learning-an associative analysis. Q J Exp Psychol 54B:3–25

Dobritsa AA, van der Goes van Naters W, Warr CG, Steinbrecht RA, Carlson JR (2003) Integrating the molecular and cellular basis of odor coding in the *Drosophila* antenna. Neuron 37:827–841

Dubnau J, Grady L, Kitamoto T, Tully T (2001) Disruption of neurotransmission in *Drosophila* mushroom body blocks retrieval but not acquisition of memory. Nature 411:476–480

Dudai Y, Corfas G, Hazvi S (1988) What is the possible contribution of Ca^{2+}-stimulated adenylate cyclase to acquisition, consolidation and retention of an associative olfactory memory in *Drosophila*. J Comp Physiol [A] 162:101–109

Dudai Y, Jan YN, Byers D, Quinn WG, Benzer S (1976) dunce, a mutant of *Drosophila* deficient in learning. Proc Natl Acad Sci USA 73:1684–1688

Duerr JS, Quinn WG (1982) Three *Drosophila* mutations that block associative learning also affect habituation and sensitization. Proc Natl Acad Sci USA 79:3646–3650

Elsner B, Hommel B (2001) Effect anticipation and action control. J Exp Psychol Hum Percept Perform 27:229–240

Erber J, Masuhr TH, Menzel R (1980) Localization of short-term memory in the brain of the bee, *Apis mellifera*. Physiol Entomol 5:343–358

Faber T, Joerges J, Menzel R (1999) Associative learning modifies neural representations of odors in the insect brain. Nat Neurosci 2:74–78

Falk R, Bleiseravivi N, Atidia J (1976) Labellar taste organs of *Drosophila melanogaster*. J Morphol 150:327–341

Farooqui T, Robinson K, Vaessin H, Smith BH (2003) Modulation of early olfactory processing by an octopaminergic reinforcement pathway in the honeybee. J Neurosci 23:5370–5380

Faucher C, Forstreuter M, Hilker M, de Bruyne M (2006) Behavioral responses of *Drosophila* to biogenic levels of carbon dioxide depend on life-stage, sex and olfactory context. J Exp Biol 209:2739–2748

Ferveur JF (2005) Cuticular hydrocarbons their evolution and roles in *Drosophila* pheromonal communication. Behav Genet 35:279–295

Feeny P, Rosenberg L, Carter M (1983) Chemical aspects of oviposition behavior in butterflies. In: Ahmad S (ed) Herbivorous insects: host-seeking behavior and mechanisms. Academic Press, New York, pp 27–76

Fiala A, Spall T, Diegelmann S, Eisermann B, Sachse S, Devaud JM, Buchner E, Galizia CG (2002) Genetically expressed cameleon in *Drosophila melanogaster* is used to visualize olfactory information in projection neurons. Curr Biol 12:1877–1884

Fischler W, Kong P, Marella S, Scott K (2007) The detection of carbonation by the *Drosophila* gustatory system. Nature 30:1054–1057

Fishilevich E, Domingos AI, Asahina K, Naef F, Vosshall LB, Louis M (2005) Chemotaxis behavior mediated by single larval olfactory neurons in *Drosophila*. Curr Biol 15:2086–2096

Fishilevich E, Vosshall LB (2005) Genetic and functional subdivision of the *Drosophila* antennal lobe. Curr Biol 15:1548–1553

Foelix RF, Stocker RF, Steinbrecht RA (1989) Fine structure of a sensory organ in the arista of *Drosophila melanogaster* and some other dipterans. Cell Tissue Res 258:277–287

Forbes B (1993) Larval learning and memory in *Drosophila melanogaster*. Diploma Thesis, University of Würzburg

Friggi-Grelin F, Coulom H, Meller M, Gomez D, Hirsh J, Birman S (2003) Targeted gene expression in *Drosophila* dopaminergic cells using regulatory sequences from tyrosine hydroxylase. J Neurobiol 54:618–627

Fujishiro N, Kijima H, Morita H (1984) Impulse frequency and action potential amplitude in labellar chemosensory neurones of *Drosophila melanogaster*. J Insect Physiol 30:317–325

Gao Q, Chess A (1999) Identification of candidate *Drosophila* olfactory receptors from genomic DNA sequence. Genomics 60:31–39

Gao Q, Yuan B, Chess A (2000) Convergent projections of *Drosophila* olfactory neurons to specific glomeruli in the antennal lobe. Nat Neurosci 3:780–785

Garcia CC, Blair HJ, Seager M, Coulthard A, Tennant S, Buddles M, Curtis A, Goodship JA (2004) Identification of a mutation in synapsin I, a synaptic vesicle protein, in a family with epilepsy. J Med Genet 41:183–186

Gendre N, Lüer K, Friche S, Grillenzoni N, Ramaekers A, A Technau GM, Stocker RF (2004) Integration of complex larval chemosensory organs into the adult nervous system of *Drosophila*. Development 131:83–92

Gerber B, Hendel T (2006) Outcome expectations drive learned behaviour in larval *Drosophila*. Proc R Soc Lond B 273:2965–2968

Gerber B, Scherer S, Neuser K, Michels B, Hendel T, Stocker RF, Heisenberg M (2004) Visual learning in individually assayed *Drosophila* larvae. J Exp Biol 207:179–188

Gerber B, Stocker RF (2007) The *Drosophila* larva as a model for studying chemosensation and chemosensory learning: a review. Chem Senses 32:65–89

Gerber B, Tanimoto H, Heisenberg M (2004) An engram found? Evaluating the evidence from fruit flies. Curr Opin Neurobiol 14:737–744

Gitler D, Takagishi Y, Feng J, Ren Y, Rodriguiz RM, Wetsel WC, Greengard P, Augustine GJ (2004) Different presynaptic roles of synapsins at excitatory and inhibitory synapses. J Neurosci 24:11368–11380

Godenschwege TA, Reisch D, Diegelmann S, Eberle K, Funk N, Heisenberg M, Hoppe V, Hoppe J, Klagges BRE, Martin JR, Nikitina EA, Putz G, Reifegerste R, Reisch N, Rister J, Schaupp M, Scholz H, Schwärzel M, Werner U, Zars T, Buchner S, Buchner E (2004) Flies lacking all synapsins are unexpectedly healthy but are impaired in complex behaviour. Eur J Neurosci 20:611–622

Goldman AL, van der Goes van Naters W, Lessing D, Warr CG, Carlson JR (2005) Coexpression of two functional odor receptors in one neuron. Neuron 45:661–666

Grillenzoni N, de Vaux V, Meuwly J, Vuichard S, Gendre N, Stocker RF (2007) Role of proneural genes in the formation of the larval olfactory organ of *Drosophila*. Dev Genes Evol 217:209–219

Ha TS, Smith DP (2006) A pheromone receptor mediates 11-*cis*-vaccenyl acetate-induced responses in *Drosophila*. J Neurosci 26:8727–8733

Hallem EA, Carlson JR (2006) Coding of odors by a receptor repertoire. Cell 125:143–160

Hallem EA, Ho MG, Carlson JR (2004) The molecular basis of odor coding in the *Drosophila* antenna. Cell 117:965–979

Hammer M (1993) An identified neuron mediates the unconditioned stimulus in associative olfactory learning in honeybees. Nature 366:59–63

Hammer M, Menzel R (1998) Multiple sites of associative odor learning as revealed by local brain microinjections of octopamine in honeybees. Learn Mem 5:146–156

Han PL, Levin LR, Reed RR, Davis RL (1992) Preferential expression of the *Drosophila* rutabaga gene in mushroom bodies, neural centers for learning in insects. Neuron 9:619–627

Han KA, Millar NS, Grotewiel MS, Davis RL (1996) DAMB, a novel dompamine receptor expressed specifically in *Drosophila* mushroom bodies. Neuron 16(6):1127–1135

Heimbeck G, Bugnon V, Gendre N, Häberlin C, Stocker RF (1999) Smell and taste perception in *D. melanogaster* larva: toxin expression studies in chemosensory neurons. J Neurosci 19:6599–6609

Heimbeck G, Bugnon V, Gendre N, Keller A, Stocker RF (2001) A central neural circuit for experience-independent olfactory and courtship behavior in *Drosophila melanogaster*. Proc Natl Acad Sci USA 98:15336–15341

Heisenberg M (2003) Mushroom body memoir: from maps to models. Nat Rev Neurosci 4:266–275

Heisenberg M, Borst A, Wagner S, Byers D (1985) *Drosophila* mushroom body mutants are deficient in olfactory learning. J Neurogen 2:1–30

Heisenberg M, Gerber B. Behavioral Analysis of Learning and Memory in *Drosophila*. In R. Menzel (Ed.), Learning Theory and Behavior. Vol. [1] of Learning and Memory: A Comprehensive Reference, 4 vols. (J. Byrne Editor), pp. [549–560] Oxford: Elsevier.

Hendel T, Michels B, Neuser K, Schipanski A, Kaun K, Sokolowski MB, Marohn F, Michel R, Heisenberg M, Gerber B (2005) The carrot, not the stick: appetitive rather than aversive gustatory stimuli support associative olfactory learning in individually assayed *Drosophila* larvae. J Comp Physiol A 191:265–279

Hildebrand JG, Shepherd G (1997) Mechanisms of olfactory discrimination: converging evidence for common principles across phyla. Annu Rev Neurosci 20:595–631

Hilfiker S, Pieribone VA, Czernik AJ, Kao H-T, Augustine GJ, Greengard P (1999) Synapsins as regulators of neurotransmitter release. Philos Trans R Soc Lond B 354:269–279

Hiroi M, Marion-Poll F, Tanimura T (2002) Differentiated nerve response to sugars among labellar chemosensilla in *Drosophila*. Zool Sci 19:1009–1018

Hiroi M, Meunier N, Marion-Poll F, Tanimura T (2004) Two antagonistic gustatory receptor neurons responding to sweet-salty and bitter taste in *Drosophila*. J Neurobiol 61:333–342

Hoare DJ, McCrohan CR, Cobb M (2008) Precise and fuzzy coding by olfactory sensory neurons. J Neurosci 28:9710–9722

Hoffmann J (2003) Anticipatory behavioral control. In: Butz MV, Sigaud O, Gerad P (eds) Anticipatory behavior in adaptive learning systems. Springer, Heidelberg, pp 44–65

Honjo K, Furukubo-Tokunaga K (2005) Induction of cAMP response element-binding protein-dependent medium-term memory by appetitive gustatory reinforcement in *Drosophila* larvae. J Neurosci 25:7905–7913

Inoshita T, Tanimura T (2006) Cellular identification of water gustatory receptor neurons and their central projection pattern in *Drosophila*. Proc Natl Acad Sci USA 103:1094–1099

Ishimoto H, Tanimura T (2004) Molecular neurophysiology of taste in *Drosophila*. Cell Mol Life Sci 61:10–18

Ito K, Awano W, Suzuki K, Hiromi Y, Yamamoto D (1997) The *Drosophila* mushroom body is a quadruple structure of clonal units each of which contains a virtually identical set of neurones and glial cells. Development 124:761–771

Ito K, Suzuki K, Estes P, Ramaswami M, Yamamoto D, Strausfeld NJ (1998) The organization of extrinsic neurons and their implications in the functional roles of the mushroom bodies in *Drosophila melanogaster* Meigen. Learn Mem 5:52–77

Jefferis GSXE, Potter CJ, Chan AM, Marin EC, Rohlfing T, Maurer CR Jr, Luo L (2007) Comprehensive maps of *Drosophila* higher olfactory centers: spatially segregated fruit and pheromone representation. Cell 128:1187–1203

Jiao Y, Moon SJ, Montell C (2007) A *Drosophila* gustatory receptor required for the responses to sucrose glucose and maltose identified by mRNA tagging. Proc Natl Acad Sci USA 104:14110–14115

Jones WD, Cayirlioglu P, Kadow IG, Vosshall LB (2007) Two chemosensory receptors together mediate carbon dioxide detection in *Drosophila*. Nature 445:86–90

Kaas JH (2005) The future of mapping sensory cortex in primates: three of many remaining issues. Philos Trans R Soc Lond B 360:653–664

Kaun KR, Hendel T, Gerber B, Sokolowski MB (2007) Natural variation in *Drosophila* larval reward learning and memory due to a cGMP-dependent protein kinase. Learn Mem 14:342–349

Keene AC, Stratmann M, Keller A, Perrat PN, Vosshall LB, Waddell S (2004) Diverse odor-conditioned memories require uniquely timed dorsal paired medial neuron output. Neuron 44:521–533

Keene AC, Waddell S (2007) *Drosophila* olfactory memory: single genes to complex neural circuits. Nat Rev Neurosci 8:341–354

Keene AC, Krashes MJ, Leung B, Bernard JA, Waddell S (2006) *Drosophila* dorsal paired medial neurons provide a general mechanism for memory consolidation. Curr Biol 16:1524–1530

Kim YC, Lee HG, Seong CS, Han KA (2003) Expression of a D1 dopamine receptor dDA1/DmDOP1 in the central nervous system of *Drosophila melanogaster*. Gene Expr Patterns 3:237–245

Kim YC, Lee HG, Han KA (2007) D1 dopamine receptor dDA1 is required in the mushroom body neurons for aversive and appetitive learning in *Drosophila*. J Neurosci 27:7640–7647

Kittel RJ, Wichmann C, Rasse TM, Fouquet W, Schmidt M, Schmid A, Wagh DA, Pawlu C, Kellner RR, Willig KI, Hell SW, Buchner E, Heckmann M, Sigrist SJ (2006) Bruchpilot promotes active zone assembly, Ca^{2+} channel clustering, and vesicle release. Science 312:1051–1054

Klagges B, Heimbeck G, Godenschwege TA, Hofbauer A, Pflugfelder GO, Reifegerste R, Reisch D, Schaupp M, Buchner S, Buchner E (1996) Invertebrate synapsins: a single gene codes for several isoforms in *Drosophila*. J Neurosci 16:3154–3165

Knight D, Iliadi K, Charlton MP, Atwood HL, Boulianne GL (2007) Presynaptic plasticity and associative learning are impaired in a *Drosophila* presenilin null mutant. Dev Neurobiol 67:1598–1613

Kondoh Y, Kaneshiro KY, Kimura K, Yamamoto D (2003) Evolution of sexual dimorphism in the olfactory brain of Hawaiian *Drosophila*. Proc R Soc Lond B 270:1005–1013

Krashes MJ, Keene AC, Leung B, Armstrong JD, Waddell S (2007) Sequential use of mushroom body neuron subsets during *Drosophila* odor memory processing. Neuron 53:103–115

Krashes MJ, Waddell S (2008) Rapid consolidation to a radish and protein synthesis-dependent long-term memory after single-session appetitive olfactory conditioning in *Drosophila*. J Neuro Sci. 28:3103–3113

Kreher SA, Kwon AY, Carlson JR (2005) The molecular basis of odor coding in the *Drosophila* larva. Neuron 46:445–456

Kreher, S.A, Mathew, D, Kim, J, Carlson, JR, (2008) Translation of sensory input into behavioral output via an olfactory system. Neuron 59: 110–124

Kurtovic A, Widmer A, Dickson BJ (2007) A single class of olfactory neurons mediates behavioural responses to *Drosophila* sex pheromone. Nature 446:542–546

Kwon JY, Dahanukar A, Weiss LA, Carlson JR (2007) The molecular basis of CO_2 reception in *Drosophila*. Proc Natl Acad Sci USA 104:3574–3578

Laissue PP, Reiter C, Hiesinger PR, Halter S, Fischbach KF, Stocker RF (1999) Three-dimensional reconstruction of the antennal lobe in *Drosophila melanogaster*. J Comp Neurol 405:543–552

Larsson MC, Domingos AI, Jones WD, Chiappe ME, Amrein H, Vosshall LB (2004) Or83b encodes a broadly expressed odorant receptor essential for *Drosophila* olfaction. Neuron 43:703–714

Lacaille F, Hiroi M, Twele R, Inoshita T, Umemoto D, Manière G, Marion-Poll F, Ozaki M, Francke W, Everaerts C, Tanimura T, Ferveur J-F (2007) A inhibitory sex pheromone tastes bitter for males. PLoS ONE 2:e661

Laurent G (1996) Odor images and tunes. Neuron 16:473–476

Lee T, Lee A, Luo L (1999) Development of the *Drosophila* mushroom bodies: sequential generation of three distinct types of neurons from a neuroblast. Development 126:4065–4076

Levin LR, Han PL, Hwang PM, Feinstein PG, Davis RL, Reed RR (1992) The *Drosophila* learning and memory gene rutabaga encodes a Ca^{2+}/calmodulin-responsive adenylyl cyclase. Cell 68:479–489

Li J, Ashley J, Budnik V, Bhat MA (2007) Crucial role of *Drosophila* neurexin in proper active zone apposition to postsynaptic densities, synaptic growth, and synaptic transmission. Neuron 55:741–755

Lienhard MC, Stocker RF (1987) Sensory projection patterns of supernumerary legs and aristae in *D. melanogaster*. J Exp Zool 244:187–201

Lin HH, Lai JS, Chin AL, Chen YC, Chiang AS (2007) A map of olfactory representation in the *Drosophila* mushroom body. Cell 128:1205–1217

Lindemann B (2001) Receptors and transduction in taste. Nature 413:219–225

Liu L, Leonard AS, Motto DG, Feller MA, Price MP, Johnson WA, Welsh MJ (2003a) Contribution of *Drosophila* DEG/ENaC genes to salt taste. Neuron 39:133–146

Liu L, Yermolaieva O, Johnson WA, Abboud FM, Welsh MJ (2003b) Identification and function of thermosensory neurons in *Drosophila* larvae. Nat Neurosci 6:267–273

Liu G, Seiler H, Wen A, Zars T, Ito K, Wolf R, Heisenberg M, Liu L (2006) Distinct memory traces for two visual features in the *Drosophila* brain. Nature 439:551–556

Louis M, Huber T, Benton R, Sakmar TP, Vosshall LB (2008) Bilateral olfactory sensory input enhances chemotaxis behavior. Nat Neurosci 11:187–199

Manoli DS, Foss M, Villella A, Taylor BJ, Hall JC, Baker BS (2005) Male-specific fruitless specifies the neural substrates of *Drosophila* courtship behavior. Nature 436:395–400

Mao Z, Roman G, Zong L, Davis RL (2004) Pharmacogenetic rescue in time and space of the rutabaga memory impairment by using Gene-Switch. Proc Natl Acad Sci USA 101:198–203

Marella S, Fischler W, Kong P, Asgarian S, Rueckert E, Scott K (2006) Imaging taste responses in the fly brain reveals a functional map of taste category and behavior. Neuron 49:285–295

Margulies C, Tully T, Dubnau J (2005) Deconstructing memory in *Drosophila*. Curr Biol 15:R700-R713

Marin EC, Jefferis GSXE, Komiyama T, Zhu H, Luo L (2002) Representation of the glomerular olfactory map in the *Drosophila* brain. Cell 109:243–255

Marin EC, Watts RJ, Tanaka NK, Ito K, Luo L (2005) Developmentally programmed remodeling of the *Drosophila* olfactory circuit. Development 132:725–737

Masuda-Nakagawa LM, Tanaka NK, O'Kane CJ (2005) Stereotypic and random patterns of connectivity in the larval mushroom body calyx of *Drosophila*. Proc Natl Acad Sci USA 102:19027–19032

Matsuo T, Sugaya S, Yasukawa J, Aigaki T, Fuyama Y (2007) Odorant-binding proteins OBP57d and OBP57e affect taste perception and host-plant preference in *Drosophila sechellia*. PLoS Biol 5:e118

McGuire SE, Le PT, Davis RL (2001) The role of *Drosophila* mushroom body signaling in olfactory memory. Science 293:1330–1333

McGuire SE, Le PT, Osborn AJ, Matsumoto K, Davis RL (2003) Spatiotemporal rescue of memory dysfunction in *Drosophila*. Science 302:1765–1768

McGuire SE, Deshazer M, Davis RL (2005) Thirty years of olfactory learning and memory research in *Drosophila melanogaster*. Prog Neurobiol 76:328–347

Melcher C, Pankratz MJ (2005) Candidate gustatory interneurons modulating feeding behavior in the *Drosophila* brain. PLoS Biol 3:e305

Meunier N, Ferveur JF, Marion-Poll F (2000) Sex-specific non-pheromonal taste receptors in *Drosophila*. Curr Biol 10:1583–1586

Meunier N, Marion-Poll F, Rospars JP, Tanimura T (2003) Peripheral coding of bitter taste in *Drosophila*. J Neurobiol 56:139–152

Miyakawa Y (1982) Behavioral evidence for the existence of sugar, salt and amino acid recptor cells and some of their properties in *Drosophila* larvae. J Insect Physiol 28:405–410

Michels B, Diegelmann S, Tanimoto H, Schwenkert I, Buchner E, Gerber B (2005) A role of synapsin for associative learning: The *Drosophila* larva as a study case. Learn Mem 12:224–231

Mombaerts P, Wang F, Dulac C, Chao SK, Nemes A, Mendelsohn M, Edmondson J, Axel R (1996) Visualizing an olfactory sensory map. Cell 87:675–686

Moon SJ, Kottgen M, Jiao Y, Xu H, Montell C (2006) A taste receptor required for the caffeine response in vivo. Curr Biol 16:1812–1817

Murthy M, Fiete I, Laurent G (2008) Testing odor response stereotypy in the *Drosophila* mushroom body. Neuron 59:1009–1023

Nayak SV, Singh RN (1983) Sensilla on the tarsal segments and mouthparts of adult *Drosophila melanogaster* Meigen (Diptera *Drosophilidae*). Int J Insect Morphol Embryol 12:273–291

Neuser K, Husse J, Stock P, Gerber B (2005) Appetitive olfactory learning in *Drosophila* larvae: effects of repetition, reward strength, age, gender, assay type, and memory span. Anim Behav 69:891–898

Ng M, Roorda RD, Lima SQ, Zemelman BV, Morcillo P, Miesenböck G (2002) Transmission of olfactory information between three populations of neurons in the antennal lobe of the fly. Neuron 36:463–474

Niewalda T, Singhal N, Fiala A, Saumweber T, Wegener S, Gerber B (2008) Salt processing in larval *Drosophila*: Choice, feeding, and learning shift from appetitive to aversive in a concentration-dependent way. Chem Senses 33:685–692

Olsen SR, Bhandawat V, Wilson RI (2007) Excitatory interactions between olfactory processing channels in the *Drosophila* antennal lobe. Neuron 54:89–103

Osborne KA, Robicho A, Burgess E, Butland S, Shaw RA, Coulthard A, Pereira HS, Greenspan RJ, Sokolowski MB (1997) Natural behavior polymorphism due to a cGMP-dependent protein kinase of *Drosophila*. Science 277:834–836

Ozaki M, Wada-Katsumata A, Fujikawa K, Iwasaki M, Yokohari F, Satoji Y, Nisimura T, Yamaoka R (2005) Ant nestmate and non-nestmate discrimination by a chemosensory sensillum. Science 309:311–314

Park SK, Mann KJ, Lin H, Starostina E, Kolski-Andreaco A, Pikielny CW (2006) A *Drosophila* protein specific to pheromone-sensing gustatory hairs delays males' copulation attempts. Curr Biol 16:1154–1159

Peele P, Ditzen M, Menzel R, Galizia CG (2006) Appetitive odor learning does not change olfactory coding in a subpopulation of honeybee antennal lobe neurons. J Comp Physiol A Neuroethol Sens Neural Behav Physiol [A] 192:1083–1103

Pelz D, Roeske T, Syed Z, de Bruyne M, Galizia CG (2006) The molecular receptive range of an olfactory receptor in vivo (*Drosophila melanogaster* Or22a). J Neurobiol 66:1544–1563

Perez-Orive J, Mazor O, Turner GC, Cassenaer S, Wilson RI, Laurent G (2002) Oscillations and sparsening of odor representations in the mushroom body. Science 297:359–365

Perisse E, Portelli G, Le Goas S, Teste E, Le Bourg E (2007) Further characterization of an aversive learning task in *Drosophila melanogaster*: intensity of the stimulus, relearning, and use of rutabaga mutants. J Comp Physiol A Neuroethol Sens Neural Behav Physiol 193:1139–1149

Python F, Stocker RF (2002) Adult-like complexity of the larval antennal lobe of *D. melanogaster* despite markedly low numbers of odorant receptor neurons. J Comp Neurol 445:374–387

Quinn WG, Harris WA, Benzer S (1974) Conditioned behavior in *Drosophila melanogaster*. Proc Natl Acad Sci USA 71:708–712

Ramaekers A, Magnenat E, Marin EC, Gendre N, Jefferis GSXE, Luo L, Stocker RF (2005) Glomerular maps without cellular redundancy at successive levels of the *Drosophila* larval olfactory circuit. Curr Biol 15:982–992

Ressler KJ, Sullivan SL, Buck LB (1994) Information coding in the olfactory system: Evidence for a stereotyped and highly organized epitope map in the olfactory bulb. Cell 79:1245–1255

Riemensperger T, Voller T, Stock P, Buchner E, Fiala A (2005) Punishment prediction by dopaminergic neurons in *Drosophila*. Curr Biol 15:1953–1960

Robertson HM, Warr CG, Carlson JR (2003) Molecular evolution of the insect chemoreceptor gene superfamily in *Drosophila melanogaster*. Proc Natl Acad Sci USA 100:14537–14542

Rodrigues V (1980) Olfactory behavior of *Drosophila melanogaster*. In: Siddiqi O, Babu P, Hall LM, Hall JC (eds) Development and Neurobiology of *Drosophila*. New York, London: Plenum, pp 361–371

Rodrigues V, Siddiqi O (1978) Genetic-analysis of chemosensory pathway. Proc Indian Acad Sci Sect B Biol Sci 87:147–160

Root CM, Semmelhack JL, Wong AM, Flores J, Wang JW (2007) Propagation of olfactory information in *Drosophila*. Proc Natl Acad Sci USA 104:11826–11831

Sachse S, Rueckert E, Keller A, Okada R, Tanaka NK, Ito K, and Vosshall L.B. (2007). Activity-dependent plasticity in an olfactory circuit. Neuron 56, 838–850

Sato K, Pellegrino M, Nakagawa T, Vosshall LB, Touhara K (2008) Insect olfactory receptors are heteromeric ligand-gated ion channels. Nature 452:1002–1006.

Scherer S, Stocker RF, Gerber B (2003) Olfactory learning in individually assayed *Drosophila* larvae. Learn Mem 10:217–225

Schipanski A (2007) Reinforcement processing in larval *Drosophila melanogaster*. Diploma Thesis, University of Würzburg

Schipanski A, Yarali A, Niewalda T, Gerber B (2008) Behavioral analyses of sugar processing in choice, feeding, and learning in larval *Drosophila*. Chem Senses 33:563–573

Schroll C, Riemensperger T, Bucher D, Ehmer J, Völler T, Erbgut K, Gerber B, Hendel T, Nagel G, Buchner E, Fiala A (2006) Light-induced activation of distinct modulatory neurons substitutes for appetitive or aversive reinforcement during associative learning in larval *Drosophila*. Curr Biol 16:1741–1747

Schlief, M.L., and Wilson, R.I. (2007). Olfactory processing and behavior downstream from highly selective receptor neurons. Nat Neurosci 10, 623–630

Schwaerzel M, Heisenberg M, Zars T (2002) Extinction antagonizes olfactory memory at the subcellular level. Neuron 35:951–960

Schwaerzel M (2003) Localizing engrams of olfactory memories in *Drosophila*. PhD Thesis, University of Würzburg

Schwaerzel M, Monastirioti M, Scholz H, Friggi-Grelin F, Birman S, Heisenberg M (2003) Dopamine and octopamine differentiate between aversive and appetitive olfactory memories in *Drosophila*. J Neurosci 23:10495–10502

Schwaerzel M, Jaeckel A, Mueller U (2007) Signaling at A-kinase anchoring proteins organizes anesthesia-sensitive memory in *Drosophila*. J Neurosci 27:1229–1233

Scott K, Brady R Jr, Cravchik A, Morozov P, Rzhetsky A, Zuker C, Axel R (2001) A chemosensory gene family encoding candidate gustatory and olfactory receptors in *Drosophila*. Cell 104:661–673

Shanbhag SR, Müller B, Steinbrecht RA (1999) Atlas of olfactory organs of *Drosophila melanogaster*. 1. Types, external organization, innervation and distribution of olfactory sensilla. Int J Insect Morphol Embryol 28:377–397

Shanbhag SR, Park SK, Pikielny CW, Steinbrecht RA (2001) Gustatory organs of *Drosophila melanogaster* fine structure and expression of the putative odorant-binding protein PBPRP2. Cell Tissue Res 304:423–437

Shang Y, Claridge-Chang A, Sjulson L, Pypaert M, Miesenböck G (2007) Excitatory local circuits and their implications for olfactory processing in the fly antennal lobe. Cell 128:601–612

Sinakevitch I, Strausfeld NJ (2006) Comparison of octopamine-like immunoreactivity in the brains of the fruit fly and blow fly. J Comp Neurol 494:460–475

Singh RN (1998) Neurobiology of the gustatory systems of *Drosophila* and some terrestrial insects. Microsc Res Techn 39:547–563

Singh RN, Singh K (1984) Fine structure of the sensory organs of *Drosophila melanogaster* Meigen larva (Diptera: *Drosophilidae*). Int J Insect Morphol Embryol 13:255–273

Slone J, Daniels J, Amrein H (2007) Sugar receptors in *Drosophila*. Curr Biol 17:1809–1816

Sone M, Suzuki E, Hoshino M, Hou D, Kuromi H, Fukata M, Kuroda S, Kaibuchi K, Nabeshima Y, Hama C (2000) Synaptic development is controlled in the periactive zones of *Drosophila* synapses. Development 127:4157–4168

Stocker RF (1994) The organization of the chemosensory system in *Drosophila melanogaster*: a review. Cell Tiss Res 275:3–26

Stocker RF (2001) *Drosophila* as a focus in olfactory research: mapping of olfactory sensilla by fine structure, odor specificity, odorant receptor expression and central connectivity. Microsc Res Techn 55:284–296

Stocker RF, Heimbeck G, Gendre N, de Belle JS (1997) Neuroblast ablation in *Drosophila* P[GAL4] lines reveals origins of olfactory interneurons. J Neurobiol 32:443–456

Stocker RF, Singh RN, Schorderet M, Siddiqi O (1983) Projection patterns of different types of antennal sensilla in the antennal glomeruli of *Drosophila melanogaster*. Cell Tissue Res 232:237–248

Stockinger P, Kvitsiani D, Rotkopf S, Tirian L, Dickson BJ (2005) Neural circuitry that governs *Drosophila* male courtship behavior. Cell 121:795–807

Strausfeld NJ, Hildebrand JG (1999) Olfactory systems: common design, uncommon origins? Curr Opin Neurobiol 9:634–639

Strausfeld NJ, Sinakevitch I, Vilinsky I (2003) The mushroom bodies of *Drosophila melanogaster*: an immunocytological and Golgi study of Kenyon cell organization in the calyces and lobes. Microsc Res Tech 62:151–169

Suh GSB, Wong AM, Hergarden AC, Wang JW, Simon AF, Benzer S, Axel R, Anderson DJ (2004) A single population of olfactory sensory neurons mediates an innate avoidance behavior in *Drosophila*. Nature 431:854–859

Tamura T, Chiang AS, Ito N, Liu HP, Horiuchi J, Tully T, Saitoe M (2003) Aging specifically impairs *amnesiac*-dependent memory in *Drosophila*. Neuron 40:1003–1011

Tanaka NK, Awasaki T, Shimada T, Ito K (2004) Integration of chemosensory pathways in the *Drosophila* second-order olfactory centers. Curr Biol 14:449–457

Tanaka NK, Tanimoto H, and Ito, K (2008). Neuronal assemblies of the *Drosophila* mushroom body. J Comp Neurol 508, 711–755

Tanimura T, Isono K, Takamura T, Shimada I (1982) Genetic dimorphism in the taste sensitivity to trehalose in *Drosophila melanogaster*. J Comp Physiol [A] 147:433–437

Tanimura T, Isono K, Yamamoto M-T (1988) Taste sensitivity to trehalose and its alteration by gene dosage in *Drosophila melanogaster*. Genetics 119:366–406

Tempel BL, Bonini N, Dawson DR, Quinn WG (1983) Reward learning in normal and mutant *Drosophila*. Proc Natl Acad Sci USA 80:1482–1486

Thorne N, Chromey C, Bray S, Amrein H (2004) Taste perception and coding in *Drosophila*. Curr Biol 14:1065–1079

Thum AS, Jenett A, Ito K, Heisenberg M, Tanimoto H (2007) Multiple memory traces for olfactory reward learning in *Drosophila*. J Neurosci 27:11132–11138

Tully T, Quinn WG (1985) Classical conditioning and retention in normal and mutant *Drosophila melanogaster*. J Comp Physiol [A] 157:263–277

Tully T, Cambiazo V, Kruse L (1994) Memory through metamorphosis in normal and mutant *Drosophila*. J Neurosci 14:68–74

Ueno K, Ohta M, Morita H, Mikuni Y, Nakajima S, Yamamoto K, Isono K (2001) Trehalose sensitivity in *Drosophila* correlates with mutations in and expression of the gustatory receptor gene Gr5a. Curr Biol 11:1451–1455

Van der Goes van Naters W, Carlson JR (2007) Receptors and neurons for fly odors in *Drosophila*. Curr Biol 17:606–612

Vassar R, Chao SK, Sitcheran R, Nunez JM, Vosshall LB, Axel R (1994) Topographic organization of sensory projections to the olfactory bulb. Cell 79:981–991

Vosshall LB, Amrein H, Morozov PS, Rzhetsky A, Axel R (1999) A spatial map of olfactory receptor expression in the *Drosophila* antenna. Cell 96:725–736

Vosshall LB, Wong AM, Axel R (2000) An olfactory sensory map in the fly brain. Cell 102:147–159.

Vosshall LB, Stocker RF (2007) Molecular architecture of smell and taste in *Drosophila*. Annu Rev Neurosci 30:505–533

Waddell S, Armstrong JD, Kitamoto T, Kaiser K, Quinn WG (2000) The *amnesiac* gene product is expressed in two neurons in the *Drosophila* brain that are critical for memory. Cell 103:805–813

Wagh DA, Rasse TM, Asan E, Hofbauer A, Schwenkert I, Durrbeck H, Buchner S, Dabauvalle MC, Schmidt M, Qin G, Wichmann C, Kittel R, Sigrist SJ, Buchner E (2006) Bruchpilot, a protein with homology to ELKS/CAST, is required for structural integrity and function of synaptic active zones in *Drosophila*. Neuron 49:833–844

Wang Y, Guo HF, Pologruto TA, Hannan F, Hakker I, Svoboda K, Zhong Y (2004a) Stereotyped odor-evoked activity in the mushroom body of *Drosophila* revealed by green fluorescent protein-based Ca^{2+} imaging. J Neurosci 24:6507–6514

Wang Z, Singhvi A, Kong P, Scott K (2004b) Taste representations in the *Drosophila* brain. Cell 117:981–991

Wicher D, Schafer R, Bauernfeind R, Stensmyr MC, Heller R, Heinemann SH, Hansson BS (2008) *Drosophila* odorant receptors are both ligand-gated and cyclic-nucleotide-activated cation channels. Nature 452:1007–1011

Wilson RI, Laurent G (2005) Role of GABAergic inhibition in shaping odor-evoked spatiotemporal patterns in the *Drosophila* antennal lobe. J Neurosci 25:9069–9079

Wilson RI, Turner GC, Laurent G (2004) Transformation of olfactory representations in the *Drosophila* antennal lobe. Science 303:366–370

Wolfe MS (2007) When loss is gain: reduced presenilin proteolytic function leads to increased Abeta42/Abeta40. Talking point on the role of presenilin mutations in Alzheimer disease. EMBO Rep 8:136–140

Wong AM, Wang JW, Axel R (2002) Spatial representation of the glomerular map in the *Drosophila* protocerebrum. Cell 109:229–241

Wustmann G, Rein K, Wolf R, Heisenberg M (1996) A new paradigm for operant conditioning of *Drosophila melanogaster*. J Comp Physiol [A] 179:429–436

Yao CA, Ignell R, Carlson JR (2005) Chemosensory coding by neurons in the coeloconic sensilla of the *Drosophila* antenna. J Neurosci 25:8359–8367

Yarali T, Hendel B, Gerber B (2007) Olfactory learning and behaviour are 'insulated' against visual processing in larval *Drosophila*. J Comp Physiol [A] 192:1133–1145

Yasuyama K, Meinertzhagen IA, Schürmann FW (2002) Synaptic organization of the mushroom body calyx in *Drosophila melanogaster*. J Comp Neurol 445:211–226

Yu D, Ponomarev A, Davis RL (2004) Altered representation of the spatial code for odors after olfactory classical conditioning: memory trace formation by synaptic recruitment. Neuron 42:437–449

Yu D, Keene AC, Srivatsan A, Waddell S, Davis RL (2005) *Drosophila* DPM neurons form a delayed and branch-specific memory trace after olfactory classical conditioning. Cell 123:945–957

Zars T (2000) Behavioral functions of the insect mushroom bodies. Curr Opin Neurobiol 10:790–795

Zars T, Fischer M, Schulz R, Heisenberg M (2000) Localization of a short-term memory in *Drosophila*. Science 288:672–675

Zeng X, Sun M, Liu L, Chen F, Wei L, Xie W (2007) Neurexin-1 is required for synapse formation and larvae associative learning in *Drosophila*. FEBS Lett 581:2509–25169

The Receptor Basis of Sweet Taste in Mammals

S. Vigues, C.D. Dotson, and S.D. Munger

Abstract The taste of sweeteners is hedonically pleasing, suggests high caloric value in food, and contributes to increased intake. In recent years, many of the molecular mechanisms underlying the detection of sweeteners have been elucidated. Of particular note is the identification of the sweet taste receptor, the heteromeric G-protein-coupled receptor T1R2:T1R3, which responds to a vast array of chemically diverse natural and artificial sweeteners. In this chapter, we discuss some of the mechanisms underlying the detection of sweeteners by mammals, with a particular focus on the function and role of the T1R2:T1R3 receptor in these processes.

1 Introduction

Sweet taste opens a particular window to our sensory world. It can indicate the presence of key nutrients, can enhance the taste and hedonic properties of food, and can influence our choices of what to ingest. Thus, sweet taste is of immense interest to sensory scientists and neuroscientists, food scientists, dieticians, and others, not to mention the general public. While carbohydrate sweeteners – that is, sugars – are appreciated for their pleasing taste and high caloric content, overingestion of sugars is associated with obesity and obesity-related disease. To meet the ever-growing demand for low-calorie sweeteners with pleasing sensory properties and versatile uses in foods and beverages, many are seeking to illuminate the molecular mechanisms of sweet taste.

Many excellent reviews have been published in the last few years covering different aspects of sweet taste function (Bachmanov and Beauchamp 2007; Boughter and Bachmanov 2007; Chandrashekar et al. 2006; Lemon and Katz 2007; Roper

S.D. Munger (✉)
Department of Anatomy and Neurobiology, University of Maryland School of Medicine, Baltimore, MD 21201, USA
e-mail: smung001@umaryland.edu

2006; Scott 2005). Here, we will focus on the detection of sweeteners, with a special emphasis on interactions between sweeteners and the sweet taste receptor.

2 What Is Sweet?

Sweetness, like other taste qualities (e.g., bitter, salty, sour), is a human percept. At levels normally found in foods and beverages, compounds that elicit a sweet taste (i.e., sweeteners) are hedonically pleasing to humans and are preferred by most mammals (Breslin and Spector 2008). Sugars such as sucrose, glucose, and fructose are prototypical sweet-tasting compounds. However, humans perceive a diverse array of natural and synthetic compounds as having a sweet taste. Natural sweeteners come from several chemical classes, including sugars, sugar alcohols (e.g., mannitol, xylitol), proteins (e.g., thaumatin, monellin), and amino acids. Synthetic sweeteners are equivalently diverse: commonly available ones include sulfamates (e.g., sodium cyclamate), dipeptides (e.g., aspartame, neotame), halogenated sugars (e.g., sucralose), and sulfonyl amides (e.g., sodium saccharin, acesulfame potassium). Mammals exhibit strong preferences for stimuli from most of these chemical classes, emphasizing the importance of sweetener detection across species.

Do these diverse compounds give rise to a common perception of sweetness or to qualitatively different sensations? Sweetness does indeed appear to be a unitary percept (Breslin et al. 1994, 1996). However, some sweeteners may be discriminable on the basis of their activation of other sensory transduction mechanisms or differences in the temporal properties of their sensory action. For example, the sweetener sodium saccharin activates bitter receptors in some people (Kuhn et al. 2004; Pronin et al. 2007), and also inhibits sweet taste at high concentrations (Galindo-Cuspinera et al. 2006). Sweet proteins such as thaumatin and monellin can have a slow onset or evoke a prolonged sweetness compared with sugars (Faus 2000), likely owing to a relatively high affinity for the sweet taste receptor.

3 Species Differences in Response to Sweeteners

While most mammals can detect and do prefer a wide variety of sweeteners, species vary considerably in their sweetener preference. Many of the compounds that taste sweet to humans are favored by a variety of mammalian species, including mice (Bachmanov et al. 2001a; Fuller 1974; Kasahara et al. 1987; Lush 1989), hamsters (MacKinnon et al. 1999), rats (Nowlis et al. 1980), rabbits (Carpenter 1956), pigs (Nofre et al. 2002; Tinti et al. 2000), opossums (Pressman and Doolittle 1966), and primates (Fisher et al. 1965; Glaser et al. 1995, 1998; Haefeli et al. 1998; Nofre et al. 1996). However, species differences do exist. Rodents are indifferent to several artificial sweeteners (e.g., cyclamate, alitame, aspartame) (Hellekant and

Danilova 1996; Nowlis et al. 1980), though mice do prefer saccharin, acesulfame K, sucralose, and dulcin (Bachmanov et al. 2001a). Interestingly, rats show little or no preference for either sucralose or dulcin (Bello and Hajnal 2005; Fisher et al. 1965; Sclafani and Clare 2004) even though the rat sweet taste receptor responds robustly to sucralose (Li et al. 2002), thus highlighting the important contributions of nontaste factors to sweetener preference and acceptance. Interestingly, cats appear not to prefer sweeteners of any type (Beauchamp et al. 1977; Carpenter 1956; Pfaffmann 1955). Differences exist between primate species, as well. For example, New World monkeys are indifferent to the artificial sweetener aspartame, whereas Catarrhini strongly prefer its taste (Glaser et al. 1995). These species differences could be a consequence of evolutionary pressures that select for the detection of food compounds prevalent in local, species-specific, environments. In any case, researchers have taken advantage of these differences to explore the biological basis of sweet taste perception, especially as it relates to stimulus detection.

4 Interindividual Differences in Sweet Taste

In humans, data supporting the existence of individual differences in the ability to detect sweet-tasting compounds are relatively sparse and unpersuasive (Blakeslee and Salmon 1935; Kahn 1951). However, some differences in peripheral anatomy, such as the density of fungiform papillae on the tongue (and presumably the number of functioning taste receptors), have been shown to influence sensitivity to stimuli of multiple taste qualities (e.g., sweet, bitter) (Duffy 2007; Miller and Reedy 1990; Stein et al. 1994). In contrast, humans can differ strongly in the degree to which they prefer sweeteners (Reed and McDaniel 2006). As preference for sweeteners can be influenced by multiple factors, including age, sex, culture, mood, appetite, digestive ability, and intake experience (Reed and McDaniel 2006; Stevens 1996), these preference differences are likely unrelated to differences in sensitivity to sweeteners.

In contrast, inbred mouse strains differ strongly in their responsiveness to sweeteners. In general, "taster" mice have lower preference thresholds and higher afferent nerve responsiveness for sweeteners than do "nontaster" mice (Bachmanov et al. 1996; Capeless and Whitney 1995; Capretta 1970; Frank and Blizard 1999; Fuller 1974; Inoue et al. 2001; Lush 1989; Ninomiya et al. 1984; Pelz et al. 1973; Zhao et al. 2003). Genetic linkage analyses identified two chromosomal loci, *Sac* (saccharin preference) (Capeless and Whitney 1995; Fuller 1974; Lush 1989; Lush et al. 1995; Ramirez and Fuller 1976) and *dpa* (D-phenylalanine preference) (Capeless and Whitney 1995; Ninomiya et al. 1991), that influence sweet taste preference in mice. Both loci map near the distal end of mouse chromosome 4. These mapping studies, combined with in silico analyses of human and mouse genome sequences and sophisticated molecular biological approaches, were the keys to the identification of the sweet taste receptor.

5 T1R2 and T1R3 Are Subunits of a Sweet Taste Receptor

Through a combination of physical mapping and genome database mining, several groups identified a gene encoding a putative G-protein-coupled receptor (GPCR) as a candidate for *Sac* (Bachmanov et al. 2001b; Kitagawa et al. 2001; Max et al. 2001; Montmayeur et al. 2001; Nelson et al. 2001; Sainz et al. 2001). This gene, named *Tas1r3*, was a paralogue of two others (*Tas1r1 and Tas1r2*) that encoded orphan GPCRs expressed in subsets of taste cells (Hoon et al. 1999), thus suggesting a role in taste function. The protein encoded by the *Tas1r3* gene, T1R3, was also highly enriched in subsets of taste cells and coexpressed with either T1R1 or T1R2 (though a small number of taste cells may express T1R3 alone). *Tas1r3* haplotypes corresponded well with the relative sensitivity of various inbred mouse strains to prototypical sweeteners (Reed et al. 2004), while transgenic complementation of a "nontaster" strain with a "taster" variant of T1R3 conferred increased sweetener sensitivity (Nelson et al. 2001). Together, these findings provided compelling functional evidence that *Tas1r3* is equivalent to *Sac*.

T1R3 was clearly a strong candidate sweet taste receptor, but in vitro studies showed that the story was not so simple. Through use of in vitro receptor activation assays in heterologous cells, it was found that T1R3 requires coexpression with T1R2 to form a fully functional receptor responsive to a wide range of sweeteners (Li et al. 2002; Nelson et al. 2001). Comparisons of the human and rodent T1R2:T1R3 receptors revealed stimulus tuning consistent with known species differences (Li et al. 2002). For example, both the human and the rodent receptors responded to several sugars and to the artificial sweeteners saccharin and sucralose, while only the human receptor was sensitive to aspartame and cyclamate. Somewhat surprisingly, T1R3 also pairs with T1R1 to form a receptor sensitive to L-amino acids — and is thus an umami ("savory") receptor (Li et al. 2002; Nelson et al. 2002). T1R3 may also function as a homomeric receptor in some cells, where it may act as a low-efficacy receptor for sugars (Nelson et al. 2001; Zhao et al. 2003).

Deletion of the *Tas1r2* and *Tas1r3* genes through gene targeting allowed in vivo confirmation of the primary role of the T1R2:T1R3 receptor in the detection of sweeteners (Damak et al. 2003; Zhao et al. 2003). *Tas1r2* and *Tas1r3* null mice each exhibit a dramatic reduction in behavioral and nerve responses to a variety of sweeteners, while *Tas1r2/Tas1r3* double-knockout mice appear completely ageusic for sweeteners. The discovery that cats, which are indifferent to sugars, carry a pseudogenized *Tas1r2* gene (Li et al. 2005) offered additional support for the requirement of the T1R2:T1R3 receptor in sweetener detection.

6 Cellular Distribution of the Sweet Taste Receptor

The two subunits of the sweet taste receptor are differentially distributed in the gustatory epithelium. T1R2 is expressed most frequently in taste buds of the circumvallate and foliate papillae (Hoon et al. 1999), less so in palatal taste buds

(Hoon et al. 1999), and rarely in taste buds of the fungiform papillae (Hoon et al. 1999). In contrast, T1R3 is expressed in approximately 30% of taste cells within all three taste-bud-containing papillae (Nelson et al. 2001). This apparent discrepancy is at least partially explained by the coexpression of T1R3 with T1R1 in many taste cells of the fungiform papillae and palate, though some studies suggest a more complex pattern of coexpression (Kim et al. 2003; Stone et al. 2007). Additionally, the response properties of mouse chorda tympani and glossopharyngeal nerves (which innervate the fungiform and circumvallate taste buds, respectively, with some overlap in foliate papillae) do not correspond to these predicted patterns (Danilova and Hellekant 2003; Ninomiya and Funakoshi 1989; Ninomiya et al. 1993, 2000). Further work is needed to resolve inconsistencies between these molecular and electrophysiological studies. It is also clear that the sweet taste receptor is not restricted to the gustatory epithelium: T1R2 and T1R3 are expressed in nutrient sensing cells of the gastrointestinal tract, where they play important roles in nutrient detection, response, and assimilation (Sternini et al. 2008).

7 The T1R2:T1R3 Receptor Is a Class C GPCR

The T1Rs most closely resemble GPCRs of class C, which also include metabotropic glutamate receptors (mGluRs), γ-aminobutyric acid type B receptors (GABA$_B$Rs), calcium-sensing receptors, and V2R vomeronasal receptors (Pin et al. 2003). Like all GPCRs, those of class C share a membrane-spanning domain comprising seven helices, three extracellular loops, three intracellular loops, and an intracellular carboxy tail. Class C GPCRs are distinguished from other GPCRs by a large extracellular amino-terminal domain. This extracellular domain contains a Venus-flytrap domain (VFD), which includes the orthosteric binding site and which shares some homology with bacterial amino acid binding proteins (Pin et al. 2003). The VFD is linked to the seven-transmembrane domain (7TMD) by a cysteine-rich domain (CRD). The T1Rs share this basic topology (Fig. 1). Class C GPCRs function as dimers: mGluRs and calcium-sensing receptors as homodimers, and GABA$_B$Rs and T1Rs as heterodimers (though, as discussed already, T1R3 homodimers may function as low-efficacy sweet receptors in some cells).

The crystal structure of the mGluR1 VFD domain has been solved in the presence and absence of the ligand (Kunishima et al. 2000). This seminal paper revealed that the glutamate ligand binds within each VFD cleft, stabilizing a closed conformation of the VFD and contributing to receptor activation. The ability of each mGluR1 subunit to bind ligand at physiological concentrations contrasts with the heterodimeric GABA$_B$R, where ligand binding to just one of the subunits promotes activation (Kaupmann et al. 1998; Pin et al. 2003). These distinct models for class C GPCR activation beg the question as to what contributions each subunit of the T1R2:T1R3 receptor makes to the detection of and response to sweeteners.

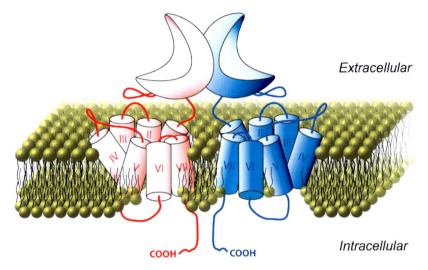

Fig. 1 Topography of the sweet taste receptor. The sweet taste receptor is an integral membrane protein complex composed of two subunits, T1R2 (*red*) and T1R3 (*blue*). Each subunit has three main domains: a large, extracellular Venus-flytrap domain (VFD) at the amino end of the protein; a seven-transmembrane helical domain typical of G-protein-coupled receptors on the carboxyl end; and a cysteine-rich linker domain that connects the other two domains

8 Structure–Function Studies of the T1R2:T1R3 Receptor

8.1 Using Receptor Chimeras To Map Functional Domains

As described already, humans and rodents differ in their sensitivity to certain artificial and natural sweeteners. For example, while humans find aspartame, neotame, cyclamate, neohesperidin dihydrochalcone, and the sweet proteins (e.g., monellin, brazzein, thaumatin) to be sweet tasting, rodents are indifferent to all of them. Several laboratories took advantage of these species differences to design an elegant series of experiments using human/rodent chimeric receptors to map regions of the T1R2:T1R3 dimer that are required for responsiveness to some of these sweeteners. These experiments, largely performed using heterologous cells, revealed that the extracellular domain of human T1R2 is required for receptor responses to aspartame and neotame (Xu et al. 2004) (Fig. 2a), the CRD of human T1R3 is required for responses to the protein brazzein (Jiang et al. 2004), and the transmembrane domain (TMD) of human T1R3 is necessary for responses to cyclamate (Jiang et al. 2005b; Xu et al. 2004) (Fig. 2b) and neohesperidin dihydrochalcone (Winnig et al. 2007). A creative in vivo experiment in which a mouse was "humanized" with the transgenic expression of the human T1R2 subunit showed that this protein

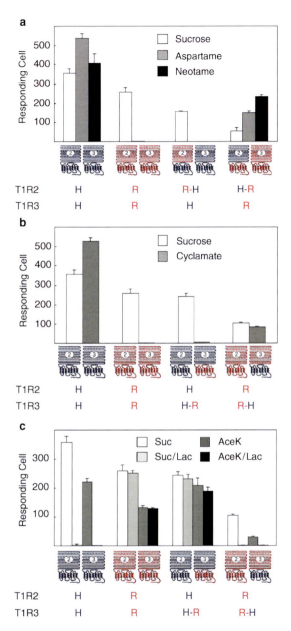

Fig. 2 Use of chimeric T1R2:T1R3 receptors to determine regions critical for sweetener response. Responses to human (*H, blue*), rat (*R, red*) or human–rat chimeric T1R2:T1R3 receptors to **a** sucrose, aspartame, or neotame, **b** sucrose or cyclamate, and **c** sucrose (*Suc*) or acesulfame potassium (*AceK*) in the absence or presence of lactisole (*Lac*). Equivalent numbers of cells were imaged for each receptor–ligand pair. (Adapted from Xu et al. 2004, with permission. Copyright 2004 National Academy of Sciences, USA)

confers human-like taste sensitivity to aspartame, monellin, and other sweeteners (Zhao et al. 2003) (though not neohesperidin dihydrochalcone, an observation explained by its interaction with the T1R3 subunit).

A similar strategy revealed the basis for sweet taste inhibition by lactisole, which also requires the TMD of human T1R3 (Winnig et al. 2005; Xu et al. 2004) (Fig. 2c). Interestingly, this inhibitor as well as high concentrations of the sweeteners saccharin and acesulfame K act as inverse agonists for the human T1R2:T1R3 (Galindo-Cuspinera et al. 2006). Release from this allosteric inhibition, such as by rinsing away the inhibitors with water, underlies the curious phenomenon of "sweet water taste" (Bartoshuk et al. 1972).

8.2 Homology Modeling

The conserved primary structure between T1Rs and mGluRs suggested that these glutamate receptors could serve as an appropriate basis for homology modeling of the sweet receptor. The crystal structure of the mGluR1 VFD with and without bound ligand (Kunishima et al. 2000) proved to be a reasonable template for modeling this domain of the T1Rs (Cui et al. 2006; Max et al. 2001; Morini et al. 2005), while the crystal structure of the class A GPCR rhodopsin (Palczewski et al. 2000) served a similar role for modeling the T1R 7TMD (Winnig et al. 2007). Systematic mutagenesis based on these models revealed an even greater diversity of sweetener binding sites than was previously thought. For example, mutations of two residues in the VFD of human T1R2, the mGluR homologues of which play key roles in glutamate binding, abolishes responses to dipeptide sweeteners but only partially affects responses to sucrose (Xu et al. 2004). These findings suggest that these two classes of small-molecule sweeteners bind to somewhat different sites in the VFD of T1R2. Extensive mutagenesis of the 7TMD of human T1R3 has revealed that two sweeteners, cyclamate and neohesperidin dihydrochalcone, as well as the sweet taste inhibitor lactisole, bind to overlapping sites that include contributions from the third, fifth, and sixth transmembrane helices (Jiang et al. 2005a, b; Winnig et al. 2005; Xu et al. 2004).

8.3 Spectroscopic Measurements of Sweetener Binding

While experiments that focus on compounds that are sweet to humans but not preferred by lower mammals have proven quite informative, they cannot examine the molecular basis for T1R2:T1R3 recognition of a class of sweeteners preferred by all sweetener-sensitive mammals: sugars. To address this issue, our laboratory devised a novel spectroscopic approach to measure ligand interactions with purified T1R2 and T1R3 VFD proteins (Nie et al. 2005, 2006). We expressed the VFDs as fusion proteins in bacteria, and purified them by affinity and ion-exchange chromatography. Binding of sugars (glucose, sucrose, and sucralose) to each domain was quantified by measuring changes in the peak intrinsic tryptophan fluorescence upon titration of ligand. Surprisingly, both subunits bound all three sugars, though with distinct affinities for each subunit. The affinities for sucrose and glucose ranged from approximately 2.5 to 15mM. These values are slightly lower than behavioral EC_{50} values for these sugars in mice (Bachmanov et al. 2001a), and may reflect the absence of the 7TMDs in these purified VFDs or the loss of cooperative interactions between the T1R2 and T1R3 subunits. However, mutation of a single residue in the T1R3 VFD associated with reduced sweet taste sensitivity in *Sac* "nontaster" mice (Reed et al. 2004) (Fig. 3a) decreased the affinity of this protein for all three ligands tested (Nie et al. 2005) (Fig. 3d), validating the approach and indicating that the mechanistic basis for the nontaster phenotype is at least partially dependent on a T1R3 subunit with reduced affinity for sweeteners.

8.4 Structure–Activity Relationships for Sweeteners: Coming Full Circle?

Prior to the identification of the sweet taste receptor, many groups tried to understand the molecular basis of sweetness by examining structure–activity relationships for different classes of sweeteners. Through a comparison the chemical structures of known sweet-tasting compounds, it was hoped that a common "sweet" motif – a so-called glucophore – could be identified and used to predict new sweeteners. While several models were developed on the basis of these approaches, all fell short of the goal of predicting sweetness (DuBois 2004). In light of the recent studies on the T1R2:T1R3 receptor, one reason for this failure is clear: none sufficiently accounted for allosteric binding sites that are selective for particular chemical classes of sweeteners but which all contribute to activation of the receptor. However, it would now seem that the time is ripe to revisit the structure–activity analysis of sweeteners in the context of receptor binding and activation. Extensive libraries of sweeteners (e.g., Spillane et al. 2006) could now be examined for their ability to bind and activate human or cross-species chimeric receptors, elucidating key determinants for sweetener–receptor interactions. The relatively straightforward structural analysis of sweet protein mutants (Assadi-Porter et al. 2003; De Simone et al. 2006; Esposito et al. 2006; Hobbs et al. 2008; Spadaccini et al. 2003) suggests that they may be particularly informative for understanding how changes in sweetener structure impact sweetener efficacy.

9 Future Directions

Though great strides have been made in the last 10 years towards understanding the molecular basis of sweet taste, much remains unknown. Our knowledge of the sweet receptor is a prime example. The absence of T1R2:T1R3 crystal structures

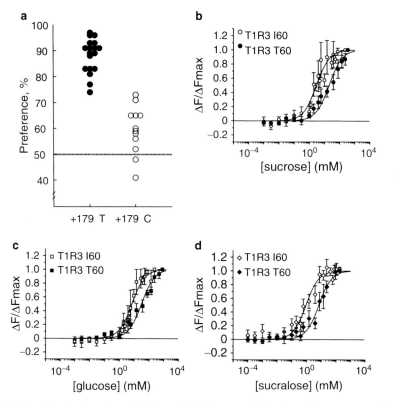

Fig. 3 Effect of a single amino acid change in T1R3 on taste behavior and taste receptor–ligand interactions. **a** Preferences for 1.6 mM saccharin in 30 inbred mouse strains that vary in their genotype at base pair 179 of the *Tas1r3* coding sequence. This is a nonsynonymous polymorphism that results in a change from isoleucine (179T) to threonine (179C) at amino acid position 60 of T1R3. Mice with the T allele at this position exhibit a significantly greater preference for saccharin. **b–d** Changes in peak intrinsic tryptophan fluorescence of the T1R3 isoleucine (I60) and threonine (T60) variants (VFD protein only) upon titration of three saccharide sweeteners. The T1R3 T60 variant showed reduced affinity for each sweetener (sucrose, $K_d = 2.9 \pm 0.4$ mM; glucose, $K_d = 7.3 \pm 0.7$ mM; sucralose, $K_d = 6.9 \pm 0.9$ mM) compared with the I60 variant (sucrose, $K_d = 20 \pm 3$ mM; glucose, $K_d = 32 \pm 5$ mM; sucralose, $K_d = 0.91 \pm 0.015$ mM). (**a** Adapted from Boughter and Bachmanov 2007, with permission. **b–d** Adapted from Nie et al. 2005, with permission)

(preferably in the presence and absence of different classes of ligands) is perhaps the biggest impediment to systematic studies of how sweeteners bind to and activate the T1R2:T1R3 heterodimer. Such knowledge would permit the precise localization of the molecular determinants critical for ligand binding and ligand-dependent receptor activation, and would facilitate the rational design of sweeteners. As informative as mutagenesis studies based on homology modeling have been, they are inadequate to resolve the specific contributions of individual residues to particular receptor functions (e.g., ligand binding, G-protein coupling) because they cannot clearly distinguish local effects from more global changes to the proteins. Unfortunately, methods for the efficient heterologous expression and purification of T1Rs (or any other mammalian chemosensory receptors) at levels suitable for crystallographic studies are still lacking, and may await a better understanding of the factors that contribute to efficient taste receptor expression, folding, and trafficking in native cells. Encouragingly, the advent of high-throughput screening approaches and recent successes in obtaining high-quality crystals of GPCRs (Cherezov et al. 2007; Palczewski et al. 2000) and receptor extracellular domains (Dellisanti et al. 2007; Kunishima et al. 2000) suggests that taste receptor crystal structures may not be far off once the critical barrier of protein production is breached.

Our understanding of the separate contributions of the T1R2 and T1R3 subunits to the function of the intact sweet taste receptor remains superficial. Chimeric and fluorescence spectroscopy studies of these proteins suggest that individual sweeteners can bind to one subunit (e.g., cyclamate) or both subunits (e.g., sugars). Using circular dichroism spectroscopy, we also showed that the VFDs of both T1R2 and T1R3 undergo distinct ligand-dependent conformational changes (Nie et al. 2005), suggesting that each subunit makes a unique contribution to overall ligand response and receptor activation. Thus, it may not be surprising that dissociated VFDs or homomeric T1R3 receptors display affinities and efficacies that differ somewhat from those of the T1R2:T1R3 heterodimer (Nie et al. 2005; Zhao et al. 2003). Class C GPCRs display intersubunit cooperativity. For example, the two subunits of mGluR1 display negative cooperativity in the presence of bound ligand (Suzuki et al. 2004). The $GABA_BR2$ subunit, which has a low affinity for γ-aminobutyric acid, nevertheless modifies the ligand sensitivity of the $GABA_BR1$ subunit (Kaupmann et al. 1998). Similar mechanisms may be at play in the T1R2:T1R3 receptor.

The conformational changes and intramolecular and intermolecular interactions that couple sweetener binding to receptor activation are also unclear. For example, how can the binding of sugars to the VFD (Nie et al. 2005, 2006), of sweet proteins to the VFD and/or the CRD (Jiang et al. 2004; Temussi 2002), and of cyclamate to the TMD (Jiang et al. 2005b; Xu et al. 2004) effectively elicit the same receptor activation? Again, crystal structures of sweet receptors bound to various sweetener ligands could offer invaluable insights into these questions. And what of the specific contributions of the T1R3 subunit to the sweet receptor, in contrast to its common roles in both the sweet (T1R2:T1R3) and the umami (T1R1:T1R3) receptor? Sweeteners such as sucrose, aspartame, and saccharin that are thought to bind the VFDs have no impact on the umami receptor (Xu et al. 2004). In contrast, cyclamate and lactisole, which both bind the TMD of T1R3, enhance or inhibit the umami receptor's response to glutamate, respectively (Xu et al. 2004). The divergent efficacies of VFD- and TMD-interacting sweeteners on umami receptor function could indicate that the functions of these two T1R3 domains generalize to different extents between the sweet and the umami taste receptor.

Other genes, as yet unidentified, have been implicated in sweet taste sensation though their specific roles are unknown. For example, the *dpa* quantitative trait locus, which appears to specifically influence sensitivity to the sweet-tasting amino acid D-phenylalanine is distinct from *Sac* (and independent of both *Tas1r2* and *Tas1r3*). Indeed, nontaster *dpa* mice generalize the taste of D-phenylalanine to bitter-tasting substances, not to other sweeteners (Shigemura et al. 2005). Receptor-activation assays and studies with *Tas1r3* null mice indicate that taste responses to D-phenylalanine are mediated by the T1R2:T1R3 receptor (Nelson et al. 2002; Zhao et al. 2003). Thus, the *dpa* locus may influence the transduction of a specific sweetener downstream of the T1R2:T1R3 receptor; however, the mechanism by which this could occur is not clear.

Even with the implication of T1R2, T1R3, the effector enzyme phospholipase C β2, and the ion channel TRPM5 in the transduction of sweeteners (Chandrashekar et al. 2006), the details of these mechanisms remain remarkably unclear. For example, though the G proteins α-gustducin, Gβ3, and Gγ13 are all implicated in the transduction of sweeteners by some T1R2:T1R3-expressing cells, other G proteins must be involved in sweet taste (Margolskee 2002). How sweet taste information is processed between cells within the taste bud is an area of intensive investigation, as are the synaptic mechanisms used for signaling of this information to afferent nerves (Roper 2006). Sweet taste information may also be modulated at the level of the taste bud by hormones, e.g., leptin (Kawai et al. 2000), glucagon-like peptide-1 (Shin et al. 2008), and neuropeptide Y (Kawai et al. 2000; Shigemura et al. 2003, 2004; Zhao et al. 2005), which may afford the opportunity to regulate sweet taste function in the context of an animal's metabolic or developmental state. Clearly, there is a great deal more to know about sweet taste, as well as much more to learn about the physiological properties of the sweet taste receptor.

Acknowledgements This work was supported by grants from the National Institute on Deafness and Other Communication Disorders (DC005786 to S.D.M.; DC007317 to S.V.). C.D.D. is supported by a National Institute of Dental and Craniofacial Research training grant (DE007309).

References

Assadi-Porter FM, Abildgaard F, Blad H, Markley JL (2003) Correlation of the sweetness of variants of the protein brazzein with patterns of hydrogen bonds detected by NMR spectroscopy. J Biol Chem 278:31331–31339

Bachmanov AA, Beauchamp GK (2007) Taste receptor genes. Annu Rev Nutr 27:389–414

Bachmanov AA, Reed DR, Tordoff MG, Price RA, Beauchamp GK (1996) Intake of ethanol, sodium chloride, sucrose, citric acid, and quinine hydrochloride solutions by mice: a genetic analysis. Behav Genet 26:563–573

Bachmanov AA, Tordoff MG, Beauchamp GK (2001a) Sweetener preference of C57BL/6ByJ and 129P3/J mice. Chem Senses 26:905–913

Bachmanov AA, Li X, Reed DR, Ohmen JD, Li S, Chen Z, Tordoff MG, de Jong PJ, Wu C, West DB, et al. (2001b) Positional cloning of the mouse saccharin preference (Sac) locus. Chem Senses 26:925–933

Bartoshuk LM, Lee CH, Scarpellino R (1972) Sweet taste of water induced by artichoke (Cynara scolymus). Science 178:988–989

Beauchamp GK, Maller O, Rogers JG (1977) Flavor preferences in cats (Felis catus and Panthera sp.). J Comp Physiol Psychol 91:1118

Bello NT, Hajnal A (2005) Male rats show an indifference-avoidance response for increasing concentrations of the artificial sweetener sucralose. Nutr Res 25:693–699

Blakeslee AF, Salmon TN (1935) Genetics of sensory thresholds: individual taste reactions for different substances. Proc Natl Acad Sci USA 21:84

Boughter JD Jr Bachmanov AA (2007) Behavioral genetics and taste. BMC Neurosci 8(Suppl 3):S3

Breslin PA, Spector AC (2008) Mammalian taste perception. Curr Biol 18:R148–155

Breslin PAS, Kemp S, Beauchamp GK (1994) Single sweetness signal. Nature 369:447

Breslin PA, Beauchamp GK, Pugh EN Jr (1996) Monogeusia for fructose, glucose, sucrose, and maltose. Percept Psychophys 58:327–341

Capeless CG, Whitney G (1995) The genetic basis of preference for sweet substances among inbred strains of mice: preference ratio phenotypes and the alleles of the Sac and dpa loci. Chem Senses 20:291–298

Capretta PJ (1970) Saccharin and saccharin-glucose ingestion in 2 inbred strains of mus-musculus. Psychonom Sci 21:133–135

Carpenter JA (1956) Species differences in taste preferences. J Comp Physiol Psychol 49:139

Chandrashekar J, Hoon MA, Ryba NJ, Zuker CS (2006) The receptors and cells for mammalian taste. Nature 444:288–294

Cherezov V, Rosenbaum DM, Hanson MA, Rasmussen SG, Thian FS, Kobilka TS, Choi HJ, Kuhn P, Weis WI, Kobilka BK, Stevens RC (2007) High-resolution crystal structure of an engineered human beta2-adrenergic G protein-coupled receptor. Science 318:1258–1265

Cui M, Jiang P, Maillet E, Max M, Margolskee RF, Osman R (2006) The heterodimeric sweet taste receptor has multiple potential ligand binding sites. Curr Pharm Des 12:4591–4600

Damak S, Rong M, Yasumatsu K, Kokrashvili Z, Varadarajan V, Zou S, Jiang P, Ninomiya Y, Margolskee RF (2003) Detection of sweet and umami taste in the absence of taste receptor T1r3. Science 301:850–853

Danilova V, Hellekant G (2003) Comparison of the responses of the chorda tympani and glossopharyngeal nerves to taste stimuli in C57BL/6J mice. BMC Neurosci 4:5

Dellisanti CD, Yao Y, Stroud JC, Wang ZZ, Chen L (2007) Crystal structure of the extracellular domain of nAChR alpha1 bound to alpha-bungarotoxin at 1.94 A resolution. Nat Neurosci 10:953–962

De Simone A, Spadaccini R, Temussi PA, Fraternali F (2006) Toward the understanding of MNEI sweetness from hydration map surfaces. Biophys J 90:3052–3061

DuBois GE (2004) Unraveling the biochemistry of sweet and umami tastes. Proc Natl Acad Sci USA 101:13972–13973

Duffy VB (2007) Variation in oral sensation: implications for diet and health. Curr Opin Gastroenterol 23:171–177

Esposito V, Gallucci R, Picone D, Saviano G, Tancredi T, Temussi PA (2006) The importance of electrostatic potential in the interaction of sweet proteins with the sweet taste receptor. J Mol Biol 360:448–456

Faus I (2000) Recent developments in the characterization and biotechnological production of sweet-tasting proteins. Appl Microbiol Biotechnol 53:145–151

Fisher GL, Pfaffmann C, Brown E. (1965) Dulcin and saccharin taste in squirrel monkeys, rats, and men. Science 150:506

Frank ME, Blizard DA (1999) Chorda tympani responses in two inbred strains of mice with different taste preferences. Physiol Behav 67:287–297

Fuller JL (1974) single-locus control of saccharin preference in mice. J Heredity 65:33–36

Galindo-Cuspinera V, Winnig M, Bufe B, Meyerhof W, Breslin PA (2006) A TAS1R receptor-based explanation of sweet 'water-taste'. Nature 441:354–357

Glaser D, Tinti JM, Nofre C (1995) Evolution of the sweetness receptor in primates. I. Why does alitame taste sweet in all prosimians and simians, and aspartame only in Old World simians? Chem Senses 20:573–584

Glaser D, Tinti JM, Nofre C (1998) Taste preference in nonhuman primates to compounds sweet in man. Ann N Y Acad Sci 855:169

Haefeli RJ, Solms J, Glaser D (1998) Taste responses to amino acids in common marmosets (Callithrix jacchus jacchus, Callitrichidae) a non-human primate in comparison to humans. Lebensm Wiss Technol 31:371–376

Hellekant G, Danilova V (1996) Species differences toward sweeteners. Food Chem 56:323

Hobbs JR, Munger SD, Conn GL (2008) Crystal structures of the sweet protein MNEI: insights into sweet protein-receptor interactions. In Weerasinghe DK, DuBois GE (eds.) Sweetness and sweeteners. American Chemical Society, Washington

Hoon MA, Adler E, Lindemeier J, Battey JF, Ryba NJ, Zuker CS (1999) Putative mammalian taste receptors: a class of taste-specific GPCRs with distinct topographic selectivity. Cell 96:541–551

Inoue M, McCaughey SA, Bachmanov AA, Beauchamp GK (2001) Whole nerve chorda tympani responses to sweeteners in C57BL/6ByJ and 129P3/J mice. Chem Senses 26:915–923

Jiang P, Ji Q, Liu Z, Snyder LA, Benard LM, Margolskee RF, Max M (2004) The cysteine-rich region of T1R3 determines responses to intensely sweet proteins. J Biol Chem 279:45068–45075

Jiang P, Cui M, Zhao B, Liu Z, Snyder LA, Benard LM, Osman R, Margolskee RF, Max M (2005a) Lactisole interacts with the transmembrane domains of human T1R3 to inhibit sweet taste. J Biol Chem 280:15238–15246

Jiang P, Cui M, Zhao B, Snyder LA, Benard LM, Osman R, Max M, Margolskee RF (2005b) Identification of the cyclamate interaction site within the transmembrane domain of the human sweet taste receptor subunit T1R3. J Biol Chem 280:34296–34305

Kahn SG (1951) Taste perception—individual reactions to different substances. Trans Ill State Acad Sci 44:263

Kasahara T, Iwasaki K, Sato M (1987) Taste effectiveness of some d- and l-amino acids in mice. Physiol Behav 39:619–624

Kaupmann K, Malitschek B, Schuler V, Heid J, Froestl W, Beck P, Mosbacher J, Bischoff S, Kulik A, Shigemoto R et al.(1998) GABA(B)-receptor subtypes assemble into functional heteromeric complexes. Nature 396:683–687

Kawai K, Sugimoto K, Nakashima K, Miura H, Ninomiya Y (2000) Leptin as a modulator of sweet taste sensitivities in mice. Proc Natl Acad Sci USA 97:11044–11049

Kim MR, Kusakabe Y, Miura H, Shindo Y, Ninomiya Y, Hino A (2003) Regional expression patterns of taste receptors and gustducin in the mouse tongue. Biochem Biophys Res Commun 312:500–506

Kitagawa M, Kusakabe Y, Miura H, Ninomiya Y, Hino A (2001) Molecular genetic identification of a candidate receptor gene for sweet taste. Biochem Biophys Res Commun 283:236–242

Kuhn C, Bufe B, Winnig M, Hofmann T, Frank O, Behrens M, Lewtschenko T, Slack JP, Ward CD, Meyerhof W (2004) Bitter taste receptors for saccharin and acesulfame K. J Neurosci 24:10260–10265

Kunishima N, Shimada Y, Tsuji Y, Sato T, Yamamoto M, Kumasaka T, Nakanishi S, Jingami H, Morikawa K (2000) Structural basis of glutamate recognition by a dimeric metabotropic glutamate receptor. Nature 407:971–977

Lemon CH, Katz DB (2007) The neural processing of taste. BMC Neurosci 8(Suppl 3):S5

Li X, Staszewski L, Xu H, Durick K, Zoller M, Adler E (2002) Human receptors for sweet and umami taste. Proc Natl Acad Sci USA 99:4692–4696

Li X, Li W, Wang H, Cao J, Maehashi K, Huang L, Bachmanov AA, Reed DR, Legrand-Defretin V, Beauchamp GK, Brand JG (2005) Pseudogenization of a sweet-receptor gene accounts for cats' indifference toward sugar. PLoS Genet 1:27–35

Lush IE (1989) The genetics of tasting in mice. VI. Saccharin, acesulfame, dulcin and sucrose. Genet Res 53:95–99

Lush IE, Hornigold N, King P, Stoye JP (1995) The genetics of tasting in mice. VII. Glycine revisited, and the chromosomal location of Sac and Soa. Genet Res 66:167–174

MacKinnon BI, Frank ME, Hettinger TP, Rehnberg BG (1999) Taste qualities of solutions preferred by hamsters. Chem Senses 24:23–35

Margolskee RF (2002) Molecular mechanisms of bitter and sweet taste transduction. J Biol Chem 277:1–4

Max M, Shanker YG, Huang L, Rong M, Liu Z, Campagne F, Weinstein H, Damak S, Margolskee RF (2001) Tas1r3, encoding a new candidate taste receptor, is allelic to the sweet responsiveness locus Sac. Nat Genet 28:58–63

Miller IJ Jr, Reedy FE Jr (1990) Variations in human taste bud density and taste intensity perception. Physiol Behav 47:1213–1219

Montmayeur JP, Liberles SD, Matsunami H, Buck LB (2001) A candidate taste receptor gene near a sweet taste locus. Nat Neurosci 4:492–498

Morini G, Bassoli A, Temussi PA (2005) From small sweeteners to sweet proteins: anatomy of the binding sites of the human T1R2_T1R3 receptor. J Med Chem 48:5520–5529

Nelson G, Hoon MA, Chandrashekar J, Zhang Y, Ryba NJ, Zuker CS (2001) Mammalian sweet taste receptors. Cell 106:381–390

Nelson G, Chandrashekar J, Hoon MA, Feng L, Zhao G, Ryba NJ, Zuker CS (2002) An amino-acid taste receptor. Nature 416:199–202

Nie Y, Vigues S, Hobbs JR, Conn GL, Munger SD (2005) Distinct contributions of T1R2 and T1R3 taste receptor subunits to the detection of sweet stimuli. Curr Biol 15:1948–1952

Nie Y, Hobbs JR, Vigues S, Olson WJ, Conn GL, Munger SD (2006) Expression and purification of functional ligand-binding domains of T1R3 taste receptors. Chem Senses 31:505–513

Ninomiya Y, Funakoshi M (1989) Peripheral neural basis for behavioural discrimination between glutamate and the four basic taste substances in mice. Comp Biochem Physiol A 92:371–376

Ninomiya Y, Higashi T, Katsukawa H, Mizukoshi T, Funakoshi M (1984) Qualitative discrimination of gustatory stimuli in three different strains of mice. Brain Res 322:83–92

Ninomiya Y, Sako N, Katsukawa H, Funakoshi M (1991) Taste receptor mechanisms influenced by a gene on chromosome 4 in mice, In: Wysocki CJ, Kare MR (eds) Genetics of perception and communication. Decker, New York, pp 267–278

Ninomiya Y, Kajiura H, Mochizuki K (1993) Differential taste responses of mouse chorda tympani and glossopharyngeal nerves to sugars and amino acids. Neurosci Lett 163:197–200

Ninomiya Y, Nakashima K, Fukuda A, Nishino H, Sugimura T, Hino A, Danilova V, Hellekant G (2000) Responses to umami substances in taste bud cells innervated by the chorda tympani and glossopharyngeal nerves. J Nutr 130:950S–953S

Nofre C, Tinti JM, Glaser D (1996) Evolution of the sweetness receptor in primates. II. Gustatory responses of non-human primates to nine compounds known to be sweet in man. Chem Senses 21:747–762

Nofre C, Glaser D, Tinti JM, Wanner M (2002) Gustatory responses of pigs to sixty compounds tasting sweet to humans. J Anim Physiol Anim Nutr (Berl) 86:90–96

Nowlis GH, Frank ME, Pfaffmann C (1980) Specificity of acquired aversions to taste qualities in hamsters and rats. J Comp Physiol Psychol 94:932–942

Palczewski K, Kumasaka T, Hori T, Behnke CA, Motoshima H, Fox BA, Le Trong I, Teller D C, Okada T, Stenkamp RE et al., (2000) Crystal structure of rhodopsin: A G protein-coupled receptor. Science 289:739–745

Pelz WE, Whitney G, Smith JC (1973) Genetic influences on saccharin preference of mice. Physiol Behav 10:263–265

Pfaffmann C (1955) Gustatory nerve impulses in rat, cat and rabbit. J Neurophysiol 18:429–440

Pin JP, Galvez T, Prezeau L (2003) Evolution, structure, and activation mechanism of family 3/C G-protein-coupled receptors. Pharmacol Ther 98:325–354

Pressman TG, Doolittle JH (1966) Taste preferences in the Virginia opossum. Psychol Rep 18:875

Pronin AN, Xu H, Tang H, Zhang L, Li Q, Li X (2007) Specific alleles of bitter receptor genes influence human sensitivity to the bitterness of aloin and saccharin. Curr Biol 17:1403–1408

Ramirez I, Fuller JL (1976) Genetic influence on water and sweetened water consumption in mice. Physiol Behav 16:163–168

Reed DR, McDaniel AH (2006) The human sweet tooth. BMC Oral Health 6(Suppl 1):S17

Reed DR, Li S, Li X, Huang L, Tordoff MG, Starling-Roney R, Taniguchi K, West DB, Ohmen JD, Beauchamp GK, Bachmanov AA (2004) Polymorphisms in the taste receptor gene (Tas1r3) region are associated with saccharin preference in 30 mouse strains. J Neurosci 24:938–946

Roper SD (2006) Cell communication in taste buds. Cell Mol Life Sci 63:1494–1500

Sainz E, Korley JN, Battey JF, Sullivan SL (2001) Identification of a novel member of the T1R family of putative taste receptors. J Neurochem 77:896–903

Sclafani A, Clare RA (2004) Female rats show a bimodal preference response to the artificial sweetener sucralose. Chem Senses 29:523–528

Scott K (2005) Taste recognition: food for thought. Neuron 48:455–464

Shigemura N, Miura H, Kusakabe Y, Hino A, Ninomiya Y (2003) Expression of leptin receptor (Ob-R) isoforms and signal transducers and activators of transcription (STATs) mRNAs in the mouse taste buds. Arch Histol Cytol 66:253–260

Shigemura N, Ohta R, Kusakabe Y, Miura H, Hino A, Koyano K, Nakashima K, Ninomiya Y (2004) Leptin modulates behavioral responses to sweet substances by influencing peripheral taste structures. Endocrinology 145:839–847

Shigemura N, Yasumatsu K, Yoshida R, Sako N, Katsukawa H, Nakashima K, Imoto T, Ninomiya Y (2005) The role of the dpa locus in mice. Chem Senses 30(Suppl 1):i84–i85

Shin YK, Martin B, Golden E, Dotson CD, Maudsley S, Kim W, Jang HJ, Mattson MP, Drucker DJ, Egan JM, Munger SD (2008) Modulation of taste sensitivity by GLP-1 signaling. J Neurochem 106:455–463

Spadaccini R, Trabucco F, Saviano G, Picone D, Crescenzi O, Tancredi T, Temussi PA (2003) The mechanism of interaction of sweet proteins with the T1R2-T1R3 receptor: evidence from the solution structure of G16A-MNEI. J Mol Biol 328:683–692

Spillane WJ, Kelly DP, Curran PJ, Feeney BG (2006) Structure–taste relationships for disubstituted phenylsulfamate tastants using classification and regression tree (CART) analysis. J Agric Food Chem 54:5996–6004

Stein N, Laing DG, Hutchinson I (1994) Topographical differences in sweetness sensitivity in the peripheral gustatory system of adults and children. Brain Res Dev Brain Res 82:286–292

Sternini C, Anselmi L, Rozengurt E (2008) Enteroendocrine cells: a site of "taste" in gastrointestinal chemosensing. Curr Opin Endocrinol Diabetes Obes 15:73–78

Stevens DA (1996) Individual differences in taste perception. Food Chem 56:303

Stone LM, Barrows J, Finger TE, Kinnamon SC (2007) Expression of T1Rs and gustducin in palatal taste buds of mice. Chem Senses 32:255–262

Suzuki Y, Moriyoshi E, Tsuchiya D, Jingami H (2004) Negative cooperativity of glutamate binding in the dimeric metabotropic glutamate receptor subtype 1. J Biol Chem 279:35526–35534

Temussi PA (2002) Why are sweet proteins sweet? Interaction of brazzein, monellin and thaumatin with the T1R2-T1R3 receptor. FEBS Lett 526:1–4

Tinti JM, Glaser D, Wanner M, Nofre C (2000) Comparison of gustatory responses to amino acids in pigs and in humans. Lebensm Wiss Technol 33:578

Winnig M, Bufe B, Meyerhof W (2005) Valine 738 and lysine 735 in the fifth transmembrane domain of rTas1r3 mediate insensitivity towards lactisole of the rat sweet taste receptor. BMC Neurosci 6:22

Winnig M, Bufe B, Kratochwil NA, Slack JP, Meyerhof W (2007) The binding site for neohesperidin dihydrochalcone at the human sweet taste receptor. BMC Struct Biol 7:66

Xu H, Staszewski L, Tang H, Adler E, Zoller M, Li X (2004) Different functional roles of T1R subunits in the heteromeric taste receptors. Proc Natl Acad Sci USA 101:14258–14263

Zhao GQ, Zhang Y, Hoon MA, Chandrashekar J, Erlenbach I, Ryba NJ, Zuker CS (2003) The receptors for mammalian sweet and umami taste. Cell 115:255–266

Zhao FL, Shen T, Kaya N, Lu SG, Cao Y, Herness S (2005) Expression, physiological action, and coexpression patterns of neuropeptide Y in rat taste-bud cells. Proc Natl Acad Sci USA 102:11100–11105

Mammalian Bitter Taste Perception

M. Behrens and W. Meyerhof

Abstract Bitter taste in mammals is achieved by a family of approximately 30 bitter taste receptor genes. The main function of bitter taste is to protect the organism against the ingestion of, frequently bitter, toxic food metabolites. The field of taste research has advanced rapidly during the last several years. This is especially true for the G-protein-coupled-receptor-mediated taste qualities, sweet, umami, and bitter. This review summarizes current knowledge of bitter taste receptor gene expression, signal transduction, the structure—activity relationship of bitter taste receptor proteins, as well as their variability leading to a high degree of individualization of this taste quality in mammals.

1 Introduction

Mammals distinguish the five basic taste qualities, sweet, sour, umami, salty, and bitter, to assess the quality of consumed food. The sensory cells that allow taste perception, the taste receptor cells, are distributed in the oral cavity, with the tongue being the main taste organ. On the tongue, taste receptor cells are found in specialized structures, the gustatory papillae, where they occur in groups of 60–100 cells called taste buds. Each taste bud exhibits a single apical porus exposed to tastants present in the oral cavity that contains the microvilli of the taste receptor cells. Although the ability to detect the five basic taste qualities is in general invariant, some exceptions related to nutritional habits are possible (e.g. the lack of a functional sweet taste receptor in the domestic cat and related species; Li et al. 2005). Whereas the complete absence of a taste quality is exceptional, the ability of mammalian species to detect particular substances varies

M. Behrens (✉)
German Institute of Human Nutrition Potsdam-Rehbruecke, Arthur-Scheunert-Allee 114–116, 14558 Nuthetal, Germany
e-mail: behrens@dife.de

considerably. Some examples for observed taste differences among mammals are the inability of rodents to taste artificial sweeteners and sweet plant proteins (Brouwer et al. 1973; Jiang et al. 2004; Li et al. 2002; Sclafani and Abrams 1986; Xu et al. 2004; Zhao et al. 2003), the relative insensitivity of cats and dogs to sodium chloride, as well as selective amino acid taste (Boudreau et al. 1985). The relative low level of amino acid sequence homology of bitter taste receptors of different species, e.g. of human and mice (Shi et al. 2003), implies already that bitter taste differences among species should be the rule rather than the exception. The observed differences in bitter tastes of mammals likely reflect the very different nutritional habits of mammalian species and their evolutionary histories, leading them to encounter different bitter plant toxins. The contact with similar as well as with completely different poisonous bitter substances shaped a bitter taste receptor repertoire unique to each species to serve as warning sensors against the ingestion of toxic food compounds (for more details see the chapter by Shi and Zhang in this volume). Moreover, the approximately 30 bitter taste receptor genes of mammals not only lead to variability of bitter taste perception across species, but also show a considerable degree of intraspecies variability observed in mouse strains (Boughter et al. 2005) and humans (Bartoshuk 2000; Bufe et al. 2005; Kim et al. 2003, 2005, Pronin et al. 2007; Soranzo et al. 2005; Wooding et al. 2004).

After a brief introduction to the anatomy of the gustatory system this review will discuss recent findings concerning bitter taste receptor gene expression and its implications for the recognition of bitter substances. Further, the components involved in the signal transduction of bitter taste will be summarized. A large part of this review is devoted to the molecular characterization of bitter taste receptor proteins and the conclusions for structure and function that can be drawn by the continuously advancing deorphanization process of these receptors. Finally, the enormous variability of bitter taste receptor genes and its consequences for the individuality of bitter taste is summarized in the last part of this review.

2 Anatomy/Morphology

The morphological structures underlying taste perception in mammals are located in the oral cavity (Miller 1995). Taste receptor cells are, in general, organized in groups of 60–100 cells, called taste buds, and are found throughout the oral cavity, including tongue, soft palate, epiglottis, larynx, and pharynx, where they might occur concentrated in specialized organs, the taste papillae. The three types of taste papillae are non-uniformly distributed over the tongue surface. On the frontal part of the tongue the morphologically simplest taste papillae carrying only one to three taste buds, the fungiform papillae, are located, whereas the posterior part of the tongue contains foliate and circumvallate papillae. Unlike the fungiform papillae, which are rather simple epithelial protrusions, the foliate and circumvallate papillae are more complex invaginations into the tongue surface forming a trench filled with saliva secreted from minor salivary glands located at the bottom of the papillae. A large number of taste buds are positioned in the epithelial walls of the trenches facing the oral cavity. Whereas foliate papillae are found symmetrically arranged on both sides of the posterior tongue, the circumvallate papillae that occur in variable numbers are localized centrally and further backwards. In contrast to the tongue surface, the other taste sensitive areas within the oral cavity contain isolated taste buds, which differ considerably in size, ranging from taste buds resembling those found on the tongue surface, e.g. on the soft palate, to smaller laryngeal taste buds containing fewer cells (Sbarbati et al. 2004). Each taste bud is exposed to tastants present in the oral cavity by a single porus where the apical microvilli of taste receptor cells are localized. Taste receptor cells are secondary sensory cells of epidermal origin with an average life span of approximately 10 days in rats (Beidler and Smallman 1965; Farbman 1980). In order to transmit taste information from the oral cavity to the brain, taste receptor cells receive afferent input by different cranial nerves. Whereas the fungiform papillae of the anterior tongue are innervated by fibres of the chorda tympani, the glossopharyngeal nerve makes contacts to taste buds of foliate and circumvallate papillae. Palatal gustatory information is transmitted via cranial nerve VII and taste buds of the epiglottis and larynx receive nerve contacts by cranial nerve X. Peripheral taste information converges on the gustatory division of the nucleus tractus solitarius (NTS), the first relay-station for taste information in the brain (Smith and Scott 2003). Whereas in rodents ascending fibres first target the medial parabrachial nucleus before taste information is then transmitted into the thalamus, in humans the NTS neurons directly project into the thalamic region (Rolls 1995). In both cases the final destination of gustatory information is the orbitofrontal cortex (secondary gustatory cortex), which is reached after passage of the insular-opercular cortex (primary gustatory cortex) (Scott and Verhagen 2000).

3 Gene Expression

A number of studies have been devoted to the analyses of TAS2R gene expression in gustatory and, more recently, in extragustatory tissues of rodents and humans. The first reports on TAS2R gene expression showed that the expression is restricted to taste receptor cells of taste buds in the oral cavity (Adler et al. 2000; Matsunami et al. 2000). Although both studies indicated already limited expression in non-gustatory tissues, a predominant function as taste receptors was concluded and demonstrated by Chandrashekar et al. (2000), who identified the first TAS2Rs as bitter taste receptors. It was also shown that only subsets of taste receptor cells within each taste bud hybridize to TAS2R probes in in situ hybridization experiments (Fig. 1). Compared with the number of α-gustducin-expressing taste receptor cells, which account for approximately 30–40% of all intragemmal cells (McLaughlin et al. 1992; Wong et al. 1999), clearly fewer taste receptor cells contain TAS2R messenger RNAs (mRNAs) (Adler et al. 2000; Matsunami et al. 2000). Double-staining analyses using TAS2R and α-gustducin probes demonstrated that TAS2R-expressing taste receptor cells represent a subset of α-gustducin-positive cells (Adler et al. 2000), which is in marked contrast to the observation that TAS1Rs rarely colocalize with α-gustducin (Hoon et al. 1999). Careful quantification of TAS2R-positive cells in rodent lingual and palatal taste buds revealed that approximately 15% of all cells within a taste bud express TAS2R genes (Adler et al. 2000). In taste buds of human circumvallate papillae, the fraction of TAS2R-expressing cells is somewhat smaller, ranging from approximately 1% to approximately 10%, depending on the receptor (Behrens et al. 2007). Interestingly, the distribution of bitter taste receptor cells is not uniform, but rather inhomogeneous among the taste buds of the three types of taste papillae. Whereas all taste buds of rodent circumvallate and foliate papillae seem to express bitter taste receptor genes (Adler et al. 2000; Matsunami et al. 2000), only approximately 10% of fungiform taste buds contain TAS2R mRNAs (Adler et al. 2000). Although only circumstantial evidence was provided, the presence of bitter-taste-receptor-negative fungiform taste buds seems to apply also to humans (Behrens et al. 2007). Since the detection of bitter tastants at the tip of the tongue, where exclusively fungiform papillae are located, is firmly established (Collings 1974), a reduced number of TAS2R-expressing taste buds appear to be sufficient for bitter compound recognition.

The bulk of electrophysiological studies done in mammals demonstrated that afferent fibres transmit information on more than one taste quality, although usually one modality predominates (Smith et al. 2000). Such fibres are termed, S-best (sucrose, i.e. sweet), H-best (HCl, i.e. sour), N-best (NaCl, i.e. salty), or Q-best (quinine, i.e. bitter) according to the taste stimulus evoking the most robust activation. On the basis of these findings, the formulation of the "across-fibre pattern theory" for taste transduction from the periphery into the brain was developed, which was seemingly incompatible with the "labelled-line theory" of strictly linear and separate information coding favoured by other researchers (for a review, see Smith et al. 2000). After the revolutionary finding of broadly tuned presynaptic cells that govern collection and transmission of diverse taste stimuli within taste buds, a more unifying mechanistic explanation of taste information processing appears to be coming in reach. On the level of single taste receptor cells, the observations leading to the across-fibre pattern theory might result from the expression of taste receptor genes specific for multiple taste qualities. A large number of studies demonstrated convincingly, however, that each taste receptor cell is devoted to detection of only a single taste quality. It was shown that TAS2Rs are not coexpressed with TAS1Rs (Adler et al. 2000), that the TAS1R subunits specifying the sweet and umami taste receptor heteromers TAS1R2 and TAS1R1 are coexpressed with the common subunit TAS1R3 but not with each other (Nelson et al. 2001), and also the sour taste receptor candidate molecule PKD2L1 is present in a population of cells distinct from bitter, sweet, and umami taste receptor cells (Huang et al. 2006).

A central question in bitter taste research is if bitter taste receptor cells are broadly tuned sensors for most, if not all, bitter tastants, or if bitter taste receptor cells are rather heterogeneous, thus providing a potential cellular basis for a discrimination

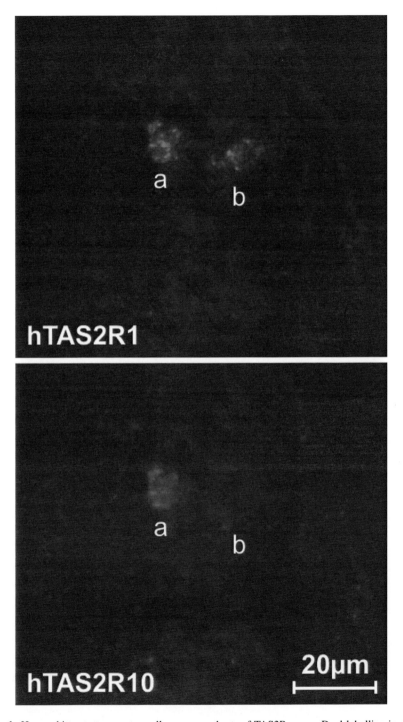

Fig. 1 Human bitter taste receptor cells express subsets of TAS2R genes. Dual labelling in situ hybridization of one taste bud in human circumvallate papillae reveals that cells **a** and **b** express hTAS2R1 (*upper panel*). Of the two cells, cell a contains also hTAS2R10 messenger RNA (mRNA), whereas cell **b** tests negative for hTAS2R10 mRNA

of bitter compounds. Whereas the majority of functional data clearly argue for a heterogeneous population of bitter taste receptor cells as shown by calcium imaging of single rodent taste receptor cells (Caicedo and Roper 2001), nerve fibre recordings (Dahl et al. 1997), and recordings of NTS neurons (Geran and Travers 2006), the findings of gene expression studies are less unanimous. Undoubtedly, all studies attempting to clarify this question utilizing in situ hybridizations of TAS2R mRNAs demonstrate that multiple TAS2R genes are coexpressed in individual bitter taste receptor cells, thus creating taste receptor cells with broader agonist spectra than any given TAS2R might respond to. However, the extent of the observed coexpression differs considerably between almost complete (Adler et al. 2000) and a limited number of receptors per bitter taste receptor cells (Behrens et al. 2007). The ultimate answer to the question of whether mammals can discriminate between bitter stimuli has to come from sensoric analyses, which, so far, have resulted in somewhat controversial findings (Brasser et al. 2005; Delwiche et al. 2001; Keast et al. 2003; Spector and Kopka 2002; Yokomukai et al. 1993), since even individual responses of bitter taste receptor cells may eventually converge upon transmission of information into higher-order brain areas.

The number of reports on the extragustatory expression of taste-signalling molecules and taste receptors itself is increasing steadily. Expression of such molecules is usually associated with one of two morphologically different cell types, solitary chemosensory cells and brush cells, respectively (for a review on brush cells, see Sbarbati and Osculati 2005). On one hand, cells that express signalling components related to the taste system have been found in nasal respiratory epithelium (Finger et al. 2003), the vomeronasal organ (Zancanaro et al. 1999), and airways (Merigo et al. 2005, 2007); on the other hand, such cells extend into the gastrointestinal tract (Sternini et al. 2008). Despite the fact that taste-related signalling compounds have been identified in the organ systems specified above as well as in additional organs, like testis (Fehr et al. 2007), suggesting that also taste receptor molecules should be found in those organs, direct evidence for bitter taste receptor gene expression in these organs is scarce. However, within the nasal respiratory system of the mouse two bitter taste receptors, mT2R8 and mT2R19, have been localized to solitary chemosensory cells by in situ hybridization, whereas a third bitter taste receptor, mT2R5, could not be detected. Intranasal irrigation with different bitter substances, including denatonium, the known agonist for mT2R8 (Chandrashekar et al. 2000), resulted in clear respiratory responses, indicating that bitter taste receptors in the respiratory system might serve a function in protecting the organism from noxious substances (Finger et al. 2003). Within the gastrointestinal tract bitter taste receptor gene expression was detected by reverse transcriptase PCR of several tissues; the cellular origin of T2R mRNAs, however, remains to be determined. As the cell line STC-1 was shown to express T2R genes and even responded to stimulation with bitter compounds, T2R mRNAs detected by reverse transcriptase PCR of gastrointestinal tissue likely originate from enteroendocrine cells (Wu et al. 2002).

4 Signal Transduction

It is assumed that bitter taste receptors accumulate on the cell surface of bitter taste receptor cell microvilli. Here, the TAS2R proteins come into contact with their agonists present in the oral cavity, resulting in the activation of receptors. As for the other G-protein-coupled-receptor-mediated taste qualities sweet and umami, activated bitter taste receptors utilize the G-protein α-subunit α-gustducin for signal transduction (McLaughlin et al. 1992). The crucial role of α-gustducin for bitter taste transduction was demonstrated by the generation of knockout mice, which exhibit dramatic reduction of their bitter tasting abilities (Wong et al. 1996). The bitter taste phenotype of α-gustducin knockout mice was rescued by the reintroduction of rat α-gustducin, further confirming its importance (Wong et al. 1999). The direct activation of α-gustducin by the heterologously expressed mouse cycloheximide receptor, mT2R5, was shown by GTP-γ-S assay after stimulation of mT2R5-expressing cells with cycloheximide (Chandrashekar et al. 2000). Mapping analyses of the carboxy terminus of α-gustducin revealed that the last 37 amino acids of α-gustducin are crucial for its recruitment by activated recombinantly expressed bitter taste receptors (Ueda et al. 2003). Although definitely a key element in bitter taste signal transduction, the role of α-gustducin is supplemented by other G-protein α-subunits present in taste receptor cells. Evidence for this comes from the observation that responses to bitter stimulation in α-gustducin knockout mice are severely reduced but not completely abolished, indicating the presence of additional Gα proteins in taste receptor cells (Wong et al. 1999). Consequently, rod transducin, a Gα protein closely related to α-gustducin, was detected in taste receptor cells and subsequently used to rescue, at least partially, the bitter taste ability of α-gustducin knockout mice (Ruiz-Avila et al. 1995). As bitter taste receptors can also couple in vitro to other Gi/Go proteins, the signal transduction cascade might be even more complex (Ozeck et al. 2004; Sainz et al. 2007). It remains to be determined if Gα protein recruitment by activated bitter taste receptor proteins is simply directed by availability/relative abundance, if different TAS2Rs exhibit variable selectivity in their G-protein coupling, or if other factors such as cellular or spatiotemporal segregation may play a role as well.

Also the β- and γ-subunits necessary to form the functional heterotrimeric G-protein complex were identified from taste cells. The differential screening of complementary DNA of α-gustducin-positive single taste cells resulted in the identification β1, β3, and γ13 subunits (Huang et al. 1999). The involvement of these G-proteins was demonstrated by functional experiments. Antibodies raised against Gβ1, Gβ3, and Gγ13 were able to reduce the responses of mouse taste tissue to stimulation with denatonium benzoate, a substance classically used as a bitter test substance (Huang et al. 1999; Rossler et al. 2000). As the blocking effect of the anti-Gβ3 antibody was more pronounced compared with that of the anti-Gβ1 antibody (Rossler et al. 2000), it seems that the composition of the heterotrimeric G-protein used in bitter taste transduction is Gα-gustducin, Gβ3, Gγ13, perhaps with a minor fraction of Gβ1-containing complexes (Fig. 2).

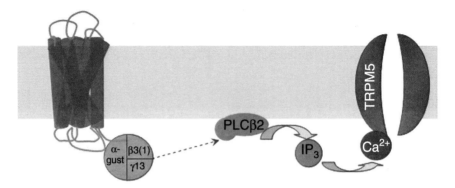

Fig. 2 The bitter taste signal transduction cascade. Bitter taste receptors are G-protein-coupled receptors. Activation of TAS2Rs results in the activation of the heterotrimeric G-protein complex α-gustducin (α-*gust*), 3 or β1, and γ13. The βγ-subunits activate phospholipase C β2, (*PLCβ2*) resulting in the production of inositol 1,4,5-trisphosphate (*IP₃*). The IP₃-mediated increase of intracellular calcium activates transient receptor potential channel 5 (*TRPM5*)

After activation of the heterotrimeric G-protein, the βγ-subunits are able to induce PLCβ2 activation. The central role of PLCβ2 was established in functional experiments as well as by the generation of knockout models. The use of PLCβ2-specific antibodies and inhibitors blocked the denatonium-induced increase in intracellular inositol 1,4,5-trisphosphate (IP₃) (Rossler et al. 1998) and calcium (Ogura et al. 1997). Consequently, genetic ablation of the PLCβ2 gene in mice led to a complete (Zhang et al. 2003) or at least a substantial loss of their bitter responsiveness (Dotson et al. 2005). Immunohistochemical analyses revealed that taste receptor cells positive for α-gustducin, Gγ13, and PLCβ2, express the type-III IP₃ receptor, filling the gap between IP₃ increases and an elevation of intracellular calcium levels (Clapp et al. 2001). The most recently identified component involved in bitter taste transduction is the transient receptor potential channel M5 (TRPM5). TRPM5 is expressed in a subset of taste receptor cells also expressing all components of the heterotrimeric G-protein (α-gustducin, Gβ3, Gγ13) as well as PLCβ2 (Perez et al. 2002). The activation of this channel is triggered by rapid changes in intracellular calcium levels (Hofmann et al. 2003; Liu and Liman 2003; Perez et al. 2003; Prawitt et al. 2003). Although independently generated TRPM5 knockout models differ in the extent of their taste phenotype, ranging from a complete loss of responsiveness upon stimulation with various bitter compounds (Zhang et al. 2003) to residual responses for bitter, sweet, and umami stimuli (Damak et al. 2006; Talavera et al. 2005), the importance of this channel for taste signal transduction is undisputed.

Since α-gustducin belongs to the Gαi protein subfamily of Gα proteins, its activation should lead to a reduction of cyclic nucleotide levels via the stimulation of phosphodiesterases. Indeed, two phosphodiesterases were identified in gustatory tissue (McLaughlin et al. 1994) and a rapid decrease in cyclic AMP after stimulation with bitter substances was observed (Yan et al. 2001). Despite the fact that even a cyclic-nucleotide-gated channel has been reported in taste receptor cells (Misaka et al. 1997), the exact role of cyclic-nucleotide signalling in taste transduction remains elusive.

5 Structure/Function

Approximately similar numbers, ranging from 12 in dog to 37 in rat (Shi and Zhang 2006), of bitter taste receptor genes were identified in a variety of mammalian species. Therefore, the intriguing question of how a comparably low number of approximately 25 human (Behrens and Meyerhof 2006) or approximately 37 rodent (Shi and Zhang 2006; Wu et al. 2005) bitter taste receptor genes might suffice for the detection of thousands of structurally diverse bitter compounds present in nature applies to all mammalian species. Most TAS2R genes exhibit high sequence diversity among each other and across species (Shi et al. 2003). Obviously, the need for the detection of so many bitter compounds without increasing the number of receptors has forced nature to create a very distinct group of receptor proteins. Although considerable energy has been directed towards the identification of agonists of bitter taste receptor genes, the majority of TAS2Rs still await deorphanization, which is required for detailed structure—function analyses. To date, almost half of the 25 human TAS2Rs have been deorphanized, whereas only three rodent T2R genes have had some of their ligands identified. At this point it is important to note that no laboratory is able to test all bitter compounds for their activation of the corresponding receptor genes; therefore, the identification of agonist spectra for bitter taste receptors is, and perhaps will remain for a long time, preliminary. However, on the basis of current literature several conclusions about the agonist specificity of human bitter taste receptors can be drawn:

1. Several bitter taste receptors are broadly tuned to detect bitter compounds belonging to different chemical classes (Fig. 3). This applies to the receptors hTAS2R7 (Sainz et al. 2007), hTAS2R14 (Behrens et al. 2004; Sainz et al. 2007), and hTAS2R46 (Brockhoff et al. 2007), indicating the need for rather non-selective hTAS2Rs during human evolution.
2. Some receptors seem to recognize complex structures present in chemically related families of bitter compounds. For the two hTAS2Rs belonging to this group, hTAS2R16 (Bufe et al. 2002) and hTAS2R38 (Bufe et al. 2005; Kim et al. 2003), β-D-glucopyranoside and N—C=S moieties, respectively, have been characterized as agonistic features each present in a number of bitter substances. These two receptors therefore appear to be tuned to detect entire families of bitter compounds. Interestingly, also hTAS2R46 recognizes chemically distinct families of bitter terpenoids, but its specificity is not restricted to these substance classes.

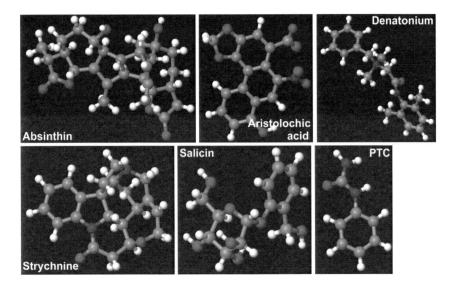

Fig. 3 Structural diversity of bitter compounds. The chemical structures of absinthin, aristolochic acid, denatonium, strychnine, D-(-)-salicin, and phenylthiocarbamate (*PTC*) are depicted. Note the different sizes, charges, and three-dimensional architectures of these compounds that all activate at least one of the human bitter taste receptors

3. At present, the largest group of hTAS2Rs recognize only a few bitter agonists. This group includes hTAS2R4, which detects denatonium benzoate and high concentrations of 6-n-propylthiouracil (PROP) (Chandrashekar et al. 2000), hTAS2R10, responding to strychnine (Bufe et al. 2002), as well as hTAS2R43 and hTAS2R44 (Kuhn et al. 2004; Pronin et al. 2007), which are activated by saccharin, acesulfame K, and aristolochic acid, and the denatonium and 6-nitrosaccharin receptor hTAS2R47 (Pronin et al. 2004). Recently, hTAS2R8 was shown to respond to saccharin, although its sensitivity for this substance is much lower than that observed for the other known saccharin receptors, hTAS2R43 and hTAS2R44 (Pronin et al. 2007). If the apparently higher selectivity observed for this receptor group persists, or if future agonist screening projects will expand the agonist spectra further remains to be seen.

Until now, only few non-human TAS2Rs have been deorphanized. The mouse receptors mT2R5 and mT2R8 recognize cycloheximide and denatonium benzoate, respectively (Chandrashekar et al. 2000), whereas rT2R9 is activated by strychnine (Bufe et al. 2002). The fact that only single agonists were identified for each rodent bitter taste receptor might indicate a higher selectivity of rodent TAS2Rs in general. However, only detailed agonist screening studies utilizing large panels of bitter compounds could substantiate this, otherwise surprising, observation.

Despite considerable screening successes the question of how so few receptors are able to detect so many bitter compounds remains unanswered. In order to elucidate this phenomenon in more detail, it is necessary to study the chemical determinants of compounds resulting in the activation of bitter taste receptors, as well as the underlying receptor structure facilitating specific agonist interaction extensively. The best studied agonist group for a bitter taste receptor are N—C=S-containing substances activating hTAS2R38. Although the receptor underlying the bitter taste of compounds including the N—C=S moiety was not identified until recently (Kim et al. 2003), chemical characterization of such substances dates back decades (Barnicot et al. 1951; Fox 1932) because a non-taster allele for the prototypical N—C=S group containing bitter agonists, PROP and phenylthiocarbamate (PTC), of hTAS2R38 is frequently found in the human population. More recently, it was reported that the hTAS2R38 haplotype predicts the perceived bitterness of PTC better than for PROP (Bufe et al. 2005), perhaps indicating additional bitter epitopes present in PROP leading to the activation of an additional receptor by this compound. Another well-characterized bitter epitope is the β-d-glycopyranose moiety activating the receptor hTAS2R16. It was shown that the steric orientations of the C1 and C4 positions of the pyranose ring are of equal importance for the activation of this receptor. The impact of substitutions at other carbon atoms of the pyranose ring differed, indicating that interactions between bitter taste receptors and agonists are complex (Bufe et al. 2002). The mixture of pronounced selectivity for certain chemical features within agonists and the variability for other residues of a chemical core structure allow the two receptors hTAS2R16 and hTAS2R38 to recognize chemically related families of bitter compounds. A variation of this theme was reported just recently for the receptor hTAS2R46. This receptor is activated by numerous naturally occurring bitter sesquiterpene lactones and diterpenoids. Careful analyses of agonistic and non-agonistic sesquiterpene lactones and diterpenoids indicated that agonists are required to contain γ-lactone or δ-lactone groups, but the exact chemical composition of the terpenoid skeleton is crucial for the agonistic properties of such compounds. Surprisingly, a number of chemically unrelated agonists, including the alkaloid strychnine, the antibiotic chloramphenicol, and the two synthetic bitter compounds denatonium and sucrose octaacetate, activated hTAS2R46 as well (Brockhoff et al. 2007). These observations place hTAS2R46 between the receptors recognizing particular chemical families of bitter compounds, prototypically represented by hTAS2R16 and hTAS2R38, and receptors such as hTAS2R14 (Behrens et al. 2004; Sainz et al. 2007), which detects a large number of structurally diverse bitter compounds. In the case of hTAS2R14, even the identification of many agonists did not allow the definition of a chemical core structure common to all agonists (Behrens et al. 2004). The recently deorphanized receptor hTAS2R7 might be even more broadly tuned than hTAS2R14 as membrane preparations of hTAS2R7 producing insect cells respond to most bitter substances tested (Sainz et al. 2007). A completely different group of bitter compounds are peptides that are generated by hydrolyses of proteins, e.g. during cheese production. Recently, it was shown that products of casein hydrolyses as well as synthetic dipeptides activate an array of human bitter taste receptors, including hTAS2R1, hTAS2R4, hTAS2R14, and hTAS2R16 (Maehashi et al. 2007).

To date, only a single study has investigated structural requirements of receptors necessary for agonist activation. Using the closely related receptors hTAS2R43, hTAS2R44, and hTAS2R47 as templates for domain swapping and mutagenesis analyses, Pronin et al. (2004) identified extracellular loops 1 and 2 as being involved in agonist activation. Further, it was noticed that transmembrane segments of TAS2Rs contribute to agonist—receptor interactions.

6 Variability of Bitter Taste Receptor Genes

Evolutionary forces shaped the mammalian bitter taste receptor gene repertoire, resulting in a variable number of functional TAS2R genes and pseudogenes. A detailed review on the evolution of taste receptor genes can be found in Shi and Zhang (2008). The fact that bitter taste receptor genes are very heterogeneous among each other, within species, and between species, as well as the large number of observed TAS2R polymorphisms reflects their evolutionary history. An important question is how much of the large variability in taste responsiveness between human individuals (Bartoshuk 2000) and also interstrain differences in mice is caused by TAS2R gene polymorphisms (Boughter et al. 2005; Chandrashekar et al. 2000) or other factors such as anatomical differences (Miller and Reedy 1990). The first and thus far only functional polymorphism identified in a rodent TAS2R gene was found in the mouse T2R5 gene, which codes for a cycloheximide responsive bitter taste receptor. It was shown that heterologously expressed mT2R5 from cycloheximide-sensitive mouse (taster) strains (DBA/2J, CBA/ca, BALB/c, C3H/He) responded to much lower concentrations of cycloheximide than mT2R5 variants from cycloheximide-insensitive (non-taster) strains (C57BL/6J, 129/Sv) (Chandrashekar et al. 2000). Whereas all complementary DNAs of taster strains corresponded to the cycloheximide-sensitive allele of the mT2R5 gene, the less-sensitive mT2R5 allele differing in several amino acid positions was found in all non-taster strains. Strikingly, experimentally determined pharmacological properties of mT2R5 receptor variants correlated closely with behavioural data, suggesting that the receptor proteins themselves determine the taste physiological properties of an organism and that heterologous expression of bitter taste receptors is a useful tool to study bitter taste physiological properties. Although numerous additional interstrain differences for the detection of bitter compounds have been reported (Boughter and Bachmanov 2007; Boughter et al. 2005), suggesting a number of additional functionally polymorphic bitter taste receptor genes in mice, no additional mT2R gene variants have been identified to date. The lack of additional functional polymorphisms in mouse bitter taste receptors is most likely caused by the small number of deorphanized mT2Rs, although genetic mapping studies, which have already identified some loci corresponding to the observed phenotypical differences, should help to isolate variant receptor genes more rapidly. In humans, where 11 hTAS2Rs have been deorphanized to date, a considerably higher number of functionally relevant bitter taste receptor gene polymorphisms have been identified. The first and so far the most pronounced polymorphism affects the PROP/PTC receptor, hTAS2R38. Although differences in the perception of PROP and PTC, separating the human population into PROP/PTC-taster and non-taster, were reported more than 70 years ago, polymorphisms in the hTAS2R38 gene underlying the observed phenotype were identified only recently by a positional cloning approach (Kim et al. 2003). Three common non-synonymous polymorphisms are found within hTAS2R38. The two predominant haplotypes found in the human population code for either the amino acids PAV (P^{49}, A^{262}, I^{296}) constituting the taster variant of hTAS2R38 or AVI in the corresponding positions for the non-taster variant. Later it was shown by recombinant expression studies that P49A and A262V are most important for receptor function, while V296I had little effect (Bufe et al. 2005), confirming the strong association of these two positions with the PROP/PTC-taster status observed earlier (Kim et al. 2003). Moreover, pharmacological features observed for hTAS2R38 expressed in HEK 293 cells correlated well with human psychophysical studies (Bufe et al. 2005), suggesting that, as already observed for mT2R5 (Chandrashekar et al. 2000), the receptor itself predominantly determines human taste perception for these bitter compounds. Another functional polymorphic hTAS2R gene is the β-d-glucopyranoside receptor, hTAS2R16 (Bufe et al. 2002). In the case of hTAS2R16 a single amino acid substitution at position 172, located in extracellular loop 2, is responsible for the observed functional differences of the two resulting receptor variants. Unlike hTAS2R38 variants, the changes in receptor sensitivity are more subtle, with the N172 variant being more sensitive than the K172 allele. Although the observed difference between both hTAS2R16 variants is only twofold, it affected human evolution profoundly. The ancient allele N172 is found with high frequencies in only restricted populations of the African continent, whereas the more sensitive derived allele K172 has spread throughout the world. It is believed that positive selection for the more sensitive hTAS2R16 variant occurred because it protects its carriers better than the less sensitive allele against intoxication by a number of poisonous cyanogenic bitter β-d-glucopyranosides frequently found in nature,. On the other hand, limited cyanoses caused by chronic ingestion of small quantities of such compounds may protect against malaria infection, which is likely the reason for the persistence of the less sensitive N172 allele in high-risk malaria areas within Africa (Soranzo et al. 2005). Recently, the different sensitivity to the bitter off-taste of an artificial sweetener, saccharin, and another bitter compound, aloin, has been connected with frequently occurring polymorphisms in the two closely related bitter taste receptors hTAS2R43 and hTAS2R44 (Pronin et al. 2007). In a previous report this pair of receptors was demonstrated to be activated by the purely bitter substance aristolochic acid, but also by the two artificial sweeteners saccharin and acesulfame K (Kuhn et al. 2004). Since both sweeteners exhibit in higher concentrations a bitter off-taste and are widely used as calorie-free supplements in the food and beverage industry, the now reported functional polymorphisms are likely to affect food choice and nutritional behaviour in a part of the population. The study by Pronin et al. showed that the critical amino acid for the sensitive variant of hTAS2R43 is located in intracellular loop 1 at position 35. It was reported that a tryptophan residue at this position is conserved in the majority of human bitter taste receptors, among them hTAS2R44. Direct comparison between the nucleic acid sequences of hTAS2R43 and hTAS2R44 revealed, however, that the nucleotide position leading to an exchange of tryptophan is different between both receptors, pointing to an independent evolutionary origin of this polymorphism, thus paralleling observations of hTAS2R38 mutations in human and chimpanzee, which led independently to the development of a non-taster allele for PROP/PTC tasting (Wooding et al. 2006). Since both receptors, hTAS2R43 and hTAS2R44, exhibit similar EC_{50} values in the low millimolar range for their responsiveness to saccharin, the taster status of individuals depends on the presence of either one or both of the sensitive alleles of hTAS2R43 or hTAS2R44. Consequently, only those individuals carrying the less sensitive alleles of both receptors exhibited a significantly reduced perception of the bitter taste of saccharin (Pronin et al. 2007). This is in contrast to the tasting ability for the substances aloin and aristolochic acid, for which the receptor hTAS2R43 is much more sensitive than hTAS2R44. In this case, the less sensitive allele of hTAS2R43 directly affects the taster status for these substances. The complex taste behaviour for this receptor pair is further exemplified by the fact that in the absence of the sensitive hTAS2R43 allele the hTAS2R44 genotype becomes visible, resulting again in a dichotomy of the observed taste phenotypes (Pronin et al. 2007).

These, at present, few examples of functional polymorphic bitter taste receptor genes indicate already that if only some of the haplotypes reported so far for hTAS2R genes (Kim et al. 2005; Wooding et al. 2004) exhibit similar functional differences, human bitter taste is highly individual. An even higher level of variability for bitter taste perception may arise from the fact that in some individuals entire hTAS2R genes or at least parts of the coding regions are deleted from the genome. Until now, this new observation has been restricted to only two receptors, hTAS2R43 and hTAS2R45 (Pronin et al. 2007), but careful genomic analyses might reveal additional receptors affected by this phenomenon, which is not necessarily restricted to humans.

7 Outlook

The past few years have been an exciting time for bitter taste research and the coming years will most likely remain exciting. While in the past progress was evident in signal transduction, receptor deorphanization, and a number of additional physiological processes, in the future we might start to understand the structure—function relationship of bitter taste receptors and their agonists, the bitter signal integration in the taste buds, and information transmission towards the brain.

References

Adler E, Hoon MA, Mueller KL, Chandrashekar J, Ryba NJ, Zuker CS (2000) A novel family of mammalian taste receptors. Cell 100:693–702

Barnicot NA, Harris H, Kalmus H (1951) Taste thresholds of further eighteen compounds and their correlation with P.T.C thresholds. Ann Eugen 16:119–128

Bartoshuk LM (2000) Comparing sensory experiences across individuals: recent psychophysical advances illuminate genetic variation in taste perception. Chem Senses 25:447–460

Behrens M, Brockhoff A, Kuhn C, Bufe B, Winnig M, Meyerhof W (2004) The human taste receptor hTAS2R14 responds to a variety of different bitter compounds. Biochem Biophys Res Commun 319:479–485

Behrens M, Foerster S, Staehler F, Raguse JD, Meyerhof W (2007) Gustatory expression pattern of the human TAS2R bitter receptor gene family reveals a heterogenous population of bitter responsive taste receptor cells. J Neurosci 27:12630–12640

Behrens M, Meyerhof W (2006) Bitter taste receptors and human bitter taste perception. Cell Mol Life Sci 63:1501–1509

Beidler LM, Smallman RL (1965) Renewal of cells within taste buds. J Cell Biol 27:263–272

Boudreau JC, Sivakumar L, Do LT, White TD, Oravec J, Hoang NK (1985) Neurophysiology of geniculate ganglion (facial nerve) taste systems: species comparisons. Chem Senses 10:89–127

Boughter JD Jr, Bachmanov AA (2007) Behavioral genetics and taste. BMC Neurosci 8 (Suppl 3):S3

Boughter JD Jr, Raghow S, Nelson TM, Munger SD (2005) Inbred mouse strains C57BL/6J and DBA/2J vary in sensitivity to a subset of bitter stimuli. BMC Genet 6:36

Brasser SM, Mozhui K, Smith DV (2005) Differential covariation in taste responsiveness to bitter stimuli in rats. Chem Senses 30:793–799

Brockhoff A, Behrens M, Massarotti A, Appendino G, Meyerhof W (2007) Broad tuning of the human bitter taste receptor hTAS2R46 to various sesquiterpene lactones, clerodane and labdane diterpenoids, strychnine, and denatonium. J Agric Food Chem 55:6236–6243

Brouwer JN, Hellekant G, Kasahara Y, van der Wel H, Zotterman Y (1973) Electrophysiological study of the gustatory effects of the sweet proteins monellin and thaumatin in monkey, guinea pig and rat. Acta Physiol Scand 89:550–557

Bufe B, Hofmann T, Krautwurst D, Raguse JD, Meyerhof W (2002) The human TAS2R16 receptor mediates bitter taste in response to beta-glucopyranosides. Nat Genet 32:397–401

Bufe B, Breslin PA, Kuhn C, Reed DR, Tharp CD, Slack JP, Kim UK, Drayna D, Meyerhof W (2005) The molecular basis of individual differences in phenylthiocarbamide and propylthiouracil bitterness perception. Curr Biol 15:322–327

Caicedo A, Roper SD (2001) Taste receptor cells that discriminate between bitter stimuli. Science 291:1557–1560

Chandrashekar J, Mueller KL, Hoon MA, Adler E, Feng L, Guo W, Zuker CS, Ryba NJ (2000) T2Rs function as bitter taste receptors. Cell 100:703–711

Clapp TR, Stone LM, Margolskee RF, Kinnamon SC (2001) Immunocytochemical evidence for co-expression of type III IP3 receptor with signaling components of bitter taste transduction. BMC Neurosci 2:6

Collings VB (1974) Human taste responses as a function of locus of stimulation on the tongue and soft palate. Percept Psychophys 16:169–174

Dahl M, Erickson RP, Simon SA (1997) Neural responses to bitter compounds in rats. Brain Res 756:22–34

Damak S, Rong M, Yasumatsu K, Kokrashvili Z, Perez CA, Shigemura N, Yoshida R, Mosinger B Jr, Glendinning JI, Ninomiya Y, Margolskee RF (2006) Trpm5 null mice respond to bitter, sweet, and umami compounds. Chem Senses 31:253–264

Delwiche JF, Buletic Z, Breslin PA (2001) Covariation in individuals' sensitivities to bitter compounds: evidence supporting multiple receptor/transduction mechanisms. Percept Psychophys 63:761–776

Dotson CD, Roper SD, Spector AC (2005) PLCbeta2-independent behavioral avoidance of prototypical bitter-tasting ligands. Chem Senses 30:593–600

Farbman AI (1980) Renewal of taste bud cells in rat circumvallate papillae. Cell Tissue Kinet 13:349–357

Fehr J, Meyer D, Widmayer P, Borth HC, Ackermann F, Wilhelm B, Gudermann T, Boekhoff I (2007) Expression of the G-protein alpha-subunit gustducin in mammalian spermatozoa. J Comp Physiol A Neuroethol Sens Neural Behav Physiol 193:21–34

Finger TE, Bottger B, Hansen A, Anderson KT, Alimohammadi H, Silver WL (2003) Solitary chemoreceptor cells in the nasal cavity serve as sentinels of respiration. Proc Natl Acad Sci USA 100:8981–8986

Fox AL (1932) The relationship between chemical constitution and taste. Proc Natl Acad Sci USA 18:115–120

Geran LC, Travers SP (2006) Single neurons in the nucleus of the solitary tract respond selectively to bitter taste stimuli. J Neurophysiol 96:2513–2527

Hofmann T, Chubanov V, Gudermann T, Montell C (2003) TRPM5 is a voltage-modulated and Ca(2+)-activated monovalent selective cation channel. Curr Biol 13:1153–1158

Hoon MA, Adler E, Lindemeier J, Battey JF, Ryba NJ, Zuker CS (1999) Putative mammalian taste receptors: a class of taste-specific GPCRs with distinct topographic selectivity. Cell 96:541–551

Huang L, Shanker YG, Dubauskaite J, Zheng JZ, Yan W, Rosenzweig S, Spielman AI, Max M, Margolskee RF (1999) Ggamma13 colocalizes with gustducin in taste receptor cells and mediates IP3 responses to bitter denatonium. Nat Neurosci 2:1055–1062

Huang AL, Chen X, Hoon MA, Chandrashekar J, Guo W, Trankner D, Ryba NJ, Zuker CS (2006) The cells and logic for mammalian sour taste detection. Nature 442:934–938

Jiang P, Ji Q, Liu Z, Snyder LA, Benard LM, Margolskee RF, Max M (2004) The cysteine-rich region of T1R3 determines responses to intensely sweet proteins. J Biol Chem 279:45068–45075

Keast RS, Bournazel MM, Breslin PA (2003) A psychophysical investigation of binary bitter-compound interactions. Chem Senses 28:301–313

Kim UK, Jorgenson E, Coon H, Leppert M, Risch N, Drayna D (2003) Positional cloning of the human quantitative trait locus underlying taste sensitivity to phenylthiocarbamide. Science 299:1221–1225

Kim U, Wooding S, Ricci D, Jorde LB, Drayna D (2005) Worldwide haplotype diversity and coding sequence variation at human bitter taste receptor loci. Hum Mutat 26:199–204

Kuhn C, Bufe B, Winnig M, Hofmann T, Frank O, Behrens M, Lewtschenko T, Slack JP, Ward CD, Meyerhof W (2004) Bitter taste receptors for saccharin and acesulfame K. J Neurosci 24:10260–10265

Li X, Staszewski L, Xu H, Durick K, Zoller M, Adler E (2002) Human receptors for sweet and umami taste. Proc Natl Acad Sci USA 99:4692–4696

Li X, Li W, Wang H, Cao J, Maehashi K, Huang L, Bachmanov AA, Reed DR, Legrand-Defretin V, Beauchamp GK, Brand JG (2005) Pseudogenization of a sweet-receptor gene accounts for cats' indifference toward sugar. PLoS Genet 1:27–35

Liu D, Liman ER (2003) Intracellular Ca2+ and the phospholipid PIP2 regulate the taste transduction ion channel TRPM5. Proc Natl Acad Sci USA 100:15160–15165

Maehashi K, Matano M, Wang H, Vo LA, Yamamoto Y, Huang L (2008) Bitter peptides activate hTAS2Rs, the human bitter receptors. Biochem Biophys Res Commun 365:851–855

Matsunami H, Montmayeur JP, Buck LB (2000) A family of candidate taste receptors in human and mouse. Nature 404:601–604

McLaughlin SK, McKinnon PJ, Margolskee RF (1992) Gustducin is a taste-cell-specific G protein closely related to the transducins. Nature 357:563–569

McLaughlin SK, McKinnon PJ, Spickofsky N, Danho W, Margolskee RF (1994) Molecular cloning of G proteins and phosphodiesterases from rat taste cells. Physiol Behav 56:1157–1164

Merigo F, Benati D, Tizzano M, Osculati F, Sbarbati A (2005) alpha-Gustducin immunoreactivity in the airways. Cell Tissue Res 319:211–219

Merigo F, Benati D, Di Chio M, Osculati F, Sbarbati A (2007) Secretory cells of the airway express molecules of the chemoreceptive cascade. Cell Tissue Res 327:231–247

Miller IJ Jr (ed) (1995) Anatomy of the peripheral taste system. Dekker, New York

Miller IJ Jr, Reedy FE Jr (1990) Variations in human taste bud density and taste intensity perception. Physiol Behav 47:1213–1219

Misaka T, Kusakabe Y, Emori Y, Gonoi T, Arai S, Abe K (1997) Taste buds have a cyclic nucleotide-activated channel, CNGgust. J Biol Chem 272:22623–22629

Nelson G, Hoon MA, Chandrashekar J, Zhang Y, Ryba NJ, Zuker CS (2001) Mammalian sweet taste receptors. Cell 106:381–390

Ogura T, Mackay-Sim A, Kinnamon SC (1997) Bitter taste transduction of denatonium in the mudpuppy Necturus maculosus. J Neurosci 17:3580–3587

Ozeck M, Brust P, Xu H, Servant G (2004) Receptors for bitter, sweet and umami taste couple to inhibitory G protein signaling pathways. Eur J Pharmacol 489:139–149

Perez CA, Huang L, Rong M, Kozak JA, Preuss AK, Zhang H, Max M, Margolskee RF (2002) A transient receptor potential channel expressed in taste receptor cells. Nat Neurosci 5:1169–1176

Perez CA, Margolskee RF, Kinnamon SC, Ogura T (2003) Making sense with TRP channels: store-operated calcium entry and the ion channel Trpm5 in taste receptor cells. Cell Calcium 33:541–549

Prawitt D, Monteilh-Zoller MK, Brixel L, Spangenberg C, Zabel B, Fleig A, Penner R (2003) TRPM5 is a transient Ca^{2+}-activated cation channel responding to rapid changes in $[Ca^{2+}]i$. Proc Natl Acad Sci USA 100:15166–15171

Pronin AN, Tang H, Connor J, Keung W (2004) Identification of ligands for two human bitter T2R receptors. Chem Senses 29:583–593

Pronin AN, Xu H, Tang H, Zhang L, Li Q, Li X (2007) Specific alleles of bitter receptor genes influence human sensitivity to the bitterness of aloin and saccharin. Curr Biol 17:1403–1408

Rolls E (ed.) (1995) Central taste anatomy and neurophysiology. Dekker, New York

Rossler P, Kroner C, Freitag J, Noe J, Breer H (1998) Identification of a phospholipase C beta subtype in rat taste cells. Eur J Cell Biol 77:253–261

Rossler P, Boekhoff I, Tareilus E, Beck S, Breer H, Freitag J (2000) G protein betagamma complexes in circumvallate taste cells involved in bitter transduction. Chem Senses 25:413–421

Ruiz-Avila L, McLaughlin SK, Wildman D, McKinnon PJ, Robichon A, Spickofsky N, Margolskee RF (1995) Coupling of bitter receptor to phosphodiesterase through transducin in taste receptor cells. Nature 376:80–85

Sainz E, Cavenagh MM, Gutierrez J, Battey JF, Northup JK, Sullivan SL (2007) Functional characterization of human bitter taste receptors. Biochem J 403:537–543

Sbarbati A, Osculati F (2005) A new fate for old cells: brush cells and related elements. J Anat 206:349–358

Sbarbati A, Merigo F, Benati D, Tizzano M, Bernardi P, Osculati F (2004) Laryngeal chemosensory clusters. Chem Senses 29:683–692

Sclafani A, Abrams M (1986) Rats show only a weak preference for the artificial sweetener aspartame. Physiol Behav 37:253–256

Scott TR, Verhagen JV (2000) Taste as a factor in the management of nutrition. Nutrition 16:874–885

Shi P, Zhang J (2006) Contrasting modes of evolution between vertebrate sweet/umami receptor genes and bitter receptor genes. Mol Biol Evol 23:292–300

Shi P, Zhang J (2008) Extraordinary diversity of chemosensory receptor gene repertoires among vertebrates. Results Probl Cell Differ doi: 400_2008_4

Shi P, Zhang J, Yang H, Zhang YP (2003) Adaptive diversification of bitter taste receptor genes in Mammalian evolution. Mol Biol Evol 20:805–814

Smith DV, Scott TR (2003) Gustatory neural coding. In: Doty RL (ed) Handbook of olfaction and gustation. Dekker, New York, pp 731–758

Smith DV, John SJ, Boughter JD (2000) Neuronal cell types and taste quality coding. Physiol Behav 69:77–85

Soranzo N, Bufe B, Sabeti PC, Wilson JF, Weale ME, Marguerie R, Meyerhof W, Goldstein DB (2005) Positive selection on a high-sensitivity allele of the human bitter-taste receptor TAS2R16. Curr Biol 15:1257–1265

Spector AC, Kopka SL (2002) Rats fail to discriminate quinine from denatonium: implications for the neural coding of bitter-tasting compounds. J Neurosci 22:1937–1941

Sternini C, Anselmi L, Rozengurt E (2008) Enteroendocrine cells: a site of 'taste' in gastrointestinal chemosensing. Curr Opin Endocrinol Diabetes Obes 15:73–78

Talavera K, Yasumatsu K, Voets T, Droogmans G, Shigemura N, Ninomiya Y, Margolskee, RF, Nilius B (2005) Heat activation of TRPM5 underlies thermal sensitivity of sweet taste. Nature 438:1022–1025

Ueda T, Ugawa S, Yamamura H, Imaizumi Y, Shimada S (2003) Functional interaction between T2R taste receptors and G-protein alpha subunits expressed in taste receptor cells. J Neurosci 23:7376–7380

Wong GT, Gannon KS, Margolskee RF (1996) Transduction of bitter and sweet taste by gustducin. Nature 381:796–800

Wong GT, Ruiz-Avila L, Margolskee RF (1999) Directing gene expression to gustducin-positive taste receptor cells. J Neurosci 19:5802–5809

Wooding S, Kim UK, Bamshad MJ, Larsen J, Jorde LB, Drayna D (2004) Natural selection and molecular evolution in PTC, a bitter-taste receptor gene. Am J Hum Genet 74:637–646

Wooding S, Bufe B, Grassi C, Howard MT, Stone AC, Vazquez M, Dunn DM, Meyerhof W, Weiss RB, Bamshad MJ (2006) Independent evolution of bitter-taste sensitivity in humans and chimpanzees. Nature 440:930–934

Wu SV, Rozengurt N, Yang M, Young SH, Sinnett-Smith J, Rozengurt E (2002) Expression of bitter taste receptors of the T2R family in the gastrointestinal tract and enteroendocrine STC-1 cells. Proc Natl Acad Sci USA 99:2392–2397

Wu SV, Chen MC, Rozengurt E (2005) Genomic organization, expression, and function of bitter taste receptors (T2R) in mouse and rat. Physiol Genomics 22:139–149

Xu H, Staszewski L, Tang H, Adler E, Zoller M, Li X (2004) Different functional roles of T1R subunits in the heteromeric taste receptors. Proc Natl Acad Sci USA 101:14258–14263

Yan W, Sunavala G, Rosenzweig S, Dasso M, Brand JG, Spielman AI (2001) Bitter taste transduced by PLC-beta(2)-dependent rise in IP(3) and alpha-gustducin-dependent fall in cyclic nucleotides. Am J Physiol Cell Physiol 280:C742–C751

Yokomukai Y, Cowart BJ, Beauchamp GK (1993) Individual-differences in sensitivity to bitter-tasting substances. Chem Senses 18:669–681

Zancanaro C, Caretta CM, Merigo F, Cavaggioni A, Osculati F (1999) alpha-Gustducin expression in the vomeronasal organ of the mouse. Eur J Neurosci 11:4473–4475

Zhang Y, Hoon MA, Chandrashekar J, Mueller KL, Cook B, Wu D, Zuker CS, Ryba NJ (2003) Coding of sweet, bitter, and umami tastes: different receptor cells sharing similar signaling pathways. Cell 112:293–301

Zhao GQ, Zhang Y, Hoon MA, Chandrashekar J, Erlenbach I, Ryba NJ, Zuker CS (2003) The receptors for mammalian sweet and umami taste. Cell 115:255–266

Orosensory Perception of Dietary Lipids in Mammals

P. Passilly-Degrace, D. Gaillard, and P. Besnard

Abstract Obesity constitutes a major public health problem for the twenty-first century, with its epidemic spread worldwide, particularly in children. The overconsumption of fatty foods greatly contributes to this phenomenon. Rodents and humans display a spontaneous preference for lipid-rich foods. However, the molecular mechanisms underlying this pattern of eating behaviour in mammals remain unclear. The orosensory perception of dietary lipids was long thought to involve only textural and olfactory cues. Recent findings challenge this limited viewpoint, strongly suggesting that the sense of taste also plays a significant role in dietary lipid perception and might therefore be involved in the preference for fatty foods and obesity. This mini-review analyses recent data related to the molecular mechanisms and physiological consequences of this means of orosensory lipid perception.

Abbreviations CT, Chorda tympani; CTX, Bilateral transection of the chorda tympani nerve; DRK, Delayed-rectifying potassium; FFA, Free fatty acid; GL, Glossopharyngeal; GLX, Bilateral transection of the glossopharyngeal nerve; GPCR, G-protein-coupled receptor; LCFA, Long-chain fatty acid; PROP, 6-n-Propylthiouracil; PTK, Protein tyrosine kinase; PUFA, Polyunsaturated fatty acid; NST, Nucleus of the solitary tract; TG, Triglyceride; TRC, Taste receptor cells; TRPM5, Transient receptor potential protein 5

1 Introduction

Eating is a complex form of behaviour governed by a combination of physiological, hedonic, cultural and even philosophical factors. The profound technical and economic changes of the twentieth century had a profound effect on our way of life

P. Besnard (✉)
Physiologie de la Nutrition, UMR INSERM U 866, École Nationale Supérieure de Biologie Appliquée à la Nutrition et à l'Alimentation (ENSBANA), Université de Bourgogne, 1, Esplanade Erasme, 21000 Dijon, France
e-mail: pbesnard@u-bourgogne.fr

and, consequently, on our eating behaviour. For the first time in its history, much of the world's population no longer has to "run after calories". One direct consequence of this fundamental change has been the emergence of plethora diseases which raise major public health problems. One of the most patent examples is obesity, which has reached epidemic proportions worldwide and is a major contributor to the global burden of chronic diseases. This phenomenon affects both adults and children (Malecka-Tendera and Mazur 2006). Recent data even suggest that the increase in prevalence of obesity is associated with a decrease in life expectancy in children (Olshansky et al. 2005).

An abundance of food has obvious consequences: it promotes our specific appetites. Lipids account for about 40% of the calories ingested in Western countries, whereas nutritional recommendations are 5–10% lower. This excessive lipid intake, associated with a qualitative imbalance (excess of saturated fatty acids and cholesterol, too high $\omega 6/\omega 3$ ratio) strongly favours the development of obesity and associated diseases (atherosclerosis, non insulin-dependent diabetes, hypertension, cancer). This attraction to fatty foods is not specific to humans. Rats and mice spontaneously prefer lipid-rich foods if provided with a free choice (Tsuruta et al. 1999; Takeda et al. 2000). This attraction to lipids is so strong that mice given free access to an oil as an optional diet rapidly become obese (Takeda et al. 2001a). The origin of this preference for lipids remains unclear.

Perception of the chemical composition of foods requires the integration of early olfactory, somesthesic and gustatory cues and delayed neuroendocrine and metabolic signals induced after ingestion. All these signals converge on specific areas of the central nervous system, in which they are integrated to induce stereotyped physiological effects (food preference or aversion, digestive anticipation). Thus, when possible, mammals preferentially select foods on the basis of their physicochemical and nutritional properties (digestibility, nutrient composition, metabolic effectiveness, lack of toxicity). This complexity accounts for the concept of taste itself varying widely according to the individual concerned. In general terms, gustation is a global feeling induced by the release during chewing of various dietary molecules acting on multiple sensors (olfactory and taste receptors, mechanoreceptors, thermoreceptors, nociceptors). For the specialist, the sense of taste is reduced to the interaction between a tastant with a specific recognition structure in the taste buds, the resulting sensory signal being conveyed to the brain via the gustatory nerves. According to this definition, four primary tastes (sweet, salty, bitter and sour) were initially described, to which umami has recently been added (Gilbertson and Boughter 2003; Sugita 2006). Until recently, it was thought that oral lipid detection involved only somesthesic and olfactory cues. This restrictive view has been challenged by recent observations suggesting that gustation is also involved in spontaneous fat preference, underlying that the fatty taste might constitute a sixth gustatory modality. This review highlights recent findings in this new field of investigations in both rodents and humans.

2 Fat Perception in Rodents

Taste perception is ensured by taste receptor cells (TRC), which are clustered in specialized onion-shaped structures, the taste buds, found at high density on the tongue and at low density in the soft palate, larynx, pharynx and upper part of the oesophagus. Taste buds consist of 50–150 TRC, as a function of species. In the lingual epithelium, taste buds are located in three types of gustatory papillae with different spatial distributions. Most are the fungiform papillae, which cover the front two thirds of the tongue. These papillae are mushroom-shaped and have a small number (one to three) of taste buds on their apical surface. The circumvallate and foliate papillae are located on the central and lateral regions, respectively, of the posterior third of the tongue. The circumvallate papillae consist of a circular depression, the walls of which contain several hundred taste buds in humans (Mela and Mattes 1988) and in rodents (Oakley 1993). Humans have about ten circumvallate papillae, whereas rodents have only one, in a central position. Foliate papillae are located at the posterior lateral edge of the tongue and contain hundreds of taste buds.

2.1 Rats and Mice Display a Spontaneous Lipid Preference

There is compelling evidence to suggest that laboratory rodents have an orosensory system devoted to lipid detection. The two-bottle preference test is a classic method for studying the feeding behaviour of animals in a free-choice situation. This simple paradigm clearly shows that rats (Tsuruta et al. 1999) and mice (Takeda et al. 2000) display a strong preference for lipid-rich solutions. However, this observation is difficult to interpret, because food preference results from the integration of olfactory, somesthesic, gustatory and postingestive signals. Thus, the relative importance of each of these parameters for the spontaneous fat attraction has been systematically explored. First, spontaneous fat attraction is maintained in anosmic rats and mice, in which olfaction is blocked by chemical means, demonstrating that smell does not play a significant role in this behaviour (Takeda et al. 2001b; Fukuwatari et al. 2003). Second, the two-bottle preference test clearly reveals that mice prefer vegetable oils to texturing agents, such as xanthan gum, suggesting that texture is not a major cue in orosensory fat perception (Takeda et al. 2000). This observation is supported by findings for conditioned taste aversion, in which a naive animal learns to avoid a newly encountered tastant after suffering adverse postingestive effects triggered by an intraperitoneal injection of LiCl. Indeed, aversion to a sucrose/corn oil mixture cannot be induced in rats by replacing the corn oil with an indigestible oil of similar texture (mineral oil), suggesting that the chemical composition of the oil, rather than its textural characteristics, plays a key role in the perception of this mixture (Smith et al. 2000). Interestingly, rats previously conditioned with corn oil display a stronger aversion to the sucrose/corn oil mixture than animals conditioned with sucrose alone, indicating that corn oil is the salient feature of the mixture (Smith et al. 2000).

Third, fat preference is not abolished in very short term experiments (0.5–5 min) designed to minimize postingestive effects (Tsuruta et al. 1999; Smith et al. 2000; McCormack et al. 2006). The persistence of an attraction to lipids in rats and mice in which textural, olfactory and postingestive cues have been simultaneously minimized strongly suggests that taste influences this feeding behaviour in rodents (Takeda et al. 2001b; Fukuwatari et al. 2003).

2.2 Which Lipids Are Detected?

Although dietary lipids are mainly constituted of triglycerides (TG), long-chain fatty acids (LCFA; more than 16 carbons) seem to be responsible for oral lipid perception. In a free-choice situation, rats have a weaker preference for TG and medium-chain fatty acids (8 to 14 carbons) than for LCFA (Tsuruta et al. 1999; Fukuwatari et al. 2003). This chemical selectivity is very tight, as LCFA derivatives, such as methyl LCFA, are not recognized (Tsuruta et al. 1999). The ability of rodents to detect LCFA specifically has also been confirmed with the conditioned taste aversion paradigm. It is noteworthy that both rats and mice can be conditioned to avoid specific LCFA (McCormack et al. 2006; Gaillard et al. 2008), with a submicromolar detection threshold (McCormack et al. 2006; Yoneda et al. 2007).

Lingual lipase, which is responsible for an efficient release of LCFA from TG in rodents, seems to play a significant role in oral fat perception. Indeed, its pharmacological inhibition leads to a dramatic decrease in lipid preference in the mouse (Kawai and Fushiki 2003). This may explain why mineral oil, which is not digestible, is not as attractive as vegetable oil in a free-choice situation (Yoneda et al. 2007). Interestingly, lingual lipase is known to be released directly into the clefts of foliate and circumvallate papillae by the von Ebner glands in rodents (Kawai and Fushiki 2003). This anatomical feature appears to be ideal for the efficient hydrolysis of TG and generates high LCFA levels close to the taste buds, facilitating their subsequent detection by TRC.

2.3 How Are LCFA Detected?

Chemosensitive proteins (ion channels, metabotropic and ionotropic receptors) located on the apical side of the TRC are responsible for taste reception. Interactions between a tastant and its specific detection system lead to changes in membrane potential and/or intracellular free calcium concentration, triggering neurotransmitter release and generating, in turn, the firing of gustatory afferent nerve fibres (Gilbertson and Boughter 2003; Sugita 2006). The gustatory perception of lipids must therefore require the existence of receptors displaying a high affinity for LCFA in the TRC. Three plausible candidates for this function have recently been identified: the delayed-rectifying potassium (DRK) channel Kv1.5, a G-protein-coupled receptor (GPCR), GPR120, and the receptor-like glycoprotein CD36.

2.3.1 DRK Channels

In TRC, various voltage-activated ion channels (K^+, Na^+, Ca^{2+}) contribute to the release of neurotransmitters after chemical stimulation (Gilbertson and Boughter 2003; Sugita 2006). LCFA are known to regulate ion channels in various cell types. Gilbertson et al. (1997) from the University of Utah (USA) used patch-clamp recording to explore the putative effect of free fatty acids (FFA) on membrane potential in TRC isolated from rat fungiform papillae. They reported that FFA are able to inhibit the DRK channels, which are known to be involved in the transduction pathways for various taste stimuli. This action is direct and strictly mediated by polyunsaturated fatty acids (PUFA). PUFA inhibition is effective only if these molecules are applied extracellularly, as in the physiological context. This suggests that the responsiveness to PUFA of taste cells may contribute to fat preference, thereby indirectly affecting body mass. To explore this hypothesis, the effect of PUFA on DRK currents was explored by patch-clamp recording in isolated fungiform TRC from obesity-resistant (S5B/P1) and obesity-prone (Osborne–Mendel) rats. Unexpectedly, PUFA-mediated depolarization was greater in TRC from obesity-resistant rats, which are known to prefer carbohydrates, than in those from obesity-prone animals, which prefer fats (Gilbertson et al. 2005). This strain-specific response was attributed to a difference in the pattern of expression of DRK channel isotypes in TRC, with obesity-resistant rats having more K^+ channels responsive to PUFA than obesity-prone animals (Gilbertson et al. 2005). Various DRK channels are found in rat fungiform papillae (Liu et al. 2005), but the shaker Kv1.5 channel, specifically inhibited by PUFA in cardiac cells (Honore et al. 1994), has been shown to be strongly expressed in TRC from obesity-resistant S5B/P1 rats (Gilbertson et al. 2005). The mechanism by which PUFA inhibit Kv1.5 channels in taste bud cells remains unknown. However, a direct effect is likely, since a physical interaction between PUFA and the extracellular domain of the Kv1.5 protein has already been reported in cardiomyocytes (Honore et al. 1994). The Kv1.5 channel may therefore be considered as an ionotropic receptor in TRC. All together, these data suggest that the control of Kv1.5 channels in TRC by PUFA does not explain the spontaneous fat preference observed in rodents.

2.3.2 GPCRs: GPR120

GPR120 belongs to the GPCR family (Fredriksson et al. 2003; Rayasam et al. 2007). It is abundantly expressed in enteroendocrine cells, particularly in the distal part of the small intestine (ileum) and the colon, in both mice and humans. In these cells, GPR120 functions as a receptor for unsaturated LCFA, leading to the secretion of incretins such as glucagon-like peptide-1 (Hirasawa et al. 2005) and cholecystokinin (Tanaka et al. 2007). Preliminary studies have shown that GPR120 is also expressed in circumvallate and fungiform TRC in rats (Matsumura et al. 2007). The physiological role of this receptor in the taste buds remains unknown.

GPR120 is also found in the mouse enteroendocrine STC-1 cell line. Interestingly, STC-1 cells have several genotypic and phenotypic features in common with TRC.

For instance, they also express the genes encoding receptors for sweet taste, T1R2 and T1R3 (Dyer et al. 2005), and for bitter taste, T2R (Wu et al. 2002), and the five basic taste stimuli induce an increase in intracellular free calcium concentration (Saitoh et al. 2007). In vitro studies in STC-1 cells have shown that unsaturated LCFA/GPR120 interaction also leads to an increase in intracellular free calcium concentration (Hirasawa et al. 2005). This result is reminiscent of the tastant-mediated increase in intracellular free calcium concentration in TRC. GPR120 is a receptor for unsaturated fatty acids (Hirasawa et al. 2005) and its involvement in the orosensory perception of dietary lipids is thus plausible. However, further studies including the effect of GPR120 gene manipulation (invalidation or overexpression) on fat preference are required to validate this hypothesis.

2.3.3 CD36 as a Lipid Sensor

CD36 is a receptor-like protein that binds saturated and unsaturated LCFA with an affinity in the nanomolar range (Baillie et al. 1996). It has the structural and functional features required of a taste-based lipid receptor. First, CD36 appears to be restricted to the gustatory epithelium in rodent tongues (Fukuwatari et al. 1997; Laugerette et al. 2005). In mice, CD36 is expressed particularly strongly in circumvallate papillae, to a lesser extent in foliate papillae and only very weakly in fungiform papillae (Laugerette et al. 2005). Immunohistochemical staining has shown the CD36 protein to be present mostly on the apical side of some of the TRC lining the taste pores (Laugerette et al. 2005). This distribution of a protein with a very high affinity for LCFA is particularly suitable for the generation of a lipid signal by taste buds. Indeed, CD36-positive TRC are directly exposed to a microclimate potentially rich in LCFA, owing to the local release of lingual lipase in the clefts of circumvallate papillae (Kawai and Fushiki 2003). Second, a role of CD36 as a lipid sensor is also supported by the predicted structure of this protein. This plasma membrane protein has a hairpin structure, with a large extracellular hydrophobic pocket located between two short cytoplasmic tails (Rac et al. 2007). Existence of a physical interaction between the intracellular C-terminal tail of CD36 and Src protein tyrosine kinase (PTK) (Huang et al. 1991) results in the formation of a functional complex, allowing the transfer of an exogenous lipid signal into the TRC. Third, CD36 gene invalidation abolishes both fat preference (Laugerette et al. 2005; Sclafani et al. 2007a) and the cephalic phase of digestion triggered by oral LCFA deposition (Laugerette et al. 2005). Altogether, these findings strongly suggest that CD36 is a lipid receptor involved in the orosensory perception of dietary lipids in rodents.

2.4 Does the Gustatory Pathway Play a Role in CD36-Mediated Fat Perception?

The key question at this stage concerned the possible mediation of oral lipid detection by the gustatory pathway. Studies of the lipid transduction signal in TRC and of the afferent nerve route used to transfer the fat signal to the central nervous system were required to address this question.

2.4.1 Mechanisms of Fat Signal Transduction

The tastant-induced release of neurotransmitters towards afferent nerve fibres leads to the orosensory perception of sapid molecules. As already mentioned, this event is known to be mediated by changes in intracellular free calcium concentration in TRC (Gilbertson and Boughter 2003). If lingual CD36 acts as a lipid receptor, LCFA binding to CD36 may also affect intracellular free calcium concentration. We tested this hypothesis, by determining the intracellular free calcium concentration in CD36-positive TRC isolated from mouse circumvallate papillae by affinity purification with magnetic beads (Gaillard et al. 2008). Saturated and unsaturated LCFA triggered a rapid and robust increase in intracellular free calcium concentration in CD36-positive cells. This effect was strictly CD36-dependent, as it was not observed in CD36-negative cells. Moreover, addition of the specific CD36 binding inhibitor, sulfo-N-succinimidyl oleic acid ester (Harmon et al. 1991) to the culture medium completely abolished the linoleic acid mediated rise in intracellular free calcium concentration in CD36-positive cells. These data provided the first demonstration that LCFA increases intracellular free calcium concentration in the taste bud cells, and that this event is CD36-dependent (Gaillard et al. 2008).

The reception of tastes other than salty and sour requires the heterotrimeric gustducin complex (Wong et al. 1996) (Fig. 1a). CD36-expressing TRC also contain the α-subunit of gustducin (Laugerette et al. 2005; Gaillard et al. 2008), but this G-protein complex is not involved in fat preference. Indeed, attraction for LCFA-enriched solutions is maintained in α-gustducin-null mice (Sclafani et al. 2007b). This result was expected, because CD36 does not belong to the GPCR family, unlike the T1R and T2R receptors responsible for sweet, bitter and umami taste detection. An alternative mechanism has recently been described for transduction of the lipid signal (El-Yassimi et al. 2008). It has been shown that LCFA binding to CD36 leads to the recruitment and activation of the Src-PTK Fyn and Yes, inducing an increase in intracellular free calcium concentration in purified mouse CD36-positive TRC (Fig. 1b). This increase results from the opening of store-operated calcium channels (El-Yassimi et al. 2008). The involvement of Src-PTK in this signalling cascade is not entirely surprising, because interactions between CD36 and Src-PTK Fyn, Lyn and Yes have been reported in other cell types (Huang et al. 1991). LCFA also seems to increase inositol triphosphate (IP3) concentration in CD36-positive TRC (El-Yassimi et al. 2008) (Fig. 1b). This CD36-dependent regulation may increase intracellular free calcium concentration still further, by mobilizing the Ca^{2+} stored in endoplasmic reticulum cisternae, as reported, for instance, for sweet taste (Bernhardt et al. 1996). The mechanism underlying this CD36-dependent pathway remains unknown. Concomitant exogenous and endogenous Ca^{2+} fluxes seem therefore to account for the increase in intracellular free calcium concentration mediated by LCFA in CD36-positive TRC. Transient receptor potential protein 5 (TRPM5) is known to play an important role in taste transduction (Fig. 1). This Na⁺/K⁺ channel responds to rapid changes in intracellular free calcium concentration by inducing transient membrane depolarization (Prawitt et al. 2003), leading to neuromediator release. The much weaker attraction to LCFA observed in TRPM5$^{-/-}$ mice than in controls demonstrates the involvement of this ion channel in the fat signalling cascade found in the TRC (Sclafani et al. 2007b) (Fig. 1b).

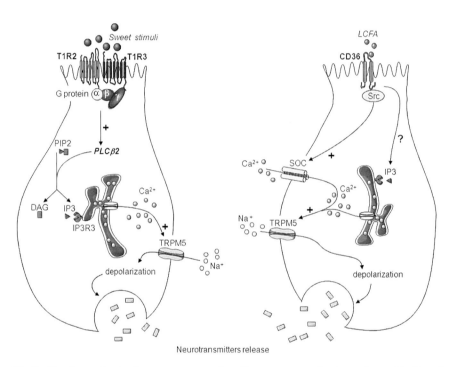

Fig. 1 Signal transduction in taste receptor cells. a Sweet taste transduction pathway. Binding of sweet molecules to the G-protein-coupled receptor T1R2 and T1R3 heteromers induces the dissociation of the G-protein. G-protein subunits then activate phospholipase C β2 (PLCβ2), which cleaves phosphatidyl inositol diphosphate (PIP2) in diacyglycerol (DAG) and inositol triphosphate (IP3). IP3 binds to isoform 3 of the IP3 receptor (IP3R3) located at the surface of the smooth endoplasmic reticulum, inducing in turn Ca^{2+} release into the cytoplasm. This increase in intracellular free calcium concentration opens transient receptor potential protein 5 (TRPM5) channels, allowing influx of Na⁺ ions and subsequent depolarization of the cell, leading to the release of neurotransmitters. b CD36-mediated fat taste transduction pathway. In CD36-positive cells immunomagnetically isolated from mouse circumvallate papillae, Src protein tyrosine kinase Yes and Fyn are activated when long-chain fatty acids (LCFA) bind to CD36, leading to a rise in intracellular free calcium concentration secondary to the opening of store-operated calcium (SOC) channels. LCFA/CD36 interaction also seem to increase IP3 concentration, suggesting that mobilization of endoplasmic reticulum Ca^{2+} could also contribute to the increase in intracellular free calcium concentration. Like sweet taste transduction, intracellular free calcium concentration rise can activate the TRPM5 channels, leading to release of neurotransmitters

Taken together, these findings provide the first evidence that LCFA have a profound effect on the function of mouse TRC, via the activation of a signalling cascade dependent on CD36.

2.4.2 The Gustatory Nerves Convey the Fat Signal

TRC from fungiform papillae and some of the anterior foliate papillae establish synaptic contacts with the chorda tympani (CT) nerve, whereas the posterior parts of the foliate and circumvallate papillae are innervated by the glossopharyngeal (GL) nerve. The possible involvement of gustatory nerves in the LCFA-mediated fat preference has recently been explored in rodents, by studying the impact of bilateral transection of the CT nerve (CTX) and/or bilateral transection of the GL nerve (GLX).

In rats, CTX decreases fat preference in a free-choice situation (Stratford et al. 2006; Pittman et al. 2007). Consistent with these data, CTX rats display much weaker conditioned aversion to linoleic acid than sham-operated control rats (Stratford et al. 2006; Pittman et al. 2007). Paradoxically, the impact of total denervation of the peripheral gustatory nerves (CTX + GLX) was not investigated in these rat studies, although such an exploration would be required for a full demonstration. In mice, the lack of functional peripheral gustatory nerves (CTX + GLX animals) fully abolishes both spontaneous linoleic acid preference and conditioned aversion (Gaillard et al. 2008). These findings demonstrate that afferent gustatory nerve fibres play a crucial role in the orosensory perception of LCFA in rodents. Although CD36 in the TRC probably acts as a lipid receptor in these mammals (Laugerette et al. 2005), its involvement in the fat signal conveyed by peripheral gustatory nerves remains to be demonstrated. However, the much lower levels of pancreatic and biliary secretions observed after oral linoleic acid deposition in CTX/GLX mice than in sham-operated control mice (Gaillard et al. 2008) provides indirect support for this hypothesis. Indeed, the cephalic phase of digestion triggered by the presence of LCFA in the oral cavity is highly CD36 dependent (Laugerette et al. 2005). A direct demonstration would require analysis of electrophysiological recordings of CT and GL nerves in $CD36^{+/+}$ and $CD36^{-/-}$ mice subjected to oral fatty acid stimulation.

2.4.3 Lipid-Induced Neuronal Activity in the Nucleus of the Solitary Tract

Lipid signals therefore appear to be transmitted by a peripheral nerve route known to be involved in the transfer of gustatory information to the brain. The nucleus of the solitary tract (NST) in the brainstem is the first synaptic relay in the nervous gustatory cascade. Immunohistochemical detection of Fos, the protein encoded by the immediate early gene c-fos, has been successfully used to identify populations of neurons activated by LCFA in the NST (Gaillard et al. 2008). In mice, oral linoleic acid deposition triggers the activation of NST areas known to receive CT and

GL afferent fibres. This activation appears to be CD36-dependent, as it is not observed in CD36-null mice subjected to oral stimulation with linoleic acid (Gaillard et al. 2008). Axons from the mandibular branch of the trigeminal nerve innervating the anterior tongue project into taste areas of the NST (Hamilton and Norgren 1984), but mechanical or textural stimulation is unlikely. Indeed, water alone or mixed with xanthan gum to mimic the texture of lipids does not affect neuronal activity in the NST (Gaillard et al. 2008). Thus, the fat signal triggered by the interaction of LCFA with CD36 in the oral cavity is transmitted through the NST. The known involvement of the lateral hypothalamus and nucleus accumbens in food intake and reward, respectively, and in the reception of synaptic inputs from the NST (Berthoud 2002) may account for the spontaneous preference for LCFA-rich food observed in mice. The digestive projections of the NST (Berthoud 2002) may also contribute to a lipid-mediated reflex, controlling pancreatobiliary secretions directly and/or indirectly, through the production of intestinal hormones.

2.5 Main Questions Posed

This brief literature review provides compelling evidence in favour of a significant role for the sense of taste in fat preference in rodents. However, the presence of several putative lipid receptors (CD36, Kv1.5 channels, GPR120) in taste buds was not expected and raises questions about the physiological roles of these receptors. Furthermore, the taste-mediated perception of LCFA by CD36 would be unusual owing to the multifunctional nature of this protein.

2.5.1 How Can DRK and CD36 Data Be Reconciled?

While lipid detection by lingual CD36 contributes to promote fat feeding (Laugerette et al. 2005), the lipid-mediated inhibition of DRK Kv1.5 channels in the TRC seems to be more efficient in rats, which prefer carbohydrates, than in animals with a marked natural preference for fats (Gilbertson et al. 1998, 2005). This apparent paradox may be accounted for by differences in the binding and structural features of these two lipid receptors. Consistent with the binding affinity of CD36 (Baillie et al. 1996), the behavioural and digestive effects of lingual CD36 are triggered by both saturated and unsaturated LCFA (Laugerette et al. 2005), whereas DRK inhibition is strictly PUFA-dependent (Gilbertson et al. 1997). Furthermore, given its receptor-like structure and its association with Src kinases, CD36 probably acts as a metabotropic receptor in the tongue, whereas Kv1.5 appears to be an ionotropic receptor. These two lipid-mediated sensory systems therefore probably coexist in the rodent tongue. Lingual CD36 may be a lipid sensor involved in fat preference, encouraging the selection of lipid-rich foods and facilitating their subsequent use by the body. By depolarizing TRC via Kv1.5 channel inhibition, PUFA may act as a taste modulator, increasing the palatability of other tastants. The preference for a subthreshold concentration of saccharin observed in the presence of linoleic acid (Gilbertson et al. 2005) is consistent with this hypothesis. Interestingly, similar results have been obtained with other sapid substances in behavioural assays, with PUFA reinforcing the attractive or aversive effects of sweet and bitter tastes, respectively (Pittman et al. 2006). Further experiments are required to validate this working hypothesis.

2.5.2 How Can We Account for the Multifunctionality of CD36?

CD36 is a plasma-membrane glycoprotein expressed in a wide variety of tissues. It is a multifunctional protein belonging to the family of class-B scavenger receptors. It increases the uptake of LCFA by cardiomyocytes and adipocytes (Coburn et al. 2000; Hajri et al. 2001) and that of oxidized low-density lipoproteins by macrophages (Febbraio et al. 2001), modifies platelet aggregation by binding to thrombospondin and collagen (Chen et al. 2000), facilitates the phagocytosis of apoptotic cells by macrophages (Ren et al. 1995) and increases the cytoadhesion of erythrocytes infected with Plasmodium falciparum (Oquendo et al. 1989). In addition, CD36 has also recently been shown to play a role in the taste reception of dietary lipids on the tongue (Laugerette et al. 2005; Gaillard et al. 2008). This new function may seem paradoxical, given the high tissue and binding specificities generally displayed by other taste receptors, such as T1R and T2R (Sugita 2006).

Despite its multifunctionality, CD36 generally plays a specific role in a given cell type. For example, it is involved in collagen-mediated cytoadhesion in platelets, whereas it mediates LCFA uptake in myocytes. This cell specificity of function probably results from both cellular context (genotype and microenvironment) and aspects of CD36 itself, as regulation of the gene encoding this receptor is unusually complex (Andersen et al. 2006) and this protein is subject to posttranslational modifications (Rac et al. 2007). The functional diversity of CD36 results partly from alternative splicing of the precursor messenger RNA molecule. Indeed, the human, rat and mouse CD36 genes contain several alternative and independent promoters and first exons (Cheung et al. 2007; Rac et al. 2007; Sato et al. 2007). The expression profiles of these alternative transcripts differ considerably between tissues (Andersen et al. 2006), resulting in the tissue-specific regulation of CD36 gene expression. For instance, differences in the CD36 promoters expressed in the liver and small intestine account for the differential regulation of the CD36 gene by peroxisome proliferator-activated receptor agonists observed in these tissues (Sato et al. 2007). The diversity of posttranslational modifications of CD36 (glycosylation, phosphorylation, acylation, palmitoylation) also contributes to the cell-type specificity of its functions. For example, the degree of glycosylation of the CD36 extracellular domain varies considerably among tissues (Greenwalt et al. 1992). This phenomenon may result in specific physiological effects due to the selection of a specific ligand in a specific cell. Thus, a role for CD36 in the chemoreception of dietary lipids by TRC remains plausible despite the multifunctional nature of this protein. The recent identification of a CD36-related receptor responsible for the olfactory detection of a fatty acid derived pheromone in Drosophila (Benton et al. 2007) suggests that the chemosensory function of CD36 previously reported in rodents may be widespread throughout the animal kingdom.

Thus, there is compelling evidence for the existence of a gustatory system devoted to LCFA detection in rodents. Does such a system exist in humans?

3 Fat Perception in Humans

Far fewer studies have been carried out on humans, but psychophysical studies carried out by Richard Mattes and co-workers strongly suggest that taste also plays a role in fat perception in humans (Mattes 2005). Sham-feeding experiments in healthy subjects have shown that oral exposure to fat is sufficient to increase postprandial plasma TG levels significantly (Mattes 1996). Interestingly, this effect, which is independent of food ingestion, is maintained even if textural (Mattes 2001a) and olfactory (Mattes 2001b) cues are minimized. It is thought that this effect results from the release from enterocytes of residual lipids from the previous meal (Mattes 2002). This event is probably dependent on the cephalic phase of digestion induced by the presence of lipids in the oral cavity. Consistent with this hypothesis, oral lipid stimulation has also been shown to increase plasma pancreatic peptide concentration in healthy subjects (Crystal and Teff 2006). This endocrine effect is very rapid, peaking only 4 min after oral exposure to fat. This finding is similar to those reported for rodents, in which the orosensory perception of dietary lipids also affects pancreatic function (Hiraoka et al. 2003; Laugerette et al. 2005). Recent studies have shown that FFA are responsible for these physiological effects (Chale-Rush et al. 2007).

3.1 Evidence for the Orosensory Detection of LCFA

Healthy adult subjects can detect saturated and unsaturated LCFA specifically. As this has been observed in experiments in which olfactory and somatosensory cues were minimized, this perception has been attributed to the sense of taste (Chale-Rush et al. 2007). The detection threshold for LCFA is much lower (0.028% w/v, on average) (Chale-Rush et al. 2007) than that for TG (5.6 and 17.3% w/v, on average, in young and old subjects, respectively) (Schiffman et al. 1998). Lipid-rich foods may contain up to 0.5% FFA (Mattes 2005), so TG hydrolysis by lingual lipase does not seem to be required for orosensory fat detection. This point is important, given continuing debate concerning whether humans have an efficient lingual lipase (Moreau et al. 1988; Hamosh 1990). LCFA may be responsible for the taste perception of dietary lipids in humans, as reported in rodents, but is there a functional basis for this assumption? Electrophysiological recordings of individual neurons in rhesus macaques and functional magnetic resonance imaging in humans can be used to address this question.

3.2 Fat-Mediated Sensory Processing in the Brain

In primates, some of the orosensory inputs of olfactory, somesthesic and gustatory origin triggered by the presence of food in the oral cavity converge on the orbitofrontal cortex (i.e. the secondary taste cortex) and amygdala (Rolls 2007). This multimodal convergence complicates evaluations of the respective roles played by the various orosensory signals in the perception of dietary lipids. However, electrophysiological recordings in macaques (Verhagen et al. 2003) and functional magnetic resonance imaging in humans (De Araujo and Rolls 2004) have shown that some of the neurons located in the orbitofrontal cortex are activated by oral fat load independently of its viscosity, suggesting that the detection of dietary lipids is not solely dependent on their textural properties. Consistently with this assumption, FFA-sensitive neurons have been found in the primary taste cortex (insula), secondary taste cortex (orbitofrontal cortex) and amygdala in macaques (Kadohisa et al. 2005). As local pH remains stable in the presence of FFA, neuronal activation due to an increase in buccal acidity is unlikely. Besides, there is only a limited functional overlap between neurons activated by oral oil deposition and neurons activated by FFA (Kadohisa et al. 2005). It therefore seems that a system devoted to an orosensory (taste?) perception of FFA exists, in addition to the textural perception of fats in primates. However, in contrast to rodents, in which only LCFA are perceived in the oral cavity, medium-chain fatty acids (e.g. lauric acid, C12:0) seem to be able to activate the neuronal pathway known to convey taste signals in macaques (Kadohisa et al. 2005). The reasons for this difference between species remain unclear. The molecular mechanism responsible for the orosensory perception of FFA in humans is currently unknown, and it remains to be demonstrated whether CD36 or other putative lipid receptors are expressed in primate TRC.

3.3 Is Fat Perception Related to 6-n-Propylthiouracil Taster Status?

Sapid molecules are not perceived similarly by all humans. For example, the avoidance of bitterness is a trait of the human species, but the extent to which a given food is perceived as bitter differs considerably among subjects. Some subjects in the general population can detect bitterness due to very low concentrations of 6-n-propylthiouracil (PROP) (Bartoshuk et al. 1994). These subjects are described as PROP-tasters. The strong response to bitter tastes of PROP-tasters is thought to result from the presence of a larger number of fungiform papillae (Bartoshuk et al. 1994). Sensitivity to PROP is also thought to be associated with a preference for sweet and fatty foods (Looy and Weingarten 1992; Tepper and Nurse 1998; Duffy et al. 1999); however, this relationship remains a matter of debate, as it has not been systematically reproduced (Drewnowski et al. 1997, 2007; Kirkmeyer and Tepper 2003). Whatever the origin of these discrepancies, it seems unlikely that sensitivity to bitter tastes, the number of fungiform papillae and fat preference are related.

4 Conclusions and Future Directions

Dietary lipid perception clearly depends on multiple factors. It has been shown that a gustatory cue devoted to fat perception operates in mice and humans, in parallel with texture and olfaction. LCFA provide the stimulus for fat perception. In mice, CD36 acts as a gustatory lipid receptor, enabling the organism to obtain sufficient energy by selecting and promoting the digestion of lipids (Fig. 2). This system, which might be considered as a sixth taste modality, would clearly be advantages in times of food scarcity. Indeed, fat-rich foods are an important source of energy, contain essential fatty acids and carry lipid-soluble vitamins (A, D, E, K) with many important, fundamental biological functions. Conversely, this "fatty" taste might contribute to increase the prevalence of obesity during periods of food abundance. Further experiments are required to explore these assumptions.

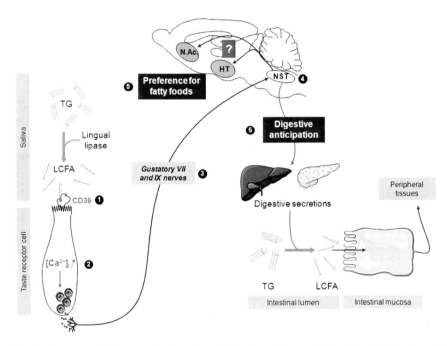

Fig. 2 Gustatory lipid perception in mice. LCFA released from triglycerides (TG) by lingual lipase bind to CD36, which acts as a gustatory lipid receptor in taste receptor cells (TRC) (1). The recognition of LCFA by CD36 induces an increase in intracellular free calcium concentration ($[Ca^{2+}]_i$) (2), an event known to generate the release of neurotransmitters by TRC. Lipid taste signal is then transmitted by the gustatory nerves (chorda tympani, nerve VII and glossopharyngeal, nerve IX) (3) to the gustatory area in the nucleus of the solitary tract (NST) in the brainstem (4). The projections of the NST to central nuclei involved in eating behaviour and to peripheral tissues, including the digestive tract, could account for the CD36-mediated attraction to lipids (5) and the cephalic phase of digestion (6) reported in mice subjected to an oral lipid stimulation. HT hypothalamus, N.Ac. nucleus accumbens

It remains unclear whether there is a similar system in humans. For example, the molecular mechanisms responsible for the chemodetection of LCFA in the oral cavity are unknown. The identification of relevant, non-invasive biological markers may make it possible to answer this question in the future. Further data concerning gustatory fat perception in mammals may lead to the development of new therapeutic approaches (nutritional and pharmacological) to decrease the risk of obesity.

References

Andersen M, Lenhard B, Whatling C, Eriksson P, Odeberg J (2006) Alternative promoter usage of the membrane glycoprotein CD36. BMC Mol Biol 7:8

Baillie AG, Coburn CT, Abumrad NA (1996) Reversible binding of long-chain fatty acids to purified FAT, the adipose CD36 homolog. J Membr Biol 153:75–81

Bartoshuk LM, Duffy VB, Miller IJ (1994) PTC/PROP tasting: anatomy, psychophysics, and sex effects. Physiol Behav 56:1165–1171

Benton R, Vannice KS, Vosshall LB (2007) An essential role for a CD36-related receptor in pheromone detection in Drosophila. Nature 450:289–293

Bernhardt SJ, Naim M, Zehavi U, Lindemann B (1996) Changes in IP3 and cytosolic Ca2+ in response to sugars and non-sugar sweeteners in transduction of sweet taste in the rat. J Physiol 490(2):325–336

Berthoud HR (2002) Multiple neural systems controlling food intake and body weight. Neurosci Biobehav Rev 26:393–428

Chale-Rush A, Burgess JR, Mattes RD (2007) Evidence for human orosensory (taste?) sensitivity to free fatty acids. Chem Senses 32:423–431

Chen H, Herndon ME, Lawler J (2000) The cell biology of thrombospondin-1. Matrix Biol 19:597–614

Cheung L, Andersen M, Gustavsson C, Odeberg J, Fernandez-Perez L, Norstedt G, Tollet-Egnell P (2007) Hormonal and nutritional regulation of alternative CD36 transcripts in rat liver--a role for growth hormone in alternative exon usage. BMC Mol Biol 8:60

Coburn CT, Knapp FF Jr, Febbraio M, Beets AL, Silverstein RL, Abumrad NA (2000) Defective uptake and utilization of long chain fatty acids in muscle and adipose tissues of CD36 knockout mice. J Biol Chem 275:32523–32529

Crystal SR, Teff KL (2006) Tasting fat: cephalic phase hormonal responses and food intake in restrained and unrestrained eaters. Physiol Behav 89:213–220

De Araujo IE, Rolls ET (2004) Representation in the human brain of food texture and oral fat. J Neurosci 24:3086–3093

Drewnowski A, Henderson SA, Shore AB (1997) Genetic sensitivity to 6-n-propylthiouracil (PROP) and hedonic responses to bitter and sweet tastes. Chem Senses 22:27–37

Drewnowski A, Henderson SA, Cockroft JE (2007) Genetic sensitivity to 6-n-propylthiouracil has no influence on dietary patterns, body mass indexes, or plasma lipid profiles of women. J Am Diet Assoc 107:1340–1348

Duffy VB, Fast K, Cohen Z, Chodos E, Bartoshuk LM (1999) Genetic taste status associates with fat food acceptance and body mass index in adults. Chem Senses 24:545–546

Dyer J, Salmon KS, Zibrik L, Shirazi-Beechey SP (2005) Expression of sweet taste receptors of the T1R family in the intestinal tract and enteroendocrine cells. Biochem Soc Trans 33:302–305

El-Yassimi A, Hichami A, Besnard P, Khan NA (2008) Linoleic acid induces calcium signaling, SRC-kinase phosphorylation and neurotransmitters release in mouse CD36-positive gustatory cells. J Biol Chem 283:12949–12959

Febbraio M, Hajjar DP, Silverstein RL (2001) CD36: a class B scavenger receptor involved in angiogenesis, atherosclerosis, inflammation, and lipid metabolism. J Clin Invest 108:785–791

Fredriksson R, Hoglund PJ, Gloriam DE, Lagerstrom MC, Schioth HB (2003) Seven evolutionarily conserved human rhodopsin G protein-coupled receptors lacking close relatives. FEBS Lett 554:381–388

Fukuwatari T, Kawada T, Tsuruta M, Hiraoka T, Iwanaga T, Sugimoto E, Fushiki T (1997) Expression of the putative membrane fatty acid transporter (FAT) in taste buds of the circumvallate papillae in rats. FEBS Lett 414:461–464

Fukuwatari T, Shibata K, Iguchi K, Saeki T, Iwata A, Tani K, Sugimoto E, Fushiki T (2003) Role of gustation in the recognition of oleate and triolein in anosmic rats. Physiol Behav 78:579–583

Gaillard D, Laugerette F, Darcel N, El-Yassimi A, Passilly-Degrace P, Hichami A, Khan NA, Montmayeur JP, Besnard P (2008) The gustatory pathway is involved in CD36-mediated orosensory perception of long-chain fatty acids in the mouse. FASEB J 22:1458–1468

Gilbertson TA, Boughter JD Jr (2003) Taste transduction: appetizing times in gustation. Neuroreport 14:905–911

Gilbertson TA, Fontenot DT, Liu L, Zhang H, Monroe WT (1997) Fatty acid modulation of K+ channels in taste receptor cells: gustatory cues for dietary fat. Am J Physiol 272:C1203–1210

Gilbertson TA, Liu L, York DA, Bray GA (1998) Dietary fat preferences are inversely correlated with peripheral gustatory fatty acid sensitivity. Ann N Y Acad Sci 855:165–168

Gilbertson TA, Liu L, Kim I, Burks CA, Hansen DR (2005) Fatty acid responses in taste cells from obesity-prone and -resistant rats. Physiol Behav 86:681–690

Greenwalt DE, Lipsky RH, Ockenhouse CF, Ikeda H, Tandon NN, Jamieson GA (1992) Membrane glycoprotein CD36: a review of its roles in adherence, signal transduction, and transfusion medicine. Blood 80:1105–1115

Hajri T, Ibrahimi A, Coburn CT, Knapp FF Jr, Kurtz T, Pravenec M, Abumrad NA (2001) Defective fatty acid uptake in the spontaneously hypertensive rat is a primary determinant of altered glucose metabolism, hyperinsulinemia, and myocardial hypertrophy. J Biol Chem 276:23661–23666

Hamilton RB, Norgren R (1984) Central projections of gustatory nerves in the rat. J Comp Neurol 222:560–577

Hamosh M (1990) Lingual and gastric lipases. Nutrition 6:421–428

Harmon CM, Luce P, Beth AH, Abumrad NA (1991) Labeling of adipocyte membranes by sulfo-N-succinimidyl derivatives of long-chain fatty acids: inhibition of fatty acid transport. J Membr Biol 121:261–268

Hiraoka T, Fukuwatari T, Imaizumi M, Fushiki T (2003) Effects of oral stimulation with fats on the cephalic phase of pancreatic enzyme secretion in esophagostomized rats. Physiol Behav 79:713–717

Hirasawa A, Tsumaya K, Awaji T, Katsuma S, Adachi T, Yamada M, Sugimoto Y, Miyazaki S, Tsujimoto G (2005) Free fatty acids regulate gut incretin glucagon-like peptide-1 secretion through GPR120. Nat Med 11:90–94

Honore E, Barhanin J, Attali B, Lesage F, Lazdunski M (1994) External blockade of the major cardiac delayed-rectifier K+ channel (Kv1.5) by polyunsaturated fatty acids. Proc Natl Acad Sci USA 91:1937–1941

Huang MM, Bolen JB, Barnwell JW, Shattil SJ, Brugge JS (1991) Membrane glycoprotein IV (CD36) is physically associated with the Fyn, Lyn, and Yes protein-tyrosine kinases in human platelets. Proc Natl Acad Sci USA 88:7844–7848

Kadohisa M, Rolls ET, Verhagen JV (2005) Neuronal representations of stimuli in the mouth: the primate insular taste cortex, orbitofrontal cortex and amygdala. Chem Senses 30:401–419

Kawai T, Fushiki T (2003) Importance of lipolysis in oral cavity for orosensory detection of fat. Am J Physiol Regul Integr Comp Physiol 285:R447–454

Kirkmeyer SV, Tepper BJ (2003) Understanding creaminess perception of dairy products using free-choice profiling and genetic responsivity to 6-n-propylthiouracil. Chem Senses 28:527–536

Laugerette F, Passilly-Degrace P, Patris B, Niot I, Febbraio M, Montmayeur JP, Besnard P (2005) CD36 involvement in orosensory detection of dietary lipids, spontaneous fat preference, and digestive secretions. J Clin Invest 115:3177–3184

Liu L, Hansen DR, Kim I, Gilbertson TA (2005) Expression and characterization of delayed rectifying K+ channels in anterior rat taste buds. Am J Physiol Cell Physiol 289:C868–C880

Looy H, Weingarten HP (1992) Facial expressions and genetic sensitivity to 6-n-propylthiouracil predict hedonic response to sweet. Physiol Behav 52:75–82

Malecka-Tendera E, Mazur A (2006) Childhood obesity: a pandemic of the twenty-first century. Int J Obes (Lond) 30(Suppl 2):S1–S3

Matsumura S, Mizushige T, Yoneda T, Iwanaga T, Tsuzuki S, Inoue K, Fushiki T (2007) GPR expression in the rat taste bud relating to fatty acid sensing. Biomed Res 28:49–55

Mattes RD (1996) Oral fat exposure alters postprandial lipid metabolism in humans. Am J Clin Nutr 63:911–917

Mattes RD (2001a) Oral exposure to butter, but not fat replacers elevates postprandial triacylglycerol concentration in humans. J Nutr 131:1491–1496

Mattes RD (2001b) The taste of fat elevates postprandial triacylglycerol. Physiol Behav 74:343–348

Mattes RD (2002) Oral fat exposure increases the first phase triacylglycerol concentration due to release of stored lipid in humans. J Nutr 132:3656–3662

Mattes RD (2005) Fat taste and lipid metabolism in humans. Physiol Behav 86:691–697

McCormack DN, Clyburn VL, Pittman DW (2006) Detection of free fatty acids following a conditioned taste aversion in rats. Physiol Behav 87:582–594

Mela DJ, Mattes RD (1988) The chemical senses and nutrition. I. Nutr Today 23:4–9

Moreau H, Laugier R, Gargouri Y, Ferrato F, Verger R (1988) Human preduodenal lipase is entirely of gastric fundic origin. Gastroenterology 95:1221–1226

Oakley B (1993) The gustatory competence of the lingual epithelium requires neonatal innervation. Brain Res Dev Brain Res 72:259–264

Olshansky SJ, Passaro DJ, Hershow RC, Layden J, Carnes BA, Brody J, Hayflick L, Butler RN, Allison DB, Ludwig DS (2005) A potential decline in life expectancy in the United States in the 21st century. N Engl J Med 352:1138–1145

Oquendo P, Hundt E, Lawler J, Seed B (1989) CD36 directly mediates cytoadherence of Plasmodium falciparum parasitized erythrocytes. Cell 58:95–101

Pittman DW, Labban CE, Anderson AA, O'Connor HE (2006) Linoleic and oleic acids alter the licking responses to sweet, salt, sour, and bitter tastants in rats. Chem Senses 31:835–843

Pittman D, Crawley ME, Corbin CH, Smith KR (2007) Chorda tympani nerve transection impairs the gustatory detection of free fatty acids in male and female rats. Brain Res 1151:74–83

Prawitt D, Monteilh-Zoller MK, Brixel L, Spangenberg C, Zabel B, Fleig A, Penner R (2003) TRPM5 is a transient Ca2+-activated cation channel responding to rapid changes in Ca2+i. Proc Natl Acad Sci USA 100:15166–15171

Rac ME, Safranow K, Poncyljusz W (2007) Molecular basis of human CD36 gene mutations. Mol Med 13:288–296

Rayasam GV, Tulasi VK, Davis JA, Bansal VS (2007) Fatty acid receptors as new therapeutic targets for diabetes. Expert Opin Ther Targets 11:661–671

Ren Y, Silverstein RL, Allen J, Savill J (1995) CD36 gene transfer confers capacity for phagocytosis of cells undergoing apoptosis. J Exp Med 181:1857–1862

Rolls ET (2007) Sensory processing in the brain related to the control of food intake. Proc Nutr Soc 66:96–112

Saitoh O, Hirano A, Nishimura Y (2007) Intestinal STC-1 cells respond to five basic taste stimuli. Neuroreport 18:1991–1995

Sato O, Takanashi N, Motojima K (2007) Third promoter and differential regulation of mouse and human fatty acid translocase/CD36 genes. Mol Cell Biochem 299:37–43

Schiffman SS, Graham BG, Sattely-Miller EA, Warwick ZS (1998) Orosensory perception of dietary fat. Curr Direct Psychol Sci:137–143

Sclafani A, Ackroff K, Abumrad NA (2007a) CD36 gene deletion reduces fat preference and intake but not post-oral fat conditioning in mice. Am J Physiol Regul Integr Comp Physiol 293:R1823–R1832

Sclafani A, Zukerman S, Glendinning JI, Margolskee RF (2007b) Fat and carbohydrate preferences in mice: the contribution of alpha-gustducin and Trpm5 taste-signaling proteins. Am J Physiol Regul Integr Comp Physiol 293:R1504–R1513

Smith JC, Fisher EM, Maleszewski V, McClain B.(2000) Orosensory factors in the ingestion of corn oil/sucrose mixtures by the rat. Physiol Behav 69:135–146

Stratford JM, Curtis KS, Contreras RJ (2006) Chorda tympani nerve transection alters linoleic acid taste discrimination by male and female rats. Physiol Behav 89:311–319

Sugita M (2006) Taste perception and coding in the periphery. Cell Mol Life Sci 63:2000–2015

Takeda M, Imaizumi M, Fushiki T (2000) Preference for vegetable oils in the two-bottle choice test in mice. Life Sci 67:197–204

Takeda M, Imaizumi M, Sawano S, Manabe Y, Fushiki T (2001a) Long-term optional ingestion of corn oil induces excessive caloric intake and obesity in mice. Nutrition 17:117–120

Takeda M, Sawano S, Imaizumi M, Fushiki T (2001b) Preference for corn oil in olfactory-blocked mice in the conditioned place preference test and the two-bottle choice test. Life Sci 69:847–854

Tanaka T, Katsuma S, Adachi T, Koshimizu TA, Hirasawa A, Tsujimoto G (2008) Free fatty acids induce cholecystokinin secretion through GPR120. Naunyn Schmiedebergs Arch Pharmacol 377:523–527

Tepper BJ, Nurse RJ (1998) PROP taster status is related to fat perception and preference. Ann N Y Acad Sci 855:802–804

Tsuruta M, Kawada T, Fukuwatari T, Fushiki T (1999) The orosensory recognition of long-chain fatty acids in rats. Physiol Behav 66:285–288

Verhagen JV, Rolls ET, Kadohisa M (2003) Neurons in the primate orbitofrontal cortex respond to fat texture independently of viscosity. J Neurophysiol 90:1514–1525

Wong GT, Gannon KS, Margolskee RF (1996) Transduction of bitter and sweet taste by gustducin. Nature 381:796–800

Wu SV, Rozengurt N, Yang M, Young SH, Sinnett-Smith J, Rozengurt E (2002) Expression of bitter taste receptors of the T2R family in the gastrointestinal tract and enteroendocrine STC-1 cells. Proc Natl Acad Sci USA 99:2392–2397

Yoneda T, Saitou K, Mizushige T, Matsumura S, Manabe Y, Tsuzuki S, Inoue K, Fushiki T (2007) The palatability of corn oil and linoleic acid to mice as measured by short-term two-bottle choice and licking tests. Physiol Behav 91:304–309

Gustation in Fish: Search for Prototype of Taste Perception

A. Yasuoka and K. Abe

Abstract Fish perceive water-soluble chemicals at the taste buds that are distributed on oropharyngeal and trunk epithelia. Recent progress in molecular analyses has revealed that teleosts and mammals share pivotal signaling components to transduce taste stimuli. The fish orthologs of taste receptors, fT1R and fT2R, show mutually exclusive expression in taste buds, and both are coexpressed with phospholipase C-β2 and the transient receptor potential M5 channel as common downstream components of taste receptor signals. Interestingly, fT1R heteromers are activated by various l-amino acids but not by sugars. This may reflects that in fish the energy metabolism depends primarily on gluconeogenesis from amino acids. fT2Rs are activated by denatonium benzoate, which is a bitter substance for mammals. It is thus likely that the preferable and aversive tastes for vertebrates, though their taste modalities somewhat vary, are transduced by the conserved sensory pathways. The comparative molecular biology of the fish taste system would lead to understanding a general logic of encoding taste modalities in vertebrates.

1 Introduction

Fish covers a wide variety of vertebrate species. These include *Teleostei, Acipenseriformes* (stargeon), *Lepisosteiformes* (gar), *Chondrichthyes* (shark), Agnatha (hagfish), etc. The model fish species *Oryzias latipes* (medaka), *Danio* rerio (zebrafish), and *Tetraodon nigroviridis* (pufferfish) belong to *Teleostei*, the most prevailing class in the modern world. Taste buds are observed in these fish species, except in *Agnatha*, whose phylogenic root is close to that of vertebrates. Instead, they possess solitary

---

K. Abe (✉)
Department of Applied Biological Chemistry, Graduate School of Agricultural and Life Sciences, The University of Tokyo, Tokyo, Japan
e-mail: aka7308@mail.ecc.u-tokyo.ac.jp

chemosensory cells (SCCs) distributed solely in epithelia and synapsed by sensory nerves. SCCs are also observed in other fish species, including *Teleostei,* as well as in mammals (Northcutt 2004; Finger 1997). SCCs have been found to share some signaling components with taste bud cells (Höfer et al. 1996; Finger et al. 2003), but in this article we will focus on the chemoreception process that starts at the taste bud cells. Fish have frequently been used for taste research, because they show higher sensitivity to tastants than mammals, and because specialized taste organs have been observed in some species. For example, catfish possess many taste buds on the barbel on which the facial nerve projects, and have served as an ideal material for analysis of taste nerve responses to amino acids or other tastants (Marui and Caprio 1982; Ogawa and Caprio 1999a). Also, fluorescent dye tracing experiments have revealed the projection of taste nerves in the hindbrain area (Finger 1976, 1978; Finger and Morita 1985). However, these studies have basically been restricted to the comparative morphology or physiological properties specific to fish. It is necessary to discover a molecular basis applicable to vertebrates in general. Recent progress in molecular biology of model fish has made an impact on taste research. During the period from 1996 to 2007, large-scale mutant screenings were performed in zebrafish and medaka, resulting in the identification of a number of mutations that affect the development of the central nervous system and sensory organs (Schier et al. 1996; Haffter and Nüsslein-Volhard 1996; Loosli et al. 2000; Naruse et al. 2004; Tanaka et al. 2007). One of the consequences of the screenings is that model fish utilize molecular mechanisms to specify brain and other sensory structures basically in the same way as mammals do, still exhibiting simplicity in cellular compositions. Especially in the case of medaka, several mutants showing abnormal trajectory of cranial nerves were isolated by staining nerve bundles in embryos (Yasuoka et al. 2004b). The results show the possibility that mechanisms underlying the formation of cranial nerves can be dissected by forward genetics. In addition, significant contributions have been made by genome sequencing projects on model fish species. The genomes of zebrafish (1,500 Mbp, http://zfin.org/cgi-bin/webdriver?MIval=aa-ZDB_home.apg), medaka (700 Mbp, http://www.shigen.nig.ac.jp/medaka/indexEn.html), and pufferfish (340 Mbp, Brenner et al. 1993) have been sequenced and annotated in comparison with vertebrate genomes (http://www.ensembl.org/index.html). These projects revealed the presence of multiple syntenic regions scattered over chromosomes and the occurrence of two major genome duplication events during the evolution of vertebrates (Nakatani et al. 2007). On the basis of these achievements, researchers began to utilize model fish as a tool to dissect the taste system from a molecular aspect. This article reviews the molecular mechanisms of the taste signal transduction in fish to figure out a general logic of taste perception in vertebrates.

2 Fish Taste Buds and Cranial Sensory Nerves

Fish taste buds, with morphological variations depending on the species (Northcutt 2004), can be observed in trunk, face, lip, oral, and pharyngeal (gill) epithelia, where sensory components of cranial nerves, facial (VII), glossopharyngeal (IX), and vagus (X) nerves project, while the distribution of mammalian taste buds is restricted to the oropharyngeal region. Figure 2 shows transgenetically labeled taste bud cells in two model fish species by green fluorescent protein expression under the control of medaka phospholipase C-β2 (mfPLC-β2) promoter (Aihara et al. 2006). Signals due to green fluorescent protein expression are observed in lip and oropharyngeal epithelia (Fig. 2a, b). The dissected tissue of transgenic medaka allows detailed observation of taste bud cells in gill rakers and pharyngeal teeth (Fig. 2c–f). In the development process, these epithelia are under the governance of segmental cell flow from neural crest and placodal lineages (Baker and Bronner-Fraser 2001). The taste buds themselves are derived from neither of these lineages, but develop autonomously from pharyngeal endoderm before the cranial sensory nerves attach their peripheral ends (Stone et al. 1995; Barlow and Northcutt 1997). However, the relationships between taste bud location and each cranial sensory nerve basically follow the anterior–posterior order, i.e., taste buds on lip and mandibular arch innervated by the facial nerve, those on pharyngeal arches by the glossopharyngeal nerve, and those on the brachial arches by the vagus nerve. In considering the phylogenic origin of tongue and the other cervical structures of mammals, it is reasonable to mention that the segmental organization of the taste system is basically conserved among teleosts and mammals. These cranial nerves are called epibranchial nerves, which consist of both neural crest and placodal lineages. Fine cell fate mapping analyses in amphibians showed that taste buds are innervated by those of placodal origin (Harlow and Barlow 2007). These taste nerves form ganglia located distal to those of crest origin, and project their afferents into rhombomeric segments in the manner facial to fourth segment, glossopharyngeal to seventh segment, and vagus to eighth segment, where secondary neurons receive sensory input. In some fish species, the hindbrain region containing these areas forms bulges called the facial lobe and the vagal lobe (Finger and Morita 1985; Kotrschal and Finger 1996). These areas are considered to correspond to the solitary nucleus in the brain stem of mammals. Part of the gustatory information is transmitted through a reticular formation to motor nuclei to regulate jaw and gill reflection (Finger and Morita 1985; Goehler and Finger 1992; Kotrschal and Finger 1996; Kanwal and Finger 1997). A similar pathway was reported in mammals (Travers and Karimnamazi 1997; Travers et al. 2000), but the pathway passing through the midbrain–hindbrain boundary is thought to be more dominant in mammals (Travers and Karimnamazi 1997; Reilly 1999; Faurion 2006). Limited information is available about the latter gustatory pathway in fish (Lamb and Finger 1996).

3 Fish Orthologs for Taste Receptors and Downstream Components

Given the organization of the fish taste system, which is comparable with that of mammals, it is natural to assume a parallelism among them in terms of signal transduction mechanisms. Conventional molecular cloning approaches were followed by gene mining in silico that benefited from the genome sequencing projects in the model fish species. The orthologs of fish phosoholipase C-β2 (fPLC-β2) were initially cloned as common effecter enzymes expressed by fish taste bud cells (Yasuoka et al. 2004a). The finding provides not only a molecular basis for fish taste signal transduction, but also a genetical tool to manipulate taste bud cells (mfPLC-β2 promoter; Fig. 1). Then the multiple sequences encoding fish T1Rs (fT1Rs) and fish T2Rs (fT2Rs) were identified in the model fish genome databases (Ishimaru et al. 2005).

The fT1Rs as well as mammalian T1Rs (Hoon et al. 1999) belong to an independent branch in the phylogenetic tree of family C G-protein-coupled receptor (GPCR), which also includes the vomeronasal receptor (V2R; Fig. 2a). One of the characteristics of fT1R is the existence of multiple T1R2 members that cannot be assigned to mammalian T1R2 group, while fT1R1 and fT1R3 are encoded by single-copy genes showing highest similarity to the mammalian counterparts. Also, the number of fT1R2 members varies depending on the fish species, indicating that vertebrate T1R2 genes experienced several gene duplication–deletion events. The gene redundancy of this kind can result in the loss of a gene as adduced by the fact that avians have no gene encoding the T1R2 member (Table 1; Lagerström et al. 2006).

Compared with fT1Rs, fT2Rs show ambiguous attributes, because the amino acid identity values between fT2Rs and mammalian T2Rs (Adler et al. 2000) (13–22%) are close to those between fT2Rs and vomeronasal receptors (13–22%; Fig. 2b). Also, it appears that the number of fT2R members in each fish (one to

Table 1 Vertebrate taste receptors

Species	T1R family members	T2R family number
human	1, 2, 3	25
chick	1, 3	~3
frog	-	~49
medaka	1, 2a, 2b, 2c, 3	~1
fugu	1, 2a, 2b, 3	~6
zebrafish	1, 2a, 2b, 3	~7

Gustation in Fish: Search for Prototype of Taste Perception 243

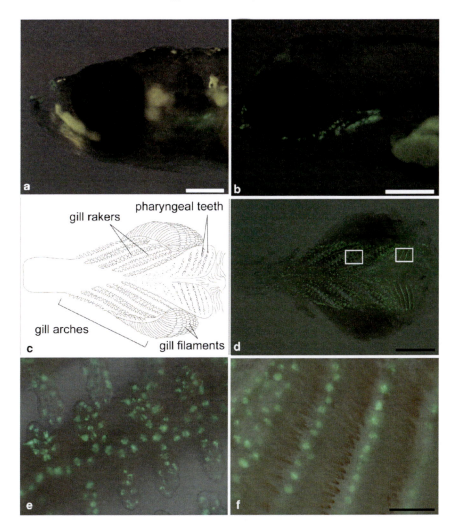

Fig. 1 Fish taste bud cells visualized by medaka *plc-β2* promoter. **a, b** Lateral views of transgenic medaka (10 days after fertilization) and transgenic zebrafish (5 days after fertilization) expressing green fluorescent protein (GFP) under the control of medaka *plc-β2* promoter. *Scale bars* 200 μm. **c** The pharyngeal region, with its dorsal side exposed. **d** Fluorescence image of the same region dissected from transgenic medaka (2 months old). *Scale bar* 1 mm. **e** Magnified image of GFP-positive cells in gill rakers in the region represented in the left open box in **d**. **f** Magnified image of GFP-positive cells along pharyngeal teeth in the region represented in the right open box in **d**. *Scale bar* 100 μm for **e, f**. Fluorescence images are overlaid on bright field images. (**a, b, d–f** Modified from Aihara et al. 2007 with permission. **c** Modified from Iwamatsu 1997)

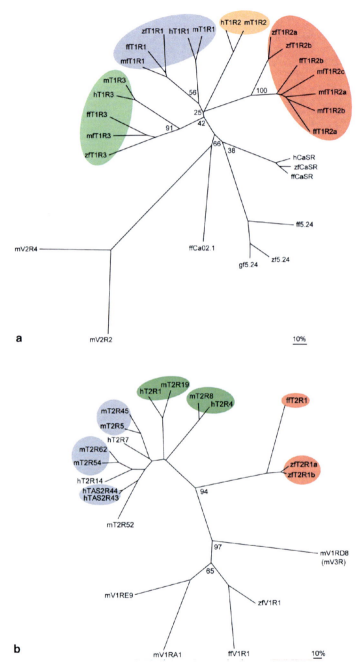

Fig. 2 Phylogenic relationships of fish and mammalian taste receptors. **a** Phylogenic tree showing fish and mammalian T1Rs. The tree was constructed by the neighbor-joining method. Fish T1Rs were categorized into the three groups: a group including mammalian T1R1 (*blue oval*), a group including mammalian T1R3 (*green oval*), and a group independent of any vertebrate T1R (*red oval*). zf5.24 and ff5.24 genes were orthologs of goldfish olfactory receptor 5.24 gene, and

seven) is smaller than that in the other vertebrates, except for avians (Fig. 3). One explanation is that there are unknown fT2Rs which cannot be identified because of their low amino acid similarities to known T2Rs. It is also possible that the fish genome has fewer T2R genes (Shi and Zhang 2007).

Besides the molecules that directly interact with G-proteins, the transient receptor potential (TRP) channel family has been of special interest in the study of sensory signal transduction, since each member of this family shows a unique expression pattern and a specialized function in certain sensory cells, for example, TRPV1 as the sensor for noxious heat and TRPC1 as the cold receptor (Venkatachalam and Montell 2007). As one of the members of this family, TRPM5 was found to be expressed by taste bud cells in mammals. Actually, the destruction of the mouse TRPM5 gene causes loss of taste responses to sweet, umami, and bitter substances (Zhang et al. 2003). The recent finding that the fish genome contains a TRPM5 ortholog which is expressed in taste bud cells (Yoshida et al. 2007) supports the idea that G-protein-coupled taste signaling pathways are conserved in higher vertebrates (Fig. 3d, left). Fish orthologs for the other components, $G\alpha_{gust}$, $G\alpha_i$, and inositol 1,4,5-triphosphate receptor 3, have not been reported yet.

4 Expression of Taste Signaling Molecules in Fish Taste Buds

The expression pattern of fish taste receptors provides primary information to predict their functional mode in taste bud cells. As to fT1R members, there are two types of fT1R-positive cells: the cells expressing both fT1R1 and fT1R3, and those expressing both fT1R2 members and fT1R3 (Fig. 3a, b). These cell types are conserved among teleosts and mammals. In addition, the partially overlapping expression of fT1R2 members may provide multimodal properties to fish taste receptor cells as proposed in the next section. In contrast to fT1Rs, fT2Rs consist of fewer members than mammalian T2Rs, but the partially overlapping expression is also observed as in the

---

Fig. 2 (continued) zfCaSR and ffCaSR genes were orthologs of human calcium sensing receptor (hCaSR) genes. ffCa02.1 is a fugu receptor distantly related to the mouse vomeronasal receptors mV2R2 and mV2R4 (Naito et al. 1998). The numbers at nodes are bootstrap values in percentages of 1,000 replicates. The value at the node where fish T1R2s, mammalian T1R2s, and the other receptors branch cannot be calculated because of its trichotomy. The value (25) indicated near the trichotomous node is for the node where T1R1s and other receptors, including vertebrate T1R3s, branch. The *scale bar* indicates 10% amino acid difference. **b** Phylogenic tree showing fish and mammalian T2Rs. The tree was constructed as in **a**. Fish T2Rs are labeled by *red ovals*. Mouse- or human-specific groups are labeled by *blue circles* and *blue oval*. Some of the mammalian T2Rs showed orthologous relationships between human and mouse (*green ovals*). The bootstrap value at the node where an orthologous group of mammalian T2R (hT2R4 and mT2R8, denatonium benzoate receptors), other mammalian T2Rs, and the other receptors branch cannot be calculated because of its trichotomy. The *scale bar* represents 10% amino acid difference. (Modified from Ishimaru et al. 2005 with permission)

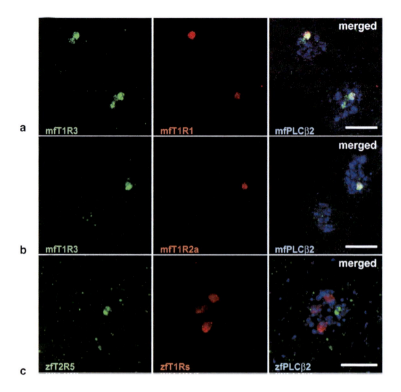

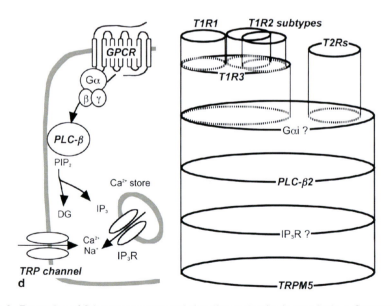

Fig. 3 Expression of fish taste receptors and signaling molecules in taste buds. **a** Coexpression of medaka mfT1R1 and mfT1R3. **b** Coexpression of medaka mfT1R1 and mfT1R2a. **c** Segregation of zebrafish zfT1R- and zfT2R-expressing cells. All taste receptors were coexpressed with fish phospholipase C-β2 (*PLCβ2*) (*merged* panels). Scale bars 20 μm. **d** Transduction pathway of

case of mammals. As far as the two closely related zebrafish T2Rs (zfT2R1a and zfT2R1b; Fig. 2b) are concerned, about 80% of the cells that test positive to one of the receptors may express the other (Okada, personal communication).

The most notable finding was the segregation of fT1R-positive cells and fT2R-positive cells (Fig. 3c). In mammals, peripheral segregation of these cell types was found to be responsible for transducing the two opposing taste modalities—preferable and aversive taste (Zhang et al. 2003; Mueller et al. 2005). Therefore, it is highly possible that fish also utilize these cell types to discriminate tastants. Finally, the expression of fPLC-β2 together with both of the fT1Rs and fT2Rs (Fig. 3a"c, merged panels) suggests that this enzyme is located at the critical step in the signaling pathways of both fish taste receptors (Fig. 3d). It was shown that in mammals the destruction of the phospholipase C-β2 (PLC-β2) gene abolishes their taste responses to both T1R and T2R ligands (Zhang et al. 2003). Knocking down of fPLC-β2 activity in vivo, for example, by means of mfPLC-β2 promoter driving inhibitor of G-protein signal can clarify this point (Aihara et al. 2008).

5 Ligands for Fish Taste Receptors

Fish taste nerves respond to various compounds, including L-amino acids, nucleic acids, fatty acids, alkaloids, organic and inorganic acids, and salts, with different sensitivities depending on the fish species (Yoshii and Kurihara 1983; Marui and Kiyohara 1987; Kiyohara and Hidaka 1991; Ogawa and Caprio 1999a, b). In the case of zebrafish, their facial nerve responds to the amino acids that exhibit sweet or umami taste to mammals, and to denatonium benzoate and quinine chloride, which taste bitter to mammals (Fig. 4a). While inosine monophosphate, when applied together with amino acids, elicits synergistic nerve responses in mammals (Yoshii et al. 1986; Hellekant and Ninomiya 1991), no such synergism was observed in zebrafish (Fig. 4b, c). Indeed these tastants were found to be received by taste buds, but their transduction mechanisms were unclear until ligands for fish taste receptors were characterized by Oike et al. (2007). In general, chemosensory GPCRs show a higher degree of interspecies diversity in terms of amino acid similarity than GPCRs receiving endogenous ligands, e.g., adrenergic receptors (Yasuoka et al. 1996), making it difficult to define their ligand reactivities on the basis of amino acid similarity. Accordingly, comprehensive screening of ligands is necessary to elucidate fish taste receptor function. Especially in the case of fT1Rs, multiple members of fT1R2 increase the possible combination of the receptor heteromers.

Fig. 3 (continued) G-protein-coupled taste receptor (*GPCR*) signal (*left*) and Venn diagrams representing the expression of signaling components (*right*). $G\alpha$, $G\alpha i$, β, γ G-protein subunits, PIP_2 phosphatidylinositol 4,5-bisphosphate, DG diacylglycerol, IP_3 inositol 1,4,5-triphosphate, IP_3R IP_3 receptor, *TRP* transient receptor potential. (a–c Modified from Oike et al. 2007 with permission). **d** Modified from Yasuoka et al. 2006 with permission)

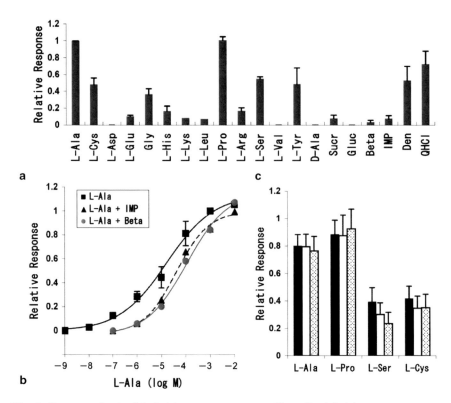

Fig. 4 Response of zebrafish facial nerve to tastants. **a** Normalized facial nerve response to tastants. The integrated neural responses were normalized to the response to 1 mM L-Ala. Each column represents the mean ± the standard error of at least three independent assays. Tastants 1 mM amino acids, 300 mM sucrose (*Sucr*), 300 mM glucose (*Gluc*), 1 mM betaine (*Beta*), 1 mM inosine monophosphate (*IMP*), 10 mM denatonium (*Den*), and 1 mM quinine HCl (*QHCl*). **b** Dose-dependent response of facial nerves to L-Ala in the absence (*squares; n = 3*) or presence of either 1 mM IMP (*triangles; n = 2*) or 1 mM betaine (*circles; n = 2*). Responses were normalized to the response to 1 mM L-Ala. Each point shows the mean ± the standard error (*triangles* and *circles*). **c** Quantification of the responses to the chemicals. Each amino acid was used at 0.1 mM in the absence (black bars) or presence of either 1 mM IMP (*white bars*) or 1 mM betaine (*hatched bars*). Responses were normalized to the mean response at 1 mM L-Ala. (Modified from Oike et al. 2007 with permission)

Fig. 5 Ligand response profiles of fish taste receptors. **a** Response of the medaka mfT1R1/mfT1R3 heteromer to amino acids. **b–d** Responses of mfT1R2 members/mfT1R3 heteromers to amino acids. Each column represents the mean ± the standard error of at least three independent measurements. **e** Response of HEK293T cells expressing zebrafish zfT2R5. All stimuli were applied for 12 s, and the starting points are indicated by arrows. The trace was derived from 13 responding cells. **f** Dose-dependent responses of zfT2R5 (*squares*) and mfT2R1 (*circles*) to denatonium benzoate. Responses were normalized to the mean response at the highest concentration. Each point represents the mean ± the standard error of at least three independent assays. Ligands were used at appropriated concentrations. Amino acids were used at 50 mM except for L-Tyr and L-Trp at 5 mM, sucrose (*Sucr*) and glucose (*Gluc*) at 150 mM, saccharin (*Sac*), trisodium citrate (*Cit*) and NaCl at 50 mM, inosine monophosphate (*IMP*) and denatonium benzoate (*Den, den*) at 10 mM, cycloheximide (*cyx*), 6-n-propylthiouracil (*PROP*), and phenylthiocarbamide (*PTC*) at 1 mM. (Modified from Oike et al. 2007 with permission)

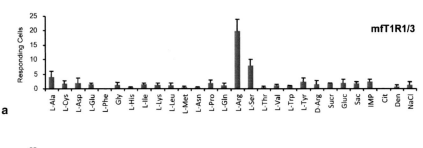
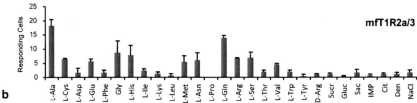
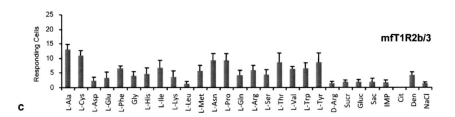
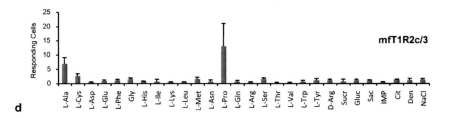
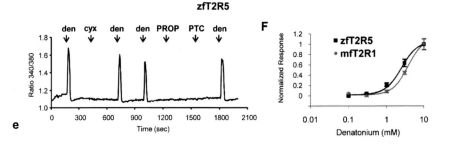

Through the extensive ligand screening using promiscuous G-protein chimera and the HEK293T expression system (Chandrashekar et al. 2000; Ueda et al. 2003), the ligand response profiles of fT1R heteromers emerged (Fig. 5a"d). In the case of medaka T1Rs (mfT1Rs), significant responses were obtained just in the combinations indicated in the Fig. 5: medaka T1R1 (mfT1R1)/medaka T1R3 (mfT1R3) and medaka T1R2 (mfT1R2) members/mfT1R3, which correlate well with the combinations deduced from the expression pattern in fish taste buds (Fig. 4d, Venn diagrams). The mfT1R1/mfT1R3 heteromer has a narrower response spectrum to amino acids (L-Arg and L-Ser) than that of the mouse T1R1/T1R3 heteromer (L-Ala, L-Ser, L-Cys, L-Gln, L-Met, L-Asn, and Gly almost equally; Nelson et al. 2002). No response was observed with betaine, which is known to elicit taste responses in several fish species (Marui and Kiyohara 1987; Kiyohara and Hidaka 1991; Valentincic and Caprio 1997). Also, the mfT1R1/mfT1R3 heteromer shows no synergism with inosine monophosphate. This suggests that fish and mammals perceive nucleotides in different ways (Nelson et al. 2002). An unexpected result was that mfT1R2/mfT1R3 heteromers did not respond to sugars and other mammalian sweeteners. Instead, these heteromers are activated by Gly and L-amino acids with different spectra depending on each mfT1R2 member (Fig. 5b"d). This is quite different from the case of the mammalian T1R2/T1R3 heteromer as the sweet taste receptor (Nelson et al. 2001). This insensitivity of mfT1R2/mfT1R3 to sugars and its widely tuned responsiveness to these amino acids were observed in the case of zebrafish T1R2/T1R3 heteromers as well (Oike et al. 2007). The results suggest that the responsiveness to sugars is lacking in teleosts and might have been acquired in mammals. The idea can be interpreted from the aspect of energy metabolism. Generally, fish living in warm water are known to have low ability to utilize carbohydrate, and their energy metabolism depends on gluconeogenesis from amino acids (Wilson 1994). Since T1Rs are located at the interface between the environment and the organism for selecting the compounds that are beneficial for survival, they should have been subjected to evolutionary pressures from both sides. It is thus likely that fT1R2/fT1R3 heteromers, as the prototype of the vertebrate taste receptor, may have evolved to sense a wide variety of L-amino acids in the environment, without gaining sensitivity to sugars that are not available for fish.

There are several T2R genes in fish genomes: at least one in medaka, six in pufferfish, and seven in zebrafish (Fig. 3). All of them are expressed in fish taste bud cells, but little information is available regarding their ligands. Among them, mfT2R1 and zfT2R5 showed relatively high amino acid identity to each other (44%), and they were supposed to receive some common ligand. We subjected them to ligand screening, and found that they were activated by denatonium benzoate as a bitter substance for mammals but not by 6-*n*-propylthiouracil, cycloheximide, or phenylthiocarbamide (Fig. 5e, f). These fT2Rs show low amino acid identities (less than 15%) to mT2R8 and hT2R4, both of which can be activated by denatonium benzoate and 6-*n*-propylthiouracil (Chandrashekar et al. 2000). Since the ligands examined were limited in number, it is possible that mfT2R1 and zfT2R5 can probably be activated by other naturally occurring ligands which are harmful to fish metabolism. However, what is important is that fish strongly avoid denatonium benzoate (Oike et al. 2007; Aihara et al. 2008), leading to the conclusion that fT2Rs are located at the initial step of aversive taste transduction in fish species.

6 Common Taste Modalities of Vertebrates

To analyze the taste discrimination process in fish, it is indispensable to quantitatively evaluate their feeding behavior towards foods. There is a contrast between taste receptor ligands; fT1R ligands as the source of energy and protein materials, and fT2R ligands as the harmful compounds. This provides a good case study. Semiquantitative analyses of the preference towards an amino acid mixture and aversion against denatonium benzoate were performed with zebrafish by observing their consumption of fluorescently labeled food containing these tastants (Oike et al. 2007). In medaka larvae as well, their uptake of the foods was quantified by analyzing fluorescence images of the gastrointestinal tract. Significant differences were observed between the food containing l-amino acids and that containing no tastant, and also between the food containing denatonium benzoate and that containing no tastant (Aihara et al. 2008). Through consideration of the ligand response profiles of fish taste receptors obtained in a cell-based assay system, it is concluded that the recognition of umami and bitter tastes is governed, respectively, by a T1R- and a T2R-mediated mechanism conserved commonly among teleosts and mammals. Also, the fact that fT1R is activated only by amino acid raises an interesting question: "Is umami taste recognized by fish as sweet taste is by mammals?" A recently reported technique using the gene for plant lectin, e.g., wheat germ agglutinin, enables us to trace this kind of sensory pathway (Ohmoto et al. 2008; Yoshihara 2002), and to solve this question. Actually, a project using mfPLC-β2 promoter and the wheat germ agglutinin gene is in motion (Aihara, personal communication). In Japan, the word "umami" is thought of as a cognate of the word "amami," which means sweetness. It is interesting whether our ancestors were primitive enough to mix up these taste modalities or had intuitive insight into the root of taste reception.

There are two basic taste modalities remaining to be elucidated, sourness and saltiness, and studies on their transduction mechanisms have had humble beginnings. Recently, two members of the polycystic kidney disease (PKD) channel, PKD1L2 and PKD2L1, were found to be expressed in the subset of taste bud cells that do not express PLC-β2; these may thus be responsible for sour taste response (Ishimaru et al. 2006; Huang et al. 2006). The PKD channels are also expressed by the neurons that may sense carbonic acid in cerebrospinal fluid. While fish orthologs for these channels have not been reported yet, their function to detect carbonic ion can be elucidated in the sensory system of fish, which possess a chemosensory organ on every gill arch to detect external and internal carbonic ion

(Perry and Gilmour 2002; David et al. 2005). It is interesting to examine the expression of fish PKD2 orthologs in this chemosensory organ rather than taste bud cells. Fish can also be a useful model animal for studying salty taste. It is predicted that there might be ion receptors in gill epithelia of euryhaline fish for rapid adaptation to the change in external osmolality (Inoue and Takei 2002; David et al. 2005). Elucidation of the molecular property of ion-sensing cells may provide a clue to understanding the whole of salt perception in vertebrate.

7 Conclusion

Amongst the five basic taste modalities, umami, sweetness, bitterness, sourness, and saltiness, the first three are now able to be interpreted in a general context among mammals and teleosts. Umami and sweetness, which inform on the presence of energy and protein sources, are not differentiated in teleosts owing to their prototypical T1Rs, and can be redefined as preferable taste. Bitter taste, as an alert for harmful compounds, is transduced by a T2R-based mechanism that is conserved among these animals. The combination of molecular genetics and behavioral biology in model fish species facilitates understanding of the molecular logic to encode the five taste modalities.

Acknowledgements We are grateful to Toshihide Nagai and Yoshiko Aihara for preparing the figures.

References

Adler E, Hoon MA, Mueller KL, Chandrashekar J, Ryba NJ, Zuker CS (2000) A novel family of mammalian taste receptors. Cell 100:693–702

Aihara Y, Yasuoka A, Yoshida Y, Ohmoto M, Shimizu-Ibuka A, Misaka T, Furutani-Seiki M, Matsumoto I, Abe K (2006) Transgenic labeling of taste receptor cells in model fish under the control of the 5′-upstream region of medaka phospholipase C-beta 2 gene. Gene Expr Pattern 7:149–157

Aihara Y, Yasuoka A, Iwamoto S, Yoshida Y, Misaka T, Abe K (2008) Construction of a taste-blind medaka fish and quantitative assay of its preference-aversion behavior. Genes Brain Behav (in press)

Baker CV, Bronner-Fraser M (2001) Vertebrate cranial placodes I. Embryonic induction. Dev Biol 232:1–61

Barlow LA, Northcutt RG (1997) Taste buds develop autonomously from endoderm without induction by cephalic neural crest or paraxial mesoderm. Development 124:949–957

Brenner S, Elgar G, Sandford R, Macrae A, Venkatesh B, Aparicio S (1993) Characterization of the pufferfish (Fugu) genome as a compact model vertebrate genome. Nature 366:265–268

Chandrashekar J, Mueller KL, Hoon MA, Adler E, Feng L, Guo W, Zucker CS, Ryba NJP (2000) T2Rs function as bitter taste receptors. Cell 100:703–711

David HE, Peter MP, Keith PC (2005) The multifunctional fish gill: dominant site of gas exchange, osmoregulation, acid—base regulation, and excretion of nitrogenous waste. Physiol Rev 85:97–177

Faurion A (2006) Sensory interactions through neural pathways. Physiol Behav 89:44–46

Finger TE (1976) Gustatory pathways in the bullhead catfish. 1. Connections of the anterior ganglion. J Comp Neurol 165:513–526

Finger TE (1978) Gustatory pathways in the bullhead catfish. II. Facial lobe connections. J Comp Neurol 180:691–705

Finger TE (1997) Evolution of taste and solitary chemoreceptor cell systems. Brain Behav Evol 50:234–243

Finger TE, Morita Y (1985) Two gustatory systems: facial and vagal gustatory nuclei have different brainstem connections. Science 227:776–778

Finger TE, Böttger B, Hansen A, Anderson KT, Alimohammadi H, Silver WL (2003) Solitary chemoreceptor cells in the nasal cavity serve as sentinels of respiration. Proc Natl Acad Sci USA 100:8981–8986

Goehler LE, Finger TE (1992) Functional organization of vagal reflex systems in the brain stem of the goldfish, Carassius auratus. J Comp Neurol 319:463–478

Haffter P, Nüsslein-Volhard C (1996) Large scale genetics in a small vertebrate, the zebrafish. Int J Dev Biol 40:221–227

Harlow DE, Barlow LA (2007) Embryonic origin of gustatory cranial sensory neurons. Dev Biol 310:317–328

Hellekant G, Ninomiya Y (1991) On the taste of umami in chimpanzee. Physiol Behav 49:927–934

Höfer D, Püschel B, Drenckhahn D (1996) Taste receptor-like cells in the rat gut identified by expression of alpha-gustducin. Proc Natl Acad Sci USA 93:6631–6634

Hoon MA, Adler E, Lindemeier J, Battey JF, Ryba NJ, Zuker CS (1999) Putative mammalian taste receptors: a class of taste-specific GPCRs with distinct topographic selectivity. Cell 96:541–551

Huang AL, Chen X, Hoon MA, Chandrashekar J, Guo W, Tränkner D, Ryba NJ, Zuker CS (2006) The cells and logic for mammalian sour taste detection. Nature 442:934–938

Inoue K, Takei Y (2002) Diverse adaptability in oryzias species to high environmental salinity. Zool Sci 19:727–734

Ishimaru Y, Okada S, Naito H, Nagai T, Yasuoka A, Matsumoto I, Abe K (2005) Two families of candidate taste receptors in fishes. Mech Dev 122:1310–1321

Ishimaru Y, Inada H, Kubota M, Zhuang H, Tominaga M, Matsunami H (2006) Transient receptor potential family members PKD1L3 and PKD2L1 form a candidate sour taste receptor. Proc Natl Acad Sci USA 103:12569–12574

Iwamatsu T (1997) The integrated book for the biology of the medaka. Daigaku Kyoiku Shuppan, Okayama (in Japanese)

Kanwal JS, Finger TE (1997) Parallel medullary gustatospinal pathways in a catfish: possible neural substrates for taste-mediated food search. J Neurosci 17:4873–4885

Kiyohara S, Hidaka I (1991) Receptor sites for alanine, proline, and betaine in the palatal taste system of the puffer, Fugu pardalis. J Comp Physiol 169:523–530

Kotrschal K, Finger TE (1996) Secondary connections of the dorsal and ventral facial lobes in a teleost fish, the rockling (Ciliata mustela). J Comp Neurol. 370:415–426

Lagerström MC, Hellström AR, Gloriam DE, Larsson TP, Schiöth HB, Fredriksson R (2006) The G protein-coupled receptor subset of the chicken genome. PLoS Comput Biol 2:e54

Lamb CF, Finger TE (1996) Axonal projection patterns of neurons in the secondary gustatory nucleus of channel catfish. J Comp Neurol 365:585–593

Loosli F, Köster RW, Carl M, Kühnlein R, Henrich T, Mücke M, Krone A, Wittbrodt J (2000) A genetic screen for mutations affecting embryonic development in medaka fish (Oryzias latipes). Mech Dev 97:133–139

Marui T, Caprio J (1982) Electrophysiological evidence for the topographical arrangement of taste and tactile neurons in the facial lobe of the channel catfish. Brain Res 231:185–190

Marui T, Kiyohara S (1987) Structure—activity relationships and response features for amino acids in fish taste. Chem Senses 12:265–275

Mueller KL, Hoon MA, Erlenbach I, Chandrashekar J, Zuker CS, Ryba NJ (2005) The receptors and coding logic for bitter taste. Nature 434:225–229. Erratum (2007) Nature 446:342

Nakatani Y, Takeda H, Kohara Y, Morishita S (2007) Reconstruction of the vertebrate ancestral genome reveals dynamic genome reorganization in early vertebrates. Genome Res 17:1254–1265

Naito T, Saito Y, Yamamoto J, Nozaki Y, Tomura K, Hazama M, Nakanishi S, Brenner S (1998) Putative pheromone receptors related to the Ca2+ sensing receptor in Fugu. Proc Natl Acad Sci USA 95:5178–5181

Naruse K, Hori H, Shimizu N, Kohara Y, Takeda H (2004) Medaka genomics: a bridge between mutant phenotype and gene function. Mech Dev 121:619–628

Nelson G, Hoon MA, Chandrashekar J, Zhang Y, Ryba NJ, Zuker CS (2001) Mammalian sweet taste receptors. Cell 106:381–390

Nelson G, Chandrashekar J, Hoon MA, Feng L, Zhao G, Ryba NJ, Zuker CS (2002) An amino-acid taste receptor. Nature 416:199–202

Northcutt RG (2004) Taste buds: development and evolution. Brain Behav Evol 64:198–206

Oike H, Nagai T, Furuyama A, Okada S, Aihara Y, Ishimaru Y, Marui T, Matsumoto I, Misaka T, Abe K (2007) Characterization of ligands for fish taste receptors. J Neurosci 27:5584–5592

Ogawa K, Caprio J (1999a) Facial taste responses of the channel catfish to binary mixtures of amino acids. J Neurophysiol 82:564–549

Ogawa K, Caprio J (1999b) Citrate ions enhance taste responses to amino acids in the largemouth bass. J Neurophysiol 81:1603–1607

Ohmoto M, Matsumoto I, Yasuoka A, Yoshihara Y, Abe K (2008) Genetic tracing of the gustatory and trigeminal neural pathways originating from T1R3-expressing taste receptor cells and solitary chemoreceptor cells. Mol Cell Neurosci 38:505–517

Perry SF, Gilmour KM (2002) Sensing and transfer of respiratory gases at the fish gill. J Exp Zool 293:249–263

Reilly S (1999) The parabrachial nucleus and conditioned taste aversion. Brain Res Bull 48:239–254

Schier AF, Neuhauss SC, Harvey M, Malicki J, Solnica-Krezel L, Stainier DY, Zwartkruis F, Abdelilah S, Stemple DL, Rangini Z, Yang H, Driever W (1996) Mutations affecting the development of the embryonic zebrafish brain. Development 123:165–178

Shi P, Zhang J (2007) Comparative genomic analysis identifies an evolutionary shift of vomeronasal receptor gene repertoires in the vertebrate transition from water to land. Genome Res 17:166–174

Stone LM, Finger TE, Tam PP, Tan SS (1995) Taste receptor cells arise from local epithelium, not neurogenic ectoderm. Proc Natl Acad Sci USA 92:1916–1920

Tanaka H, Maeda R, Shoji W, Wada H, Masai I, Shiraki T, Kobayashi M, Nakayama R, Okamoto H (2007) Novel mutations affecting axon guidance in zebrafish and a role for plexin signalling in the guidance of trigeminal and facial nerve axons. Development 134:3259–3269

Travers J, Dinardo LA, Karimnamazi H (1997) Motor and premotor mechanisms of licking. Neurosci Biobehav Rev 21:631–647

Travers JB, DiNardo LA, Karimnamazi H (2000) Medullary reticular formation activity during ingestion and rejection in the awake rat. Exp Brain Res 130:78–92

Ueda T, Ugawa S, Yamamura H, Imaizumi Y, Shimada S (2003) Functional interaction between T2R taste receptors and G-protein alpha subunits expressed in taste receptor cells. J Neurosci 23:7376–7380

Valentincic T, Caprio J (1997) Visual and chemical release of feeding behavior in adult rainbow trout. Chem Senses 22:375–382

Venkatachalam K, Montell C (2007) TRP channels. Annu Rev Biochem 76:387–417

Wilson RP (1994) Utilization of dietary carbohydrate by fish. Aquaculture 124:67–80

Yasuoka A, Abe K, Arai S, Emori Y (1996) Molecular cloning and functional expression of the alpha1A-adrenoceptor of medaka fish, Oryzias latipes. Eur J Biochem 235:501–507

Yasuoka A, Aihara Y, Matsumoto I, Abe K (2004a) Phospholipase C-beta 2 as a mammalian taste signaling marker is expressed in the multiple gustatory tissues of medaka fish, Oryzias latipes. Mech Dev 121:985–989

Yasuoka A, Hirose Y, Yoda H, Aihara Y, Suwa H, Niwa K, Sasado T, Morinaga C, Deguchi T, Henrich T, Iwanami N, Kunimatsu S, Abe K, Kondoh H, Furutani-Seiki M (2004b) Mutations affecting the formation of posterior lateral line system in Medaka, Oryzias latipes. Mech Dev 121:729–738

Yasuoka A, Okada S, Abe K (2006) General logic for taste reception. Jikken Igaku Suppl Yodo-sha 125–30 (in Japanese)

Yoshida Y, Saitoh K, Aihara Y, Okada S, Misaka T, Abe K (2007) Transient receptor potential channel M5 and phospholipaseC-beta2 colocalizing in zebrafish taste receptor cells. Neuroreport 18:1517–1520

Yoshihara Y (2002) Visualizing selective neural pathways with WGA transgene: combination of neuroanatomy with gene technology. Neurosci Res 44:133–140

Yoshii K, Kurihara K (1983) Ion dependence of the eel taste response to amino acids. Brain Res 280:63–67

Yoshii K, Yokouchi C, Kurihara K (1986) Synergistic effects of 5-nucleotides on rat taste responses to various amino acids. Brain Res 367:45–51

Zhang Y, Hoon MA, Chandrashekar J, Mueller KL, Cook B, Wu D, Zuker CS, Ryba NJ (2003) Coding of sweet, bitter, and umami tastes: different receptor cells sharing similar signaling pathways. Cell 112:293–301

Index

A

Adaptive diversification, 3, 17–19
Adenylyl cyclase type III (ACIII), 65–67, 69
Allelic exclusion, 59
Amami, 251
Amino acids, 97, 107–109, 111, 114, 239, 240, 242, 245, 247, 248, 250, 251
Antennae, 122–124, 126–134
Antennal lobe, 123, 132–134
Arista, 142–144, 148, 149
Aristolochic acid, 212, 215, 216
Artificial sweeteners, 204, 215
Asp-Arg-Tyr (DRY) motif, 65, 67
Astray, 112–114
Aversive, 239, 247, 251
Axonal projection, 57–71
Axon sorting, 57, 67–69, 71

B

Behavioral, 252
Bile acids, 109, 111, 114
Birth-and-death process, 5, 14, 19
Bitter compounds, 206, 208, 210–215
Bitter taste receptors, 3, 13, 203–216
Bombykol, 128, 131, 134
Brain
    amygdala, 233
    hypothalamus, 230, 234
    nucleus accumbens, 230, 234
    nucleus of the solitary tract (NST), 221, 229, 234
    orbitofrontal cortex, 223
Brush cells, 208

C

Calyx glomeruli, 154
Carnivores, 14, 17
CD36, 224, 226–234

Cephalic phase of digestion, 226, 229, 232, 234
Chemokines, 111
Chemosensation
    odorant receptor (OR), 2–8, 17
    olfaction, 1, 10
    taste, 1–3, 13–16
    trace amine-associated receptors (TAARs), 2, 4, 7–8
    V1Rs, 2, 4, 8–13, 16–18
    V2Rs, 2, 4, 8, 11–13, 16–18
Chemosensory cells, 240
Chemosensory receptors, 78, 84
    MHC class I, 78, 81
    ORs, 82, 85, 91
    phylogeny, 80, 81
    V1Rs, 79, 80, 82–84, 87, 91
    V2Rs, 80, 82, 84, 85
Chorda tympani, 191
Ciliated neurons, 50, 51
Ciliated OSNs, 100–106, 108–111
Circumvallate papillae, 204–207
11-cis-vaccenyl acetate, 125, 134
Class I OR, 58, 63, 64
CNGA2, 65, 67–70
Combinatorial representation, 40
Conditioned aversion, 229
Conditioned taste aversion, 223, 224
Crypt cells, 50–51, 101, 104, 105
cyclic AMP (cAMP), 57, 65–67, 69, 71
Cyclic-nucleotidegated (CNG) channels, 129, 130

D

Delayed-rectifying potassium (DRK) channel, 224–225, 230, 231
Denatonium benzoate, 209, 212, 239, 245, 247, 248, 250, 251
17 α 20b-Dihydroxy-4-pregnene-3-one-20-sulfate (17,20P-S), 110

257

Dorsal paired median (DPM) neurons, 170
dpa (D-phenylalanine), 189, 197, 198
DRK. *See* Delayed-rectifying potassium
Drosophila
  adult, 139, 141–154, 156–167, 169, 171–175
  appetitive learning, 165–166, 169–174
  aversive learning, 165, 169–171, 174–175
  labellum, 142, 143, 151, 156–159, 161–162
  larval, 139, 142, 143, 145–147, 153–154, 159, 163–166, 171–175
  olfactory organs, 141–147
  olfactory system, 139–140, 144, 148, 154, 165
  tarsi, 156–157, 162
  taste sensilla, 151, 155, 157, 162
Duplication, 240, 242

E
Energy metabolism, 239, 250
Enteroendocrine cells, 208
Epibranchial nerves, 241
Evolutionary origin, 37, 42, 46, 47
Exocrine gland peptide (ESP), 12, 13
Expression domains, 49
Expression zones, 49

F
Fatty acids, 222, 224–226, 228, 233, 234
Foliate, 223, 224, 226, 229
  papillae, 205, 206
Fruit flies, 125, 127, 134, 135
fT1Rs, 239, 242, 245–248, 250, 251
fT2Rs, 239, 242, 245, 246–248, 250, 251
Functional polymorphisms, 214, 215
Fungiform papillae, 204–206

G
Gal4/UAS, 99–100
Gene
  clusters, 42, 44–46
  duplicate, 10
  gains, 47
  loss, 38, 43, 47
  number variation, 5, 7
  turnover, 10, 12, 17
Glomeruli, 57, 58, 62, 64–71, 123, 132–134
Glossopharyngeal (IX), 191, 240, 241
GPCRs. *See* G-protein coupled receptors
G-protein, 57, 62, 65, 67, 121, 125, 129–130, 132

G-protein-coupled receptors (GPCRs), 25, 30, 31, 125, 126, 128, 130, 221, 224–227
Gr genes, 159–160, 163, 164
Gustatory receptors, 127
α-Gustducin, 205–206, 209–210

H
Harmful, 251, 252
Heterotrimeric G-protein, 209–210
Homology region (H region), 60
Hormones
  glucagon-like peptide-1, 198
  leptin, 198
  neuropeptide Y, 198
Hypothalamus, 230, 234

I
Inhibition, 188, 194, 197
In situ hybridization, 205, 207, 208
Introns, 44, 45

J
Jawless vertebrates, 47

K
Kenyon cells (KC), 142, 143, 151–152, 154, 167–168

L
Lactisole (*Lac*), 193, 194, 197
L-amino acids, 239, 247, 250, 251
Lateral olfactory tract (LOT), 109–111
Laure, 114
Lingual lipase, 224
L2 neurons, 156–158, 161
Locus control region (LCR), 59–62
LOT. *See* Lateral olfactory tract

M
Main olfactory system, 3–8
Mammals, 3–8, 10, 11, 13, 14, 16
  human, 188, 194
  mouse, 188–189–192, 196
  rats, 188–189, 193
Maxillary palps, 122–124, 127, 129, 134, 135
Medaka, 39, 42, 45, 47
Medial olfactory tract (MOT), 109–111

Index

Methyl eugenol, 135
M10 family, 12
Microvillous OSNs, 101–106, 108–111
Microvillous receptor neurons, 50–51
Monogenic expression, 40, 49
MOT. *See* Medial olfactory tract
Mutagenesis, 97, 99, 111

N
Negative-feedback regulation, 57, 60–62, 70
Negative selection, 37, 46, 48
Nerves
    chorda tympani (CT), 221, 229, 234
    glossopharyngeal (GL) nerve, 221, 229, 230, 234
Neuronal activity, 57, 68–71
Neuronal identity code, 71
N,*N*-diethyl-3-methylbenzamide (DEET), 135
Nucleotides, 97, 99, 103, 109–111
Nucleus accumbens, 230, 234

O
Obesity, 221, 222, 225, 234, 235
Octopamine, 142, 164, 167, 171, 175
Odor
    map, 98, 105, 107–111, 114
    space, 49
Odorant receptors (ORs), 2–8, 17, 18, 37, 38, 40–43, 45, 47–52, 57–71, 121, 122, 124–135, 147–149
    characteristics, 25
    genes
        choice, 57–71
        cluster, 60–62
        expressions, 29, 31–33
        sequence, 26, 29–30
    population variation, 33–34
    proteins, 26, 27, 30
    repertoire, 25–31, 33, 34
Odor receptors. *See* Odorant receptors
Odour–shock learning, 167–169, 171
Odour–sugar learning, 167, 169, 171, 172
Odysseus, 111–112
OE. *See* Olfactory epithelium
Olfactory axon convergence, 105–108
Olfactory bulb (OB), 57, 58, 62–67, 70, 71, 123, 132
Olfactory coding, 49
Olfactory cortex, 70
Olfactory epithelium (OE)
    samples, 34
    specific and ectopic expression, 32
    tissues, 31–34
Olfactory neurons, 123–125, 132, 134, 135
Olfactory receptor, 37–52
Olfactory receptor neurons (ORNs), 121–134, 141, 142, 144, 146–149, 151, 153, 154, 163
Olfactory sensory neurons (OSNs), 31, 32, 57–60, 62–71
Olfactory-system-related gene families, 2
Olfactory transduction, 122, 129, 131, 132
OlfC, 37, 38, 41, 45–48, 50, 51
Omnivorous mammals, 14
One neuron-one receptor rule, 57–61
OR. *See* Odorant receptors
ORAs, 37, 38, 41, 43–44, 47–49, 51, 52
OR83b, 144, 147–148
Or83b, 126–128, 130–133, 135
OR-instructed axonal projection, 57, 58, 65, 66, 70, 71
ORNs. *See* Olfactory receptor neurons
ORs. *See* Odorant receptors
Orthologs, 43, 51
OSN. *See* Olfactory sensory neuron

P
Papillae
    chorda tympani (CT), 229
    circumvallate, 223–229
    foliate, 223, 224, 226, 229
    fungiform, 223, 225, 226, 229, 233
Partially overlapping, 245
Particular molecular receptive range, 98
Perception, 221–235
Pharyngeal endoderm, 241
Pheromone dependent behavior
    genetic alterations, 89–91
    physical alterations, 91
Pheromones, 121, 122, 124–126, 128–134
    ESPs, 79, 81
    urine, 77–79, 81, 82, 84, 88
Phospholipase C-β2, 239, 241, 246, 247
Pickpocket (*ppk*) gene, 161, 163, 164
PKD. *See* Polycystic kidney disease
Placodal, 241
Plcβ2, 210
Polycystic kidney disease (PKD), 251
Positive selection, 37, 42, 43, 45, 46, 48
Preference, 221–227, 229–231, 233
Projection neurons, 142, 143, 149, 151–154, 165, 168–169, 171
PROP. *See* Propylthiouracil
PROP/PTC receptor, 215

PROP-tasters, 223
Propylthiouracil (PROP), 212, 213, 215, 216
Prostaglandin F2α, 110
Protein tyrosine kinase (PTK), 221, 226, 227
Pseudogenes, 40–41, 45, 214
Pseudogenization, 10, 17, 19
Pufferfish, 28, 39, 42, 43, 47

R
Receptor
  binding, 191, 194–197
  bitter receptors, 188, 189
  calcium-sensing, 191
  $GABA_B$, 191, 197
  γ-aminobutyric acid type B, 191
  G-protein-coupled receptor (GPCR), 187, 190–192, 194, 197
  inverse agonists, 194
  metabotropic glutamate receptors (mGluRs), 191
  sugars, 190, 197
  sweet taste, 187–192, 195, 197, 198
  T1R1, 190, 191, 197
  T1R2, 187, 190–198
  T1R3, 187, 190–198
  T1Rs, 191, 194, 197
  T1R2:T1R3, 187, 190–198
  V2R vomeronasal, 191
Receptotopic map, 40, 51
Refinement, 67, 70
Reproductive pheromones, 110
Roundabouts, 112
Rutabaga gene, 152, 165, 168, 169

S
Saccharin (*Sac*), 189, 190, 195, 198, 212, 215, 216
Salivary glands, 205
Salt, 247, 251, 252
Segregation, 246, 247
Selective constraints, 14, 18
Sensilla, 122–127, 135
  i-type, 158, 162
  s-type, 151, 158–160
Sequence, 1, 3, 7, 8, 9, 16–19
Sesquiterpene lactones, 213
Seven-transmembrane proteins, 26, 27, 30
Sexual dimorphism, 126, 134
Sham training, 172–174
Signal transduction, 227–229
Silkmoth, 123, 125, 128, 130, 131, 133, 134
Silkmoth bombykol pheromone, 128
Slits, 112
Solitary, 239, 241
Solitary chemosensory cells, 208
Species-specific genes, 14
Stickleback, 39, 45, 47
Sugars, 239, 250
Sweet, 245, 247, 250–252
Sweeteners
  acesulfame, 188–189, 193, 194
  aspartame, 188–190, 192–194, 197
  binding, 194–197
  brazzein, 192
  cyclamate, 188, 190, 192–194, 197
  glucose, 188, 195, 196
  monellin, 188, 192, 194
  neohesperidin dihydrochalcone, 192, 194
  neotame, 188, 192, 193
  saccharin, 188, 189, 190, 194, 196, 197
  sucralose, 188, 189, 190, 195, 196
  sucrose, 188, 193–197
  sugars, 187, 188, 190, 194–195, 197
  thaumatin, 188, 192
Sweet/umami, 2, 3, 16, 17
Synapsin, 152, 166, 172–174
Syntenic regions, 240

T
TAAR. *See* Trace amine-associated receptors
Tandem duplication, 42
Tarsi, 156–157, 162
Tas1r1, 190
Tas1r2, 190, 198
Tas1r3, 190, 196, 198
Tas2r gene expression, 205
Tas2r gene polymorphisms, 214
Taste buds, 190–191, 198, 203–206, 216, 222–226, 230
Taste neurons
  labellar sensilla, 158–159
  labellar taste pegs, 157–158
  pharyngeal taste organs, 159
  tarsi, 156–157
Taste receptor cells (TRC), 203–211, 221, 223–231, 233, 234
Tastes, 221–228, 230–234
Teleost, 37–52
Teleostei, 239, 240
TG. *See* Triglycerides
Tol2 transposon, 99
T1R, 3, 4, 16–18
T2R, 3, 4, 13–19

Index

Trace amine-associated receptors (TAAR), 2, 7–8, 37, 38, 40, 41, 44–45, 47–51
Transduction, 121–135, 225–228
Transgenesis, 99
Transient receptor potential channel C2 (TRPC2), 13
Transmembrane helices, 30
TRC. *See* Taste receptor cells
(Z)-7-Tricosene, 159
Triglycerides (TG), 221, 224, 232, 234
TRPC2. *See* Transient receptor potential channel C2
TRPM5, 210, 245
Two-bottle preference test, 223

U
Umami, 190, 197, 245, 247, 251, 252

V
Vagus (X) nerves, 241
Variation, 1, 3–5, 7–12, 17, 18

Vertebrates, 1–19, 38, 44, 45, 47, 48, 52
Vomeronasal circuitry
   central processing, 89–90
   gonadotropin releasing hormone (GnRH) neurons, 88
   primary neurons, 87–88
Vomeronasal receptor, 2, 8–13
Vomeronasal transduction, 85, 89
V1R, 2, 4, 8–13, 16–18, 37, 38, 40, 43–45, 47–51
V2R, 37, 38, 40, 45–50
V2Rs, 2, 4, 8, 11–13, 16–18
VUM_{mx1} neuron, 166–167

W
Wheat germ agglutinin, 251

Z
Zebrafish, 28, 39–42, 45–47, 50, 51, 97–114
Zone, 58, 63–65
Zone-to-zone projection, 64